Disinfection By-Product Formation and Control During Chloramination

About the Awwa Research Foundation

The Awwa Research Foundation (AwwaRF) is a member-supported, international, nonprofit organization that sponsors research to enable water utilities, public health agencies, and other professionals to provide safe and affordable drinking water to consumers.

The Foundation's mission is to advance the science of water to improve the quality of life. To achieve this mission, the Foundation sponsors studies on all aspects of drinking water, including supply and resources, treatment, monitoring and analysis, distribution, management, and health effects. Funding for research is provided primarily by subscription payments from approximately 1,000 utilities, consulting firms, and manufacturers in North America and abroad. Additional funding comes from collaborative partnerships with other national and international organizations, allowing for resources to be leveraged, expertise to be shared, and broad-based knowledge to be developed and disseminated. Government funding serves as a third source of research dollars.

From its headquarters in Denver, Colorado, the Foundation's staff directs and supports the efforts of more than 800 volunteers who serve on the board of trustees and various committees. These volunteers represent many facets of the water industry, and contribute their expertise to select and monitor research studies that benefit the entire drinking water community.

The results of research are disseminated through a number of channels, including reports, the Web site, conferences, and periodicals.

For subscribers, the Foundation serves as a cooperative program in which water suppliers unite to pool their resources. By applying Foundation research findings, these water suppliers can save substantial costs and stay on the leading edge of drinking water science and technology. Since its inception, AwwaRF has supplied the water community with more than $300 million in applied research.

More information about the Foundation and how to become a subscriber is available on the Web at **www.awwarf.org**.

Disinfection By-Product Formation and Control During Chloramination

Prepared by:
Gerald E. Speitel Jr. and **Phillip G. Pope**
University of Texas at Austin, Department of Civil Engineering
1 University Station C1700, Austin, Texas 78712-0273

and

M. Robin Collins and **Melanie Martin-Doole**
University of New Hampshire, Department of Civil Engineering
Kingsbury Hall, Durham, New Hampshire 03824-3591

Sponsored by:
Awwa Research Foundation
6666 West Quincy Avenue, Denver, CO 80235-3098

Published by:

Awwa Research Foundation
Advancing the Science of Water®

IWA Publishing

DISCLAIMER

Published by IWA Publishing, Alliance House, 12 Caxton Street, London SW1H 0QS, UK

Telephone: +44 (0) 20 7654 5500; Fax: +44 (0) 20 7654 5555; Email: publications@iwap.co.uk;
Web: **www.iwapublishing.com**

AwwaRF report number 91040F. Originally published by AwwaRF for its subscribers in 2004.
IWA Publishing version published 2005.

ISBN 1 84339 930 X

CONTENTS

TABLES

FIGURES

FOREWORD

The Awwa Research Foundation is a nonprofit corporation that is dedicated to the implementation of a research effort to help utilities respond to regulatory requirements and traditional high-priority concerns of the industry. The research agenda is developed through a process of grass-roots consultation with members, utility subscribers, and working professionals. Under the umbrella of a Strategic Research Plan, the Research Advisory Council prioritizes the suggested projects based upon current and future needs, applicability, and past work; the recommendations are forwarded to the Board of Trustees for final selection. The foundation also sponsors research through the unsolicited proposal process; the Collaborative Research, Research Applications, and Tailored Collaborative programs; and various joint research efforts with organizations such as the U.S. Environmental Protection Agency, the U.S. Bureau of Reclamation, and the Association of California Water Agencies.

This publication is a result of one of those sponsored studies, and it is hoped that its findings will be applied in communities throughout the world. The following report serves not only as a means of communicating the results of the water industry's centralized research program but also as tool to enlist the further support of the nonmember utilities and individuals.

Projects are managed closely from their inception to the final report by the foundation's staff and large cadre of volunteers who willingly contribute their time and expertise. The foundation serves a planning and management function and awards contracts to other institutions such as water utilities, universities, and engineering firms. The funding for this research effort comes primarily from the Subscription Program, through which water utilities subscribe to the research program and make an annual payment proportionate to the volume of water they deliver and consultants and manufacturers subscribe based on their annual billings. The program offers a cost-effective and fair method for funding research in the public interest.

A broad spectrum of water supply issues is addressed by the foundation's research agenda: resources, treatment and operation, distribution and storage, water quality and analysis, toxicology, economics, and management. The ultimate purpose of the coordinated effort is to assist water suppliers to provide the highest possible quality of water economically and reliably. The true benefits are realized when the results are implemented at the utility level. The foundation's trustees are pleased to offer this publication as a contribution toward that end.

Walter J. Bishop
Chair, Board of Trustees
Awwa Research Foundation

James F. Manwaring, P.E.
Executive Director
Awwa Research Foundation

ACKNOWLEDGMENTS

The authors of this report are indebted to the following water utilities and individuals for their cooperation and participation in this project:

Brick Township Municipal Utilities Authority, Brick, New Jersey
City of Austin Water and Wastewater Utility, Austin, Texas
Boca Raton Utility Services, Boca Raton, Florida
St. Paul Water Utility, St. Paul, Minnesota
Charleston Commissioners of Public Works, Charleston, SC
Manatee County Public Works Department, Bradenton, FL
St. Bernard Parish, Dept. of Public Works, Chalmette, Louisiana
City of Port Orange, Florida, Port Orange, Florida

The authors especially acknowledge the assistance of John Rissel of Brick Utilities for performing haloacetic acid and bromide analyses and for his technical support and advice throughout the project.

In addition, the advice and help of the Project Advisory Committee (PAC) and the AwwaRF project officers, Kenan Ozekin and Alice Fulmer, are sincerely appreciated. The PAC consisted of Peter Vikesland, Assistant Professor, Department of Civil and Environmental Engineering, Virginia Polytechnic Institute; Ali A Karimi, Water Quality Specialist, Los Angeles Department of Water and Power; and Lloyd D. Stowe, Chief of Plant Operations, Washington Aqueduct Division, U.S. Army Corps of Engineers

EXECUTIVE SUMMARY

INTRODUCTION

Forthcoming regulations and other developments are changing disinfection practices in the drinking water industry. In response to current and anticipated disinfection by-product (DBP) regulations, many utilities have begun to use chloramines as a secondary disinfectant (Diehl et al. 2000). Chloramination produces DBPs such as haloacetic acids (HAAs), trihalomethanes (THMs), and haloacetonitriles (HANs) in lower concentrations than chlorination. Typically, HAA formation during chloraminiation is 5 to 20% of that observed with chlorination (Speitel, 1999).

Previous research has demonstrated that dihalogenated haloacetic acids (DXAAs) are the most commonly formed HAAs during chloramination. Some utilities may have difficulty meeting the new maximum contaminant level (MCL) for HAAs because chloramination does not limit the formation of DXAAs to the same extent as it does other DBPs (Symons et al. 1998). However, DXAA formation during chloramination tends to be less than that observed in chlorination. For example, Zhang et al. (1999), working with Suwannee River fulvic acid, observed that DXAA formation after 5 days of contact with chloramines was 40 to 50% of that with chlorine. DBPs of regulatory concern, other than DXAAs, are well controlled by chloramination; thus, a better understanding of DXAA formation and control is the primary research need to ensure that chloramination remains a viable treatment option for utilities in the future.

OBJECTIVES

Many factors influence DBP formation during chloramination. Previous work indicates that pH, chlorine-to-ammonia-nitrogen (Cl_2/N) ratio, bromide, temperature, precursor type and concentration, chloramine residual concentration, and mode of chloramine addition affect the formation of HAAs. A review and assessment of the practical needs of utilities led to the following research objectives:

1. A better understanding of the reactivity of key natural organic matter (NOM) fractions (e.g., hydrophobic, transphilic, and hydrophilic) with respect to DXAA formation, as well as the effects of treatment processes on both the removal and reactivity of these NOM fractions.
2. A better delineation of the influence of pH and Cl_2/N ratio on DXAA formation.
3. Characterization of DXAA formation kinetics and the impact of treatment processes on the kinetics, in particular including the impact of significant periods of prechlorination.
4. The rate and extent of DXAA formation at elevated summer water temperatures.
5. The effect of bromide concentration on DXAA speciation and kinetics.

APPROACH

To address the above objectives, the project research program consisted of laboratory experimentation, mathematical modeling, and sampling of selected treatment plants and distribution systems. The first phase of the research consisted of batch screening experiments on two water sources of differing water quality: Lake Austin, Texas and Metedeconk River, New Jersey. The selected waters were subjected to various treatments and NOM fractionations. The Lake Austin and Metedeconk River source, treated, and NOM fractions resulted in the study of 12 different waters. These 12 waters were selected to both simulate the effect of treatment and to study the reactivity of specific natural organic matter (NOM) fractions. The key objective of the batch screening experiments was to identify the water quality and chloramination variables that are most important in DXAA formation. The screening approach used a fractional experimental design to systematically evaluate the many factors that may influence DBP formation during chloramination. Each water was exposed to sixteen separate experimental conditions that were used to evaluate the influence of Cl_2/N ratio, pH, temperature, chloramine residual concentration, and bromide concentration on DXAA formation after 48 hours of exposure to the selected chloramination conditions. Sixteen trials for these five factors, each at two levels, resulted in an experimental resolution where all the main factors and two-factor interactions could be estimated by analysis of variance (ANOVA) without confounding. The experimental design allowed the identification of the major water quality and chloramination variables that contribute to DXAA formation.

Additional batch experiments were undertaken to further study the most important variables that were identified in the initial batch screening experiments. The two factors studied were Br^-/DOC ratio and pH. The Br^-/DOC ratio was selected instead of bromide alone to normalize the DOC concentrations among the waters. The experiments were conducted on source and softened waters from Biscayne Aquifer, Florida and St. Paul, Minnesota, as well as Metedeconk River, New Jersey source water. Two experimental designs were created for the follow-up batch screening experiments: one for the source waters and one for the treated waters. Three levels of each of the two variables (pH and Br^-/DOC ratio) were selected to evaluate non-linear effects of variables. In addition, the reactivity of specific fractions of the NOM was examined in more detail than in the initial batch screening experiments.

DXAA formation kinetics were studied in four waters by measuring DXAA concentrations over chloramination contact times of 0.5 to 72 hours. The effects of treatment (*i.e.*, coagulation or softening), bromide concentration, water temperature, and prechlorination were examined. Prechlorination times of 5 and 20 minutes were tested. The laboratory kinetics measurements for Lake Austin water were confirmed by sampling the Davis Water Treatment Plant and distribution system in Austin, TX.

Mathematical models were developed to predict DXAA formation during chloramination based on water quality and chloramination conditions. The data from the batch screening experiments, plus one additional water source, were used to develop the models (179 data points). Because DXAA formation occurs rapidly, kinetic effects were ignored in the models. The performance of the models were confirmed by comparing their predictions to DXAA concentrations measured in three treatment plants and distribution systems that were not used in the development of the models.

CONCLUSIONS

Factors Affecting DXAA Formation

An initial round of batch screening experiments was conducted on two diverse water sources: Lake Austin water, a low SUVA, hard, high alkalinity source treated by lime softening, and Metedeconk River water, a high-SUVA, soft, low alkalinity source treated by alum coagulation. Source waters, treated waters, and NOM fractions were studied to determine the influence of pH, bromide concentration, Cl_2/N ratio, temperature, and chloramine residual on DXAA formation during chloramination. A fractional factorial experimental design was used to assess the significance and relative contribution of each factor, as well as water type, through ANOVA.

To complement the initial batch screening experiments, five source and treated waters were studied from a variety of locations to broaden the range of water quality and NOM characteristics studied. The two most influential factors attributed to DXAA formation in the initial batch screening experiments, the bromide concentration and pH, were studied in a fractional factorial experimental design. Three levels rather than two were investigated for each factor to assess possible non-linear effects. The findings from both sets of batch screening experiments are summarized below.

pH

The ANOVA analyses over a broad variety of experimental conditions usually pointed to pH as being the most significant factor in DXAA formation. DXAA formation decreased as the pH increased. Therefore, utilities interested in decreasing DXAA formation should first examine the possibility of raising the pH of their water.

Bromide

The evaluation of bromide's impact on DXAA formation was studied in two ways. In the initial batch screening experiments two levels of bromide concentration were considered: ambient and ambient plus 0.5 mg/L. In the follow-up batch experiments, the Br^-/DOC ratio was selected as the factor for studying the impact of bromide after accounting for varying DOC content; ratios ranged from less than 1 to 87.5 µg/mg and were grouped into three levels: <15, 40, and 87.5 µg/mg, respectively. The Br^-/DOC ratio was not a controlled factor in the initial batch screening experiments; therefore, a wide variety of Br^-/DOC ratios prevailed in these experiments, ranging from 9 to 415. The impact of bromide can be viewed from several perspectives: the DXAA mass concentration or yield, DXAA molar concentration or yield, and DXAA speciation, which can be described by the molar bromine incorporation ratio (n_D'). Generally, bromide impacts the DXAA speciation and DXAA mass concentration or yield more so than DXAA molar concentration or yield. Therefore, the results are discussed below from each of these perspectives to provide a more complete picture.

The ANOVA analyses on the initial batch screening experiments showed that pH and bromide concentration were the most significant factors in DXAA formation. DXAA formation increased and a shift in speciation to the bromine-substituted species occurred as the bromide concentration increased. It should be noted, however, that some experimental conditions in these batch experiments resulted in relatively high bromide concentrations and Br^-/DOC ratios in comparison to most natural waters. Therefore, the wide range of bromide concentrations studied may have influenced the relative significance of bromide as factor in DXAA formation.

The follow-up batch experiments used Br^-/DOC ratios that were more representative of the range seen in typical practice. Under these conditions, the Br^-/DOC ratio had a minor influence on DXAA molar yield for the waters tested, which somewhat contradicted the results found in the initial batch screening experiments. Increasing Br^-/DOC ratios caused a shift toward the bromine-substituted species without significantly influencing the DXAA molar yield. Because of the higher molecular weight of bromine in comparison to chlorine, the shift in speciation implies an increase in the DXAA mass yield or concentration. Therefore, the overall results suggest that the DXAA molar yield generally will not increase over the range of Br^-/DOC ratios typical of practice. A source water experiencing a large increase in bromide concentration, however, may exhibit an increase in the DXAA molar yield. An increase in the Br^-/DOC ratio will cause a shift toward the bromine-substituted species and increase the DXAA mass yield and concentration. An increase in the Br^-/DOC ratio is undesirable from two perspectives. First, the regulations are based on mass concentration, so the likelihood of violating the HAA MCL increases. Second, the bromine-substituted HAAs may be of greater health concern. In general, higher Br^-/DOC ratios and lower pH led to greater bromine incorporation into DXAAs. At Br^-/DOC ratios of less than 200 µg/mg (the typical range of most natural waters), increases in the Br^-/DOC ratio caused an increase in DXAA bromine incorporation. Values of n_D', which characterizes the bromine substitution of DXAAs on a molar basis, varied considerably among the water sources studied, but ranged from a low of 0 at Br^-/DOC ratios around 10 µg/mg to a high of 1.4 at a Br^-/DOC ratio of 200 µg/mg. At Br^-/DOC ratios above 200 µg/mg, further increases caused no discernable impact on bromine incorporation into DXAAs.

Other Factors

The Cl_2/N mass ratio was the next most significant contributor to DXAA formation after pH and bromide. DXAA formation decreased as the Cl_2/N ratio decreased. Temperature and chloramine residual also impacted DXAA formation, but were much less influential than the aforementioned factors. DXAA formation increased as the temperature increased, and the impact of temperature was more pronounced as the Cl_2/N ratio increased. Increasing the chloramine residual concentration caused an increase in DXAA formation in Lake Austin water, but had a minimal impact on DXAA formation in Metedeconk River water.

Impact of Treatment

Treatment can also have a significant effect on DXAA formation. The effectiveness of treatment on DXAA formation is largely related to overall DOC removal, although preferential removal of more reactive NOM fractions also may contribute to reduced DXAA formation. Comparison of DXAA yield before and after treatment helps distinguish whether decreased DXAA formation is related to differing reactivity of the waters or is just a reflection of the lower DOC concentration after treatment. Softening Lake Austin and Biscayne Aquifer source water resulted in minimal DOC removal and therefore only slightly decreased DXAA formation. Alum coagulation of Metedeconk River and Charleston source water removed a larger fraction of the DOC, resulting in a more significant decrease in DXAA formation. Treatment reduced the DXAA yield in all of the source waters studied except Biscayne Aquifer. This decrease in yield indicates that treatment generally removes the more reactive DXAA precursors. Treatment processes that remove DOC can also affect DXAA formation and speciation through the resulting increase in the Br^-/DOC ratio.

NOM Characteristics

The follow-up batch experiments showed that DXAA mass yield decreased with an increasing hydrophobic/DOC ratio or, alternatively, a decreasing hydrophilic/DOC ratio, thereby indicating that the NOM in the hydrophilic fraction was more reactive in forming DXAA than the NOM in the hydrophobic fraction. In addition, the DXAA mass yield was approximately proportional to chloramine demand; as the chloramine demand increased, the DXAA mass yield increased. Many other water quality characteristics such as DOC, UV, SUVA, SUVA x Br^-, hydrophilic DOC, and hydrophobic DOC were considered, but none yielded valuable correlations with DXAA formation when considered individually.

The most influential terms in the DXAA mass and molar yield models (discussed below) were, in order, hydrophobic DOC concentration, hydrophobic/DOC ratio, hydrophilic/DOC ratio, SUVA, and UV_{254}. The models highlighted the importance of multiple NOM characteristics in predicting DXAA formation, although it was not generally possible to correlate individual NOM characteristics with DXAA formation. The need to consider multiple NOM characteristics simultaneously illustrates the complexity of the relationship between NOM and DXAA formation.

DXAA Formation Kinetics

Experiments were conducted on both source and treated waters from four different locations to study the impact of pH, bromide concentration, temperature, and treatment on DXAA formation kinetics during chloramination. In addition, the impact of prechlorination on DXAA formation kinetics was studied because of the widespread use of short periods of chlorination prior to chloramination. The waters selected, Lake Austin, Metedeconk River, Biscayne Aquifer, and surface impoundments near Charleston, SC, spanned the typical range of bromide and DOC concentrations, as well as alkalinity and SUVA, common in source waters. The conclusions drawn from these experiments are:

1. The kinetics of DXAA formation during chloramination are rather fast. For example, approximately 73% of the 72-hour DXAA concentration formed after only 0.5 hours and almost 82% formed after 1 hour in Metedeconk River source water. Therefore, the vast majority of the DXAA formation will occur within the water treatment plant itself.
2. The bromine-substituted species (BCAA and DBAA) formed very rapidly, with the formation reactions typically being completed in less than 0.5 hours. In contrast, DCAA formation kinetics were characterized by an initial period (~ 1 hour) of rapid formation followed by slow formation over the remainder of the 72-hour incubation period.

3. More rapid DXAA formation kinetics were observed at pH 7 than at pH 9 during both the rapid and slow periods of formation. This pH trend indicates that acid catalyzed monochloramine reactions may be the cause of increased HAA formation as the pH decreases.
4. Increasing temperature caused only a minimal increase in DXAA formation over the range tested in Charleston and Lake Austin treated waters. However, additional experiments with Lake Austin softened water indicated that increased HAA formation as a result of increased temperature may be more pronounced at lower pH and higher Cl_2/N ratios. Nevertheless, the effect of temperature alone on DXAA formation appears to be relatively small; seasonal changes in NOM concentration or reactivity are also likely to be important in situations in practice where large changes in DXAA formation are observed with seasonal temperature changes.
5. During short periods of prechlorination (5 or 20 minutes), significantly more DXAA formation, as well as formation of mono- and tri-halogenated acetic acids, occurred relative to chloramination alone. Also, as expected, the longer the period of free chlorination the greater the DXAA formation; however, the increase was much less than proportional to the prechlorination time.
6. During the prechlorination period, free chlorine reacts with most of the fast-reactive NOM sites. As a result, after subsequent ammonia addition and chloramine formation very little additional DXAA is formed. At least 75% of the 72-hour DXAA concentration formed during prechlorination in all the waters studied, and greater than 90% of the 72-hour DXAA concentration typically formed during prechlorination. Therefore, the general expectation is that very little of the DXAA formation in chlorination/chloramination plants is associated with the chloramination step, despite the dramatic difference in the chloramine and chlorine contact times that is typical of such plants. Thus, HAA control strategies should be focused on the chlorination step in plants that prechlorinate prior to chloramination.
7. The absence of significant DXAA formation during chloramination after prechlorination suggests that both chlorine and monochloramine (or monochloramonium ion) react with the same fraction of the NOM to produce DXAAs, or that the DXAAs formed during chloramination do so as a result of the HOCl that is formed by monochloramine hydrolysis.

Predictive Model of DXAA Formation

The final step of this study consisted of the development of simple and practical models for predicting DXAA formation during chloramination based on water characteristics and chloramination conditions. Linear regression modeling was performed on the data generated during the initial batch screening experiments (112 points) with Lake Austin and Metedeconk River, the follow-up batch screening experiments (35 points) with Metedeconk River, Biscayne Aquifer source and treated, and St. Paul source and treated waters, and finally Charleston source water (32 points), which was not included in either of the screening experiments. Models for DXAA mass yield, DXAA molar yield, and DXAA mass concentration were developed. The major findings are summarized as follows:

1. The modeling effort was a culmination of the work accomplished throughout this study and assumed fast kinetics based on the findings from the kinetics experiments.
2. The models developed in this study, Ln DXAA Molar Yield, Ln DXAA Mass Yield, and Ln DXAA Mass Concentration adequately predicted DXAA formation (R^2 of 0.90, 0.89, and 0.84, respectively). In general, the models over-predicted the levels observed in practice; however, the full-scale system evaluation of the models was considered satisfactory, especially since the trends between model predictions and measurements were consistent.
3. The modeling effort allowed the ranking of factors through scaled-estimate analysis, which was compared to the factor ranking of the initial and follow-up batch screening experiments based ANOVA analyses. The most influential model terms for the two DXAA Yield models were, in rank order, hydrophobic DOC concentration, the hydrophobic/DOC ratio, the hydrophilic/DOC ratio, SUVA, UV_{254}, pH, bromide concentration, NH_2Cl Demand, NH_2Cl Residual, and chlorine to ammonia ratio (Cl_2/N). It is interesting to note that the first six most influential terms are all related to water/NOM characteristics. The most influential model terms for the DXAA Mass model were almost in the same order except that Ln DOC concentration was first, the influence of SUVA and UV_{254} was slightly diminished, and the influence of pH and bromide concentration was slightly increased. In general, the most influential model terms were the NOM characteristics of the water followed by pH, bromide concentration and chloramination conditions. This ranking of model factors based on scaled estimates coincided with the factor rankings from the initial and follow-up batch experiments, which were based on percent contributions from ANOVA analyses.

RECOMMENDATIONS TO UTILITIES

The use of models can be helpful for utilities to evaluate general tendencies in their distribution system and anticipate possible difficulties with respect to DBP regulation compliance. The development of simple and practical models predicting DXAA formation during chloramination based on select water characteristics and chloramination conditions was the final stage of this research. A number of kinetic models predicting monochloramine loss and/or DBP formation have been presented in the literature. The models developed in this present study were not based on kinetics considerations as most of the DXAAs were formed in less than 1 hour during chloramination. Numerous analytical measurements of water characteristics, chloramination conditions, and DXAA concentrations were gathered throughout the research to create an extensive database for model development through linear regression analysis. The water quality variables and chloramination conditions were identified as the predictors and the DXAA data (in mass, mass yield, and molar yield) were identified as the response. The multi-linear regression models are listed in Table 7.1.

The use of these models will allow utilities to qualitatively assess their distribution system DXAAs based on treated water characteristics and chloramination conditions. The DXAAs calculated from the model will tend to over-predict the actual DXAAs found in the distribution system, but the trends should still be applicable to actual treated water quality and chloramination conditions. Thus, the influence from changes in the NOM and inorganic characteristics of the treated water (due to changing conditions in the source water quality from watershed influences or treatment process variations) may be qualitatively assessed to determine any problematic issues in satisfying the HAA regulations. In addition, changes or modifications to chloramination conditions may also be qualitatively evaluated in a similar fashion.

In addition to the multi-linear regression models for assessing problematic HAA by-product issues, this research also provided guidance to utilities on an approach for systematically implementing the most influential changes to the treatment and/or chloramination process for reducing DXAAs in the distribution system. Based on the statistical experimental approach used in this study (plus the personal experiences of the authors) the corrective treatment changes to minimize distribution system DXAAs from chloramination should be implemented in the following order:

- **Increase pH**: The pH during chloramination should be raised above pH 7. Results from this study suggested that approximately a 50% reduction in DXAAs can be achieved by increasing pH from 7 to 9. This reduction was independent of all other factors and conditions.
- **Decrease free chlorination interval**: The time interval between chlorine and ammonia additions should be reduced as much as possible with preference given to simultaneous additions. As much as 95% of the 72-hour DXAA concentration can be formed with as little as 5 minutes of contact time between free chlorination and ammonia addition.
- **Decrease NOM/DOC precursors**: As expected, as much NOM/DOC precursor material as possible should be removed before chloramination. Sole dependence on softening to reduce organic precursors sufficiently should be carefully evaluated; enhanced softening, involving significant magnesium precipitation, may be needed to achieve substantial (> 20%) DOC removal (Smith et al. 2004). Both hydrophobic and hydrophilic NOM fractions should be removed, but preference should be given to the hydrophilics (XAD-8 eluate). Consequently, treatment processes that are known to favor hydrophilic DOC removal (e.g., biological treatment) may need to be considered. In addition, utilities need to be mindful that DOC removal will increase the Br^-/DOC ratio, which will shift the DXAA speciation more toward the bromine-substituted species and in some cases may increase the DXAA yield.
- **Decrease Cl_2/N ratio**: Although not as significant as the aforementioned changes, decreasing the Cl_2/N ratio will decrease DXAA formation somewhat in most cases.
- **Decrease chloramine residuals**: Although reductions in DXAAs are expected for lower chloramine residuals, the amount of reduction was small in this research. Managing DXAA reduction by limiting chloramine residuals should be used only as a last resort.

CHAPTER 1
INTRODUCTION

Forthcoming regulations and other developments are changing disinfection practices in the drinking water industry. In response to current and anticipated disinfection by-product (DBP) regulations, many utilities have begun to use chloramines as a secondary disinfectant (Diehl et al. 2000). Chloramination produces DBPs such as haloacetic acids (HAAs), trihalomethanes (THMs), and haloacetonitriles (HANs) in lower concentrations than chlorination. Typically, HAA formation during chloraminiation is 5 to 20% of that observed with chlorination (Speitel, 1999).

Previous research has demonstrated that dihalogenated haloacetic acids (DXAAs) are the most commonly formed HAAs during chloramination, and therefore, because chloramination does not limit the production of DXAAs as well as other DBPs, some utilities may have difficulty meeting the new maximum contaminant level (MCL) for HAAs (Symons et al. 1998). However, DXAA formation in chloramination tends to be less than that observed in chlorination. For example, Zhang et al. (1999), working with Suwannee River fulvic acid, observed that DXAA formation after 5 days of contact with chloramines was 40 to 50% of that with chlorine. DBPs of regulatory concern, other than DXAAs, are well controlled by chloramination; thus, a better understanding of DXAA formation and control is the primary research need to ensure that chloramination remains a viable treatment option for utilities in the future.

Many factors influence DBP formation during chloramination. Previous work indicates that pH, chlorine-to-ammonia-nitrogen (Cl_2/N) ratio, bromide, temperature, precursor type and concentration, chloramine residual concentration, and mode of chloramine addition affect the formation of HAAs. Additional research in the following areas is required to meet the practical needs of utilities:

1. A better understanding of the reactivity of key natural organic matter (NOM) fractions with respect to DXAA formation, as well as the effects of treatment processes on both the removal and reactivity of these NOM fractions.
2. A better delineation of the influence of pH and Cl_2/N ratio on DXAA formation.
3. Characterization of DXAA formation kinetics and the impact of treatment processes on the kinetics, in particular including the impact of significant periods of prechlorination.
4. The rate and extent of DXAA formation at elevated summer water temperatures.
5. The effect of bromide concentration on DXAA speciation and kinetics.

To address the above issues, the project research program consisted of laboratory experimentation, mathematical modeling, and sampling of selected treatment plants and distribution systems. An initial series of batch screening experiments identified the relative significance of the many variables that could effect DXAA formation during chloramination. Follow-up batch experiments were then conducted on several waters to study the most important variables identified in the initial batch screening experiments. DXAA formation kinetics were studied in four waters, including the effects of treatment (*i.e.*, coagulation or softening), bromide concentration, water temperature, and prechlorination. The laboratory kinetics measurements were confirmed by sampling a full-scale treatment plant distribution system. A mathematical model was developed to predict DXAA formation during chloramination based on water quality and chloramination conditions. Because DXAA formation occurs rapidly, kinetic effects were ignored in the model. The performance of the model was confirmed by comparing its predictions to DXAA concentrations measured in three treatment plants and distribution systems.

CHAPTER 2
LITERATURE REVIEW

INTRODUCTION

Two AwwaRF-sponsored projects, initiated in the mid-1990's, identified two previously unknown issues with respect to chloramination: (1) dihalogenated haloacetic acids (DXAAs) are by far the most commonly formed haloacetic acids (HAAs) (Symons et al., 1998; Singer et al., 1999), which is in contrast to chlorination where significant formation of both di- and trihaloacetic acids is commonplace, and (2) because chloramination does not limit the formation of DXAAs as well as it does other disinfection by-products (DBPs), some utilities may have difficulty meeting the then anticipated Stage 2 maximum contaminant level (MCL) of 40 μg/L for the five regulated haloacetic acids (HAA5) (Symons et al., 1998). DBPs of regulatory concern, other than DXAAs, are well-controlled by chloramination; therefore, a better understanding of DXAA formation and control is the primary research need to help ensure that all utilities are able to use chloramination in the future, should they so desire.

In setting the stage for this research, it is useful to first reiterate the relative merits of chloramination over chlorination. Even though some problems with DXAA may be encountered, Table 2.1 clearly shows that chloramination is vastly superior to chlorination with respect to HAA formation (Speitel, 1999). Typically, HAA formation with concurrent addition of chlorine and ammonia is 5 to 20% of that observed with chlorination. Even DXAA formation in chloramination tends to be less than that observed in chlorination. For example, Zhang et al. (1999), working with Suwannee River fulvic acid, observed that DXAA formation after 5 days of contact with chloramines was 40 to 50% of that with chlorine. Singer et al. (1999) found that DXAA formation after 24 hours of contact with chloramines was 10 to 30% of that with chlorine in four waters that were alum coagulated, settled, and filtered prior to disinfectant addition. The potential DXAA problems in chloramination are generally small enough to reasonably expect that an improved understanding of DXAA formation will permit utilities to meet the Stage 2 MCL through relatively minor changes in treatment practices.

Table 2.1
Relative haloacetic acid formation with chloramination versus chlorination

Water Source	Relative HAA Formation from Chloramination*	Reference
Seymour Reservoir (BC)	23% of HAASDS†	Long et al. (1992)
Capilano Reservoir (BC)	15% of HAASDS	Long et al. (1992)
Quabbin Reservoir (MA)	6 – 20% of HAASDS	Zhu and Reckhow (1993)
Mississippi River (LA)	10% of HAASDS	Lykins et al. (1994)
Palm Beach (FL) & Myrtle Beach (SC) extracts	5 – 10% of HAASDS	Cowman and Singer (1996)
Lake Austin (TX)	3 – 20% of HAAFP‡	Symons et al. (1998)
Lake Houston (TX)	7 – 17% of HAAFP	Symons et al. (1998)
CA State Project	5 – 30% HAAFP	Symons et al. (1998)
Suwannee River fulvic acid	26 – 29% HAAFP	Zhang et al. (1999)

* Chloramination HAA concentration/chlorination HAA concentration x 100.
† Haloacetic acid simulated distribution system concentration with chlorine.
‡ Haloacetic acid formation potential with chlorine.

FACTORS INFLUENCING DBP FORMATION

Many factors may influence DBP formation during chloramination. These include:

- pH and Cl_2/N mass ratio
- Bromide concentration (influences both DBP and chloramine speciation)
- Temperature
- Precursor type and concentration
- Chloramine concentration
- Mode of chloramine addition (concurrent or staggered addition of chlorine and ammonia)
- Pretreatment by other oxidants, in particular ozone, prior to chloramine addition.

With respect to HAAs, the two AwwaRF-sponsored projects cited above, as well as a few pilot and full-scale studies, largely provide the current basis for our understanding of HAA formation during chloramination. In this section, the available information is evaluated in the context of the factors influencing DBP formation, listed above.

To briefly summarize the past AwwaRF projects, Symons et al. (1998) performed detailed batch studies on three natural waters looking at pH (6, 8, 10), Cl_2/N mass ratio (3, 5, and 7/1), residual chloramine concentration (1, 2, and 4 mg/L), bromide concentration (ambient and ambient plus 0.5 mg/L), mixing intensity during chlorine and ammonia addition (low, medium, high), and delayed ammonia addition (30 sec). The work also included pilot plant runs on three waters, distribution system sampling and limited batch studies on an additional five waters, and analysis of historical data for the five waters. The primary analytical measurements were dissolved organic halogen (DOX) and trihalomethanes (THMs). HAA and cyanogen halide (CNX) concentrations were measured on a subset of the experiments run. The primary limitations of this study are the absence of a comprehensive HAA data set (because the significance of HAA formation was unknown entering the project and did not become apparent until well into the project), a focus almost solely on natural waters to consider worst case situations (*i.e.*, the effect of treatment was largely, but not totally, ignored), and detailed characterization of the natural organic matter (NOM) was not performed.

Singer et al. (1999) examined the effect of ozonation on chlorine and chloramine DBP formation, including HAAs, in a series of batch experiments. Because the project included ozonation, chlorination, and chloramination, the range of chloramination conditions had to be limited for practical considerations and included one pH (8) and one Cl_2/N mass ratio (4/1). Both source water and treated water (various combinations of ozonation and coagulation, settling, and filtration) were studied for five waters of varying characteristics, especially in terms of the nature of the NOM. Most experiments, however, were performed on model waters prepared from hydrophobic extracts of the five water sources. The model water experiments examined the effects of ozonation pH and total organic carbon (TOC), inorganic carbon, and bromide concentrations on DBP formation. Also, much more characterization of NOM was performed than on the Symons et al. project, and the utility of such measurements in providing information on reactivity and the potential for DBP formation was demonstrated. With respect to HAA formation in chloramination, the work of Singer et al. (1999) is limited by the single pH and Cl_2/N values studied and by the uncertainty associated with extrapolating results from hydrophobic extracts to natural waters. A more detailed evaluation of the two AwwaRF-sponsored projects and other relevant research is provided below for the factors that influence DBP formation during chloramination.

pH and Cl_2/N Ratio

Batch studies by Symons et al. (1998) showed that, in general, DBPs (including HAAs) decreased in concentration as the pH increased and the Cl_2/N ratio decreased. For example, at a Cl_2/N mass ratio of 3/1 the HAA_6 concentration at pH 10 was typically 30 to 50% of that at pH 6. Some significant departures from the general trend were noted, however, around neutral pH (7-8). For example, the DOX data for Lake Austin water (Figure 2.1) showed significant concentration peaks around pH 8 for a Cl_2/N mass ratio of 5/1, both at the ambient bromide concentration and with 0.5 mg/L of bromide added. These results suggest either a pH dependence of NOM reactivity or a pH dependence of haloamine speciation (and thereby reactivity). The former is more likely near neutral pH; for example, the reaction rates for chlorination of various phenols peak in the vicinity of pH 8 (Stumm and Morgan, 1996). Unfortunately, insufficient HAA data were collected to determine if similar peaking in the HAA concentrations likewise occurred. These results imply that additional work focused on pH 7-8 and Cl_2/N mass ratios commonly used in practice (3, 4, and 5/1) is warranted to obtain a more complete understanding of the role of pH and Cl_2/N ratio in HAA formation.

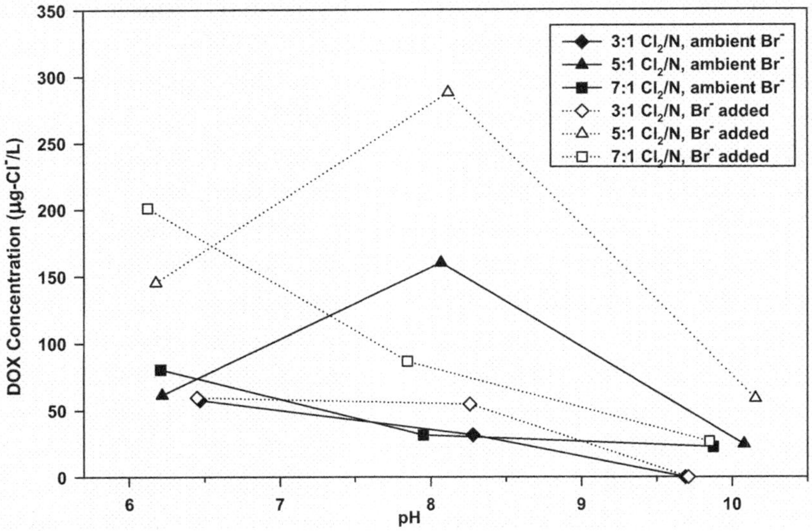

Source: Diehl et al., 2000.

Figure 2.1 DOX concentrations in Lake Austin water as a function of pH and Cl_2/N mass ratio

Bromide

Available data (Cowman and Singer, 1996; Symons et al., 1998; Singer et al., 1999; Zhang et al., 1999; Diehl et al., 2000) indicate two effects of elevated bromide concentrations on HAA formation. First, the speciation shifts toward the bromine-substituted HAAs, in particular the bromine-substituted DXAAs. Formation of bromochloroacetic acid (currently an unregulated HAA) is favored over formation of dibromoacetic acid. Second, the total HAA concentration on a mass basis increases somewhat. Cowman and Singer (1996) found that Myrtle Beach extracts provided a constant molar yield of HAAs with increasing bromide concentration, so the increase in mass concentration may be largely reflective of bromine's greater molecular weight compared to chlorine's. In general, mid-range bromide concentrations (0.05 to 0.1 mg/L) cause a small increase in HAA mass concentrations (10-30%).

Temperature

Most of the DBP formation work during chloramination has been done at constant temperature. Singer et al. (1999) incubated samples at 20°C, while Symons et al. (1998) used 22°C. Clearly, it is reasonable to expect HAA concentrations to increase with increasing temperature because of faster reaction kinetics. For example, Roth and Ozment (1999) showed an approximate doubling of HAA concentrations (14 to 28 µg/L) in the Carbondale, IL distribution system when the water temperature increased from 46°F to 76°F. The treatment plant uses both free chlorine and chloramines, so the kinetic effects on both disinfectants are reflected in the data. A review of Information Collection Rule (ICR) data (EPA, 2000) for surface water treatment plants in the southern half of the U.S. known to be practicing chloramination likewise shows elevated HAA concentrations for the July/September samples in comparison to the October/December samples. In some cases, the HAA concentrations in the summer samples were twice those in the fall samples. Clearly, additional research on HAA formation at elevated temperatures is needed based on the significant temperature effect suggested by the available data.

Precursor Type and Concentration

Only a limited amount of work has been done on the role of precursor type and reactivity in DBP formation. Symons et al. (1998) studied a series of source waters of low to medium specific ultraviolet absorbance (SUVA). Singer et al. (1999) studied five source waters with a broad range of SUVA values, as well as hydrophobic extracts from these waters. The impact of ozonation and conventional treatment (*i.e.*, coagulation, flocculation, settling, and filtration) separately and together also was examined for a subset of the source waters and extracts. For untreated hydrophobic extracts, DXAA yield (μg DXAA/mg TOC) was highly correlated to the phenolic carbon content for both chlorination and chloramination (Figure 2.2). As expected, NOM reactivity, as quantified by DXAA yield, was much greater with chlorine than with chloramines.

Phenolic carbon could only be measured on untreated extracts; correlations with SUVA must be examined for the treated extracts. Unfortunately, SUVA measurements are available for only three of the five treated extracts examined by Singer et al. (1999). DXAA yield in the hydrophobic extracts likewise was highly correlated to SUVA for both chlorination and chloramination (Figure 2.3). Ozonation prior to chloramination decreased SUVA, as would be expected, but did not significantly affect DXAA yield. Thus, the slope of the DXAA yield versus SUVA plot changed substantially (Figure 2.3), but destruction of UV-absorbing characteristics (presumably double bonds and aromaticity) had little effect on DXAA formation, suggesting that these UV-absorbing NOM sites were not DXAA precursor sites. The good correlation between DXAA yield and SUVA and the absence of an effect on DXAA yield by SUVA removal through ozonation seem somewhat contradictory, indicating that much more exploration of the reactivity of various NOM fractions is needed to better understand the role of NOM in DXAA formation.

The correlation between DXAA yield and SUVA is not as good for source waters ($R^2 < 0.7$) as for hydrophobic extracts ($R^2 > 0.85$) (Figure 2.4); nevertheless, a relationship between DXAA yield and SUVA is suggested. For the waters studied by Singer et al. (1999), conventional treatment significantly decreased the TOC concentration, as expected, but only decreased SUVA in one of the three waters for which data are available. The DXAA yield decreased by approximately 20% in two of the waters and by 45% in the third. Overall, the combination of a lower TOC concentration and a lower DXAA yield had a substantial impact on the DXAA concentration, which was 20 to 30% of that formed when the source waters were chloraminated. Thus, as with chlorination, removal of NOM in treatment processes has a positive impact on DBP formation.

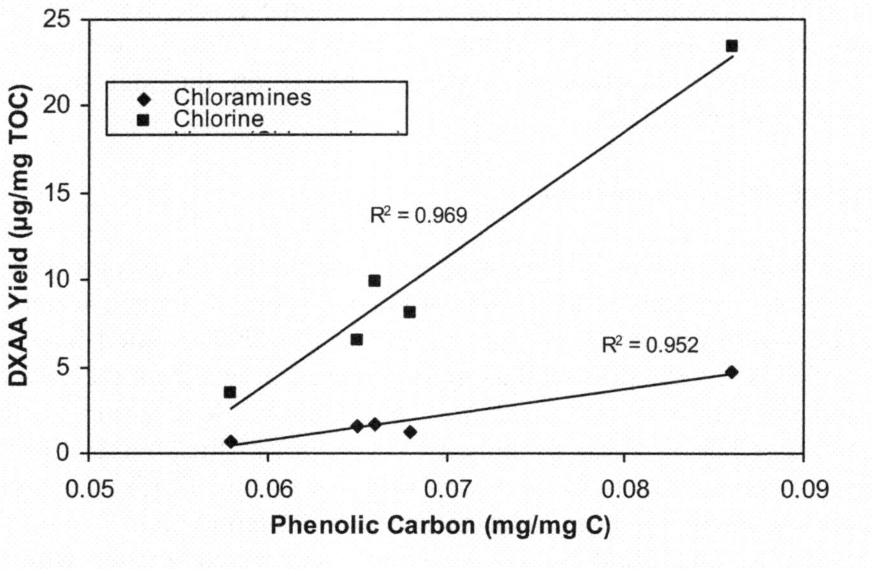

Source: Data from Singer et al., 1999.

Figure 2.2 Reactivity of extracts versus phenolic carbon. Chloramination conditions: pH 8, 20°C, 4/1 Cl_2/N, and a 2 mg/L residual as Cl_2 after 24 hours; chlorination conditions: pH 8, 20°C, and a 1 mg/L residual as Cl_2 after 24 hours.

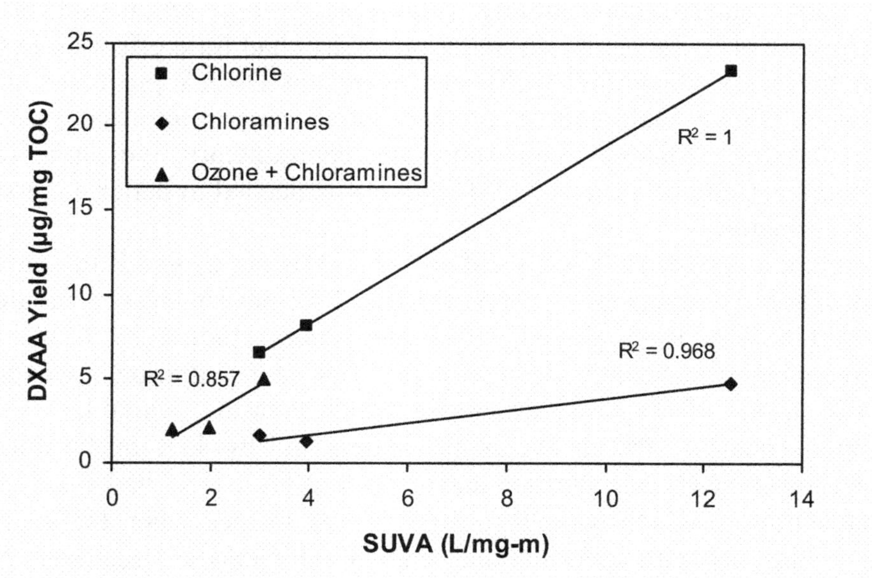

Source: Data from Singer et al., 1999.

Figure 2.3 Reactivity of extracts versus SUVA. Chloramination conditions: pH 8, 20°C, 4/1 Cl_2/N, and a 2 mg/L residual as Cl_2 after 24 hours; chlorination conditions: pH 8, 20°C, and a 1 mg/L residual as Cl_2 after 24 hours; ozone + chloramination: ozonation at pH 7 and O_3:TOC = 1:1, and chloramination at pH 8, 20°C, and a 2 mg/L residual as Cl_2 after 24 hours.

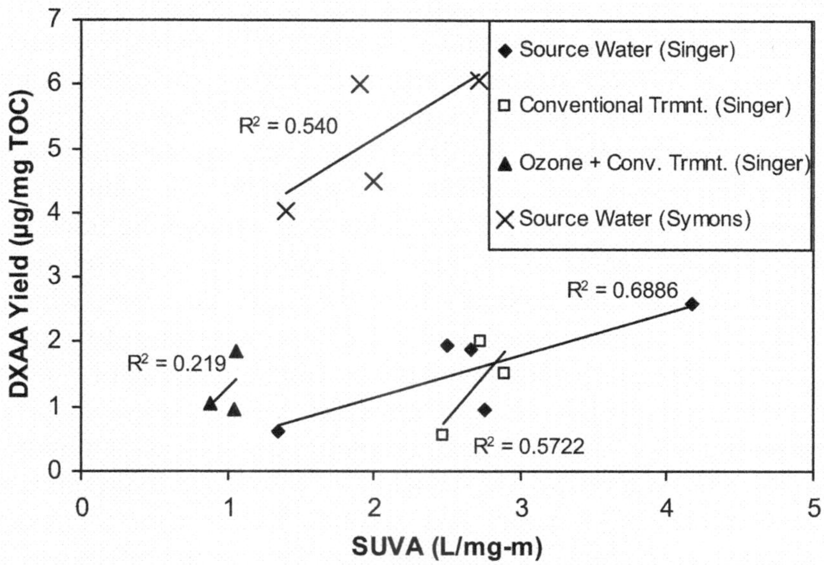

Source: Data from Singer et al. 1999 and Symons et al. 1998.

Figure 2.4 Chloramine reactivity of bulk water versus SUVA. Chloramination conditions (Singer): pH 8, 20°C, 4/1 Cl_2/N, and a 2 mg/L residual as Cl_2 after 24 hours; ozone + chloramination conditions (Singer): ozonation at pH 7 and O_3:TOC = 1:1, and chloramination at pH 8, 20 °C, and a 2 mg/L residual as Cl_2 after 24 hours; Chloramination conditions (Symons): pH 8, 20°C, 5/1 Cl_2/N, and a 2 mg/L residual as Cl_2 after 48 hours

Preozonation before conventional treatment substantially decreased SUVA relative to both that of the source water and for conventional treatment alone, which is essentially the same effect as seen with the hydrophobic extracts (Figure 2.3). The impact of ozonation on DXAA formation was not as clear cut. The DXAA yield decreased in two waters and increased in one relative to conventional treatment, while TOC removal increased in two waters and decreased in one. As a result, DXAA formation was about the same as with conventional treatment in two waters and less than that observed with conventional treatment in the third water. The benefits of ozonation are uncertain based on the work of Singer et al. (1999), but additional pilot and full-scale data presented below provide stronger evidence for ozone's positive effect on HAA formation.

The DXAA yield for four of the source waters studied by Symons et al. (1998) also is shown in Figure 2.4. The strength of the correlation between DXAA yield and SUVA was the same as for the data of Singer et al. (1999); however, the DXAA yield was much greater. The data for both projects were collected at pH 8, but both the Cl_2/N mass ratio (5/1 vs. 4/1) and the incubation times (48 vs. 24 hr) were larger in the work of Symons et al. (1998). Interpretation of the results is confounded by the different incubation periods, but the data at least suggest that the relationship between DXAA yield and SUVA may be significantly influenced by chloramination conditions, which would be expected based on the discussion above dealing with the significance of pH and Cl_2/N ratio.

Natural organic matter typically consists of hydrophobic (humic) substances and hydrophilic (nonhumic) substances. Teng and Veenstra (1996) studied the reactivity of different apparent molecular weight fractions of hydrophilic and hydrophobic organics isolated from Kaw Reservoir, Oklahoma water. Precursors in the 1,000 to 10,000 molecular weight fraction produced more HAAs than the other molecular weight fractions studied; however, the < 1,000 molecular weight fraction produced the greatest DCAA yield. In addition, chloramination of hydrophobic fractions exhibited greater DCAA yields than hydrophilic fractions. Hwang et al. (2000) fractionated and isolated five NOM fractions (hydrophobic, transphilic, hydrophilic acid plus neutral, hydrophilic bases, and colloids) from the Colorado River, which is a low-humic water. These waters were chloraminated at a 5/1 Cl_2/N mass ratio to yield a total chlorine residual of 2 mg/L after a 24-hour incubation at pH 8. Chloramination of Colorado River source water resulted in predominately DXAA formation, but yielded only 6.2 percent of the total HAAs that formed during chlorination. The hydrophilic, transphilic, and hydrophobic fractions all contributed significantly in terms of HAA yield. However, the hydrophilic acid plus neutral fraction exhibited the greatest HAA yield. This fraction was nearly 6 times more reactive than the hydrophobic fraction, accounting for approximately 49% of the total yield from the isolated NOM fractions. In addition, the hydrophilic fractions (hydrophilic acid plus neutral and hydrophilic bases) represented approximately 76% of the total yield of the isolated NOM, while the hydrophobic fraction only accounted for about 9%, indicating that treatments to minimize HAA formation in low-humic waters should address the removal of polar NOM fractions.

DBP precursors can be removed by treatment processes such as alum coagulation, softening, and filtration. However, the extent of precursor removal by these treatment processes varies considerably. Coagulation has been shown to preferentially remove hydrophobic and higher molecular weight NOM more effectively than other NOM constituents (Collins et al. 1986; Krasner and Amy 1995; White el al. 1997). In addition, White et al. (1997) determined that alum reacts preferentially to remove hydrophobic NOM, and that greater amounts of hydrophilic NOM are removed only after the coagulant demand of the hydrophobic NOM has been satisfied. Softening is also capable of achieving significant NOM removal (Liao and Randtke, 1985; Thompson et al., 1997; Smith el al., 2004), typically removing NOM with high molecular weights (Liao and Randkte, 1985; Semmens and Staples, 1986; and El-Rehaili and Weber, 1987).

Chloramine Residual Concentration

Symons et al. (1998) studied the impact of chloramine residual concentration on DBP formation (*i.e.*, DOX, THMs, HAAs, and CNX) in three source waters. Over a broad range of pH and Cl_2/N ratios no relationship between residual concentration (1, 2, and 4 mg/L) and DBP formation was apparent in 48-hour simulated distribution system (SDS) tests. More rigorous statistical analyses of this extensive data set also failed to show any statistically significant relationship between DBP formation and chloramine residual concentration (Diehl 2001). Thus, the available data suggest that residual concentration is of secondary importance in DBP formation over the range of concentrations typical in practice. Symons et al. (1998) only studied a 48-hour incubation period, however, which may have failed to capture kinetic effects of chloramine concentration that might be apparent at shorter incubation periods.

Modes of Chloramine Addition

Although chloramination seems, at first glance, to be relatively straightforward, it has been implemented in practice in a wide variety of ways, which can lead to significant differences in DBP formation among utilities practicing chloramination. The traditional approach to chloramination involves the addition of ammonia and chlorine simultaneously (or nearly so) very early in the treatment train, usually at the rapid mix basin. Mixing might be less than perfect and chemical addition might not be truly simultaneous, with the possible presence of free chlorine being an obvious concern with respect to DBP formation. Symons et al. (1998) studied the effects of delayed ammonia addition (versus preformed chloramines) and mixing intensity to simulate the less than perfect conditions that might be encountered at full scale. A 30-second delay between chlorine and ammonia addition produced no significant effect on DBP concentrations or speciation after 48 hours of incubation; in particular, no discernable impact on HAA formation was evident. Therefore, the formation kinetics appear slow enough such that very short periods of free chlorine exposure will have no effect on DXAA concentrations at the longer time frames of regulatory interest. These results also provide confidence that the common practice of using preformed chloramines in bench-scale studies does not introduce troubling artifacts into the data.

Increasingly, utilities are not practicing traditional chloramination. Many plants now have a significant period of free chlorination prior to ammonia addition for purposes of meeting CT requirements. For example, Seidel et al. (2000) evaluated ICR data for 70 plants practicing chloramination. Of these, only 9 used chloramination alone, while the remainder used a combination of chlorine and chloramination. Similarly, a recent survey of medium and large utilities found that that 42 of 58 plants added ammonia after chlorine, with an average lag time of 25.7 minutes for ammonia addition (Connell et al., 2000). In practice, the lag time between chlorine and ammonia can range from a few minutes to several hours. As might be expected, significant exposure to free chlorine will produce greater DBP formation than with chloramines alone. For example, Symons et al. (1998) studied an unfiltered lake water from the Pacific Northwest in which ammonia addition occurred 2 to 5 hours after chlorine addition. A sample from the distribution system showed an HAA_6 concentration of 38 µg/L, while bench-scale chloramination of the source water with preformed chloramines produced a concentration of only 8 µg/L. Rajbhandari et al. (1999) studied the impact of free chlorine contact time on Manatee Lake water from southwest Florida and Mississippi River water from Minneapolis, Minnesota. These settled waters were prechlorinated for 0.25, 1, or 3 hours prior to ammonia addition to achieve a 4.5/1 Cl_2/N mass ratio and a 3 mg/L 24-hour target chloramine residual. Increasing free chlorine contact times caused both an increase the chlorine required to meet the target residuals as well as an increase in HAA_9 (Table 2.2). Thus, brief exposure to free chlorine that may result from mixing inefficiencies or slightly staggered chlorine and ammonia addition points is not the major issue with respect to free chlorine exposure in chloramination. Rather, significant periods of free chlorination prior to ammonia addition necessitated by disinfection regulations are a far greater concern with respect to DBP formation. Many utilities now face a chlorine/chloramines/CT balancing act that will only get more difficult in the future. In many respects, the rate of DBP formation during chloramination is the key issue (i.e., how much additional DBP formation will occur in the treatment plant and distribution system after chlorination is terminated by ammonia addition?). Therefore, a better understanding of DXAA formation kinetics with chloramines is needed as an important input to the complicated decision-making process facing some utilities.

Unfortunately, kinetic data are very limited for DXAA formation during chloramination. Singer et al. (1999) studied the kinetics for one hydrophobic extract with and without ozonation before chloramination (Figure 2.5). A period of relatively rapid dichloroacetic acid (DCAA) formation occurred over the first hour, with a much slower rate of formation thereafter. Although the extent of formation is much less than with chlorine, the shape of the curve is very similar to that observed with chlorination DBPs: rapid initial formation followed by a decreased rate of formation. Of particular interest in this regard is whether or not an initial chlorination step would "consume" all the rapidly-reacting NOM material resulting in only a slow rate of DXAA formation during a subsequent chloramination step. Only very limited data are available on the DBP formation kinetics of chlorine and chloramine combinations. Long et al. (1992) examined DCAA formation with chlorine alone and with 1 hour of chlorination followed by chloramination (Figure 2.6). These data suggest that the rapidly-reacting NOM was "consumed" during the chlorination step, with further DCAA formation occurring at a slow rate during subsequent chloramination (*i.e.*, data after 1 hour). Carlson et al. (1996) evaluated the potential of continued HAA formation in distribution systems by dosing treated water with chlorine and allowing it to react for a short time prior to ammonia addition at a 5/1 Cl_2/N mass ratio. HAAs were measured immediately after ammonia addition and again after 13 days of storage. Approximately 86% of the total HAA formation formed prior to ammonia addition, indicating minimal additional HAA formation occurs during chloramination. A much more detailed assessment of DXAA formation kinetics during chloramination and the impact of prechlorination on these kinetics is needed, however, to truly clarify the matter.

Table 2.2
Impact of free chlorine contact time on HAA formation

Water	Treatment	FCCT (hr)	FC_0 (mg/L)	HAA_9 (µg/L)
Manatee Lake Water	Settled	0.25	6.5	50
		1	7.6	56
		3	8.2	69
Mississippi River Water	Settled	0.25	4.5	26
		1	5.0	49
		3	6.0	56

Source: Data from Rajbahandari et al. 1999.
FCCT – free chlorine contact time; FC_0 – initial chlorine dose

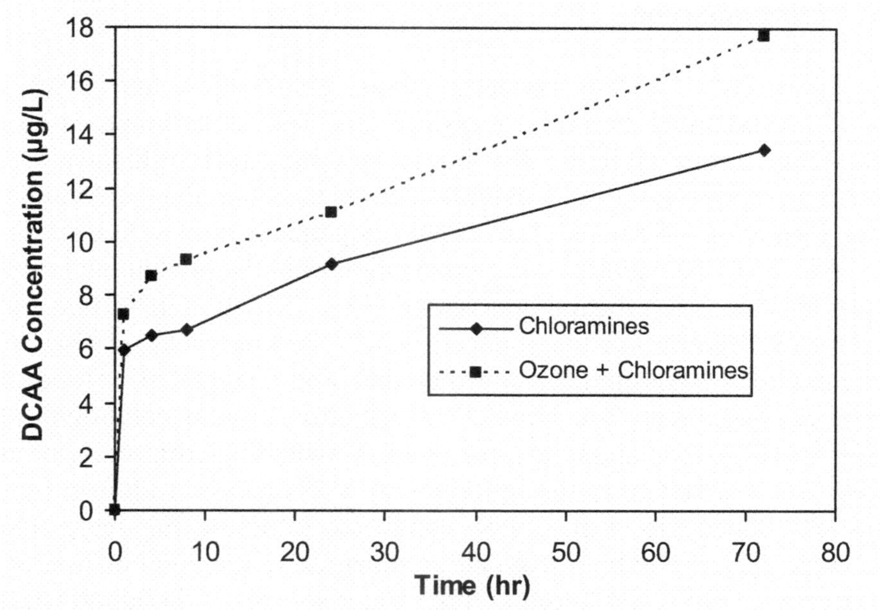

Source: Data from Singer et al. 1999.

Figure 2.5 DCAA formation kinetics. Chloramination conditions: pH 8, 20°C, 4/1 Cl_2/N, and a 2 mg/L residual as Cl_2 after 24 hours; ozone + chloramination: ozonation at pH 7 and 3.17 mg/L O_3 dose, and chloramination at pH 8, 20 °C, and a 2 mg/L residual as Cl_2 after 24 hours.

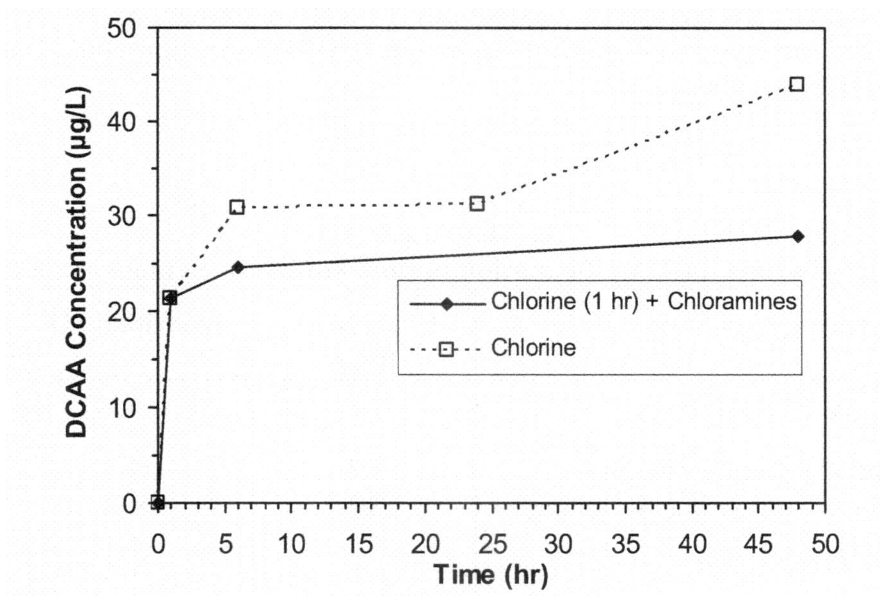

Source: Data from Long et al. 1992.

Figure 2.6 Chlorination and chloramination kinetics. Chlorine dosed in samples buffered to pH 6 to achieve a 1 mg/L residual as Cl_2; secondary disinfection by chloramines: samples buffered to pH 8.25 (after chlorination at pH 6 for 1 hour) dosed to achieve a 3/1 Cl_2/N ratio and a 2 mg/L residual as Cl_2.

Ozonation plus Chloramination

Another approach to recent concerns about disinfection and DBP control is the implementation of ozonation as the primary disinfectant, with chloramines used as a secondary disinfectant to maintain a residual in the distribution system. Much of the available data indicate that the combination of ozonation and chloramination is better than chloramination alone in controlling HAA formation. Symons et al. (1998) conducted pilot studies with ozone on two source waters (Table 2.3). The addition of preozonation to both conventional lime softening and conventional alum treatment significantly decreased HAA formation in 48-hour simulated distribution system (SDS) tests on the finished water. Lykins et al. (1994) studied different disinfection schemes for Mississippi River water that was coagulated, settled, pre-disinfected, sand filtered, and post-disinfected to simulate distribution system conditions (Table 2.4). A dramatic decrease in HAA formation was observed in switching from chlorine to chloramines, and a further decrease resulted from the introduction of ozonation, followed by chloramination after GAC filtration. Jacangelo et al. (1989) conducted pilot and full-scale studies on several surface waters (Table 2.5). Switching from chloramination to the combination of ozonation and chloramination decreased HAA formation in three water sources, although to varying extents. In the first water, the large decrease in HAA formation can be attributed to the elimination of a 4-hour period of free chlorination in the chloramination treatment scheme. The large decrease in the HAA formation in the second water may have resulted from the relatively large ozone dose (1.45 mg O_3/mg TOC). A much smaller decrease in HAA formation occurred in the third water with an ozone dose of 0.51 mg O_3/mg TOC.

Although the bench-scale results of Singer et al. (1999) are equivocal about the benefits of preozonation in decreasing HAA formation upon subsequent chloramination, the other available data consistently indicate a positive effect on HAA formation. In fact, Symons et al. (1998) recommended the introduction of ozonation if DBP formation could not be controlled through chloramination alone.

Table 2.3
Impact of ozone on HAA formation in pilot studies of Symons et al. (1998)

Operating Condition	Cl_2/N Mass Ratio	Incubation pH	HAA_6 (μg/L)
Lake Austin Water:			
Conventional Lime Softening	5/1	8	9.6
Preozonation & Conv. Lime Softening	5/1	8	1.4
California State Project Water:			
Conventional Alum	5/1	8	15
Preozonation & Conv. Alum	5/1	8	2.6
Pre-O_3, Conv. Alum, & Biofiltration	5/1	8	2.4

Table 2.4
Jefferson Parish (LA) pilot plant SDS HAAs

Disinfectant(s)	Incubation pH	TOC* (mg/L)	HAA_6†,‡ (μg/L)
Pre- & Post-Chlorine§	7-8	3.2	146
Pre- & Post-Chloramines	7-8	3.2	14
Pre-Ozone, Post Chloramines	7-8	2.9	8.7

Source: Data from Lykins et al., 1994.

* Concentration in disinfection contact chamber effluent.
† Five-day simulated distribution system incubation time.
‡ Ambient bromide concentration not reported, but HAA_6 speciation suggests relatively low bromide concentration.
§ Pre-disinfection occurred before sand filtration; post-disinfection after sand filtration.

Table 2.5
Impact of changes in disinfectants on HAA formation

Disinfectant Change	TOC (mg/L)	O_3 (mg/L)	Cl_2/N Mass Ratio	Br^- (μg/L)	pH	Decrease in HAA_5 Concentration (%)
NH_2Cl* to O_3 & NH_2Cl	2.46	2.0	5.9/1	150	8.2	80
NH_2Cl to O_3 & NH_2Cl	5.51	8.0	5.0/1	230	9.3	64
NH_2Cl to O_3 & NH_2Cl	3.91	2.0	4.2/1	320	7.8	16

Source: Data from Jacangelo et al., 1989.

* Four-hour delay between chlorine and ammonia addition; no delay with ozonation. Chlorine and ammonia added concurrently for all other cases.

HALOAMINE CHEMISTY

Chloramines

Chloramine chemistry is complicated because of the several species of combined and free chlorine that may be present and because chloramines are inherently unstable and decay through a complex set of reactions that result in the oxidation of ammonia and the formation of chloride. An understanding of this chemistry is essential to an improved understanding of DBP formation during chloramination. Jafvert and Valentine (1992) proposed a model of chloramine decomposition kinetics for a broad range of Cl_2/N ratios and concentrations. Valentine et al. (1998) and Vikesland et al. (2001) narrowed the focus of the model to address conditions typical of chloramination practice in drinking water treatment. The significant chloramine reactions under drinking water conditions are listed in Table 2.6.

Reactions 2.1 to 2.4 represent the substitution (2.1 and 2.3) and hydrolysis (2.2 and 2.4) reactions between HOCl and ammonia or the chloramines. Reactions 2.5 and 2.6 represent the disproportionation reactions of the chloramine species. Reactions 2.7 through 2.10 represent the redox reactions that occur in the absence of measurable levels of free chlorine. The overall rate of chloramine decomposition near neutral pH and above is limited by the rate of formation of dichloramine, which then rapidly decomposes. The rate of monochloramine loss increases as the solution pH decreases because of the enhanced rate at which dichloramine forms at lower pH values (Vikesland et al. 2001). Dichloramine formation occurs in two ways: (1) through monochloramine hydrolysis (Reaction 2.2) and subsequent monochloramine reaction with free chlorine (Reaction 2.3) and (2) through monochloramine disproportionation (Reaction 2.5). Reaction 2.5 is a net reaction, which consists of an equilibrium reaction producing monochlorammonium ion (NH_3Cl^+), followed by a reaction between this species and monochloramine. Reaction 2.5 is general acid catalyzed; therefore, a variety of proton-donating species can act to accelerate the reaction rate. General acid catalysts include bicarbonate, carbonic acid, phosphate, sulfate, and acetic acid (Vikesland et al., 2001). Of practical significance in drinking water treatment, the rate of chloramine decay shows a positive concentration dependence on alkalinity (*i.e.*, bicarbonate) and natural organic matter (*i.e.*, organic acids present in humic and fulvic materials) (Valentine et al., 1998). At free ammonia concentrations of less than 1 to 2 mg/L, Reactions 2.2 and 2.3 are the dominant pathway for dichloramine formation, while Reaction 2.5 dominates at higher ammonia concentrations. Under typical drinking water treatment conditions (neutral pH or above and Cl_2/N mass ratios of 3-5/1), monochloramine is the primary species present; free chlorine, dichloramine, and monochlorammonium ion are present at very low concentrations.

Table 2.6
Chloramination reactions of importance in drinking water treatment

No.	Reaction
2.1	$HOCl + NH_3 \rightarrow NH_2Cl + H_2O$
2.2	$NH_2Cl + H_2O \rightarrow NH_3 + HOCl$
2.3	$NH_2Cl + HOCl \rightarrow NHCl_2 + H_2O$
2.4	$NHCl_2 + H_2O \rightarrow NH_2Cl + HOCl$
2.5	$NH_2Cl + NH_2Cl \rightarrow NHCl_2 + NH_3$
2.6	$NHCl_2 + NH_3 \rightarrow NH_2Cl + NH_2Cl$
2.7	$NH_2Cl + H_2O \rightarrow I$
2.8	$I + NH_2Cl \rightarrow HOCl$ + products
2.9	$I + NHCl_2 \rightarrow$ products
2.10	$NH_2Cl + NHCl_2 \rightarrow$ products

"I" is an unidentified reactive intermediate; "products" refers to a combination of nitrogen gas, ammonia, chloride, nitrate and unknown products.

Brominated Species

Bromide is widespread in source waters. Amy et al. (1994) found an average concentration of 100 μg/L in the U.S. with a range of 0 to 2.3 mg/L. The presence of bromide substantially complicates haloamine chemistry because both HOCl and monochloramine react with bromide forming a variety of products. A summary of the important reactions is provided in Table 2.7. When chlorine is added to water containing ammonia and bromide, Reactions 2.1 and 2.11 proceed rapidly in parallel, with the $[NH_3]/[Br^-]$ ratio largely determining which reaction dominates (Bousher et al., 1989). Under drinking water treatment conditions, Reaction 2.1 dominates, so limited HOBr formation occurs, largely because of the very low HOCl concentration (Valentine et al., 1998). The rate constant for Reaction 2.1 is much larger than for Reaction 2.11 (Jafvert and Valentine 1992; White 1999), but NH_4^+ dominates at typical pH values, resulting in a low NH_3 concentration. The resulting ratio of reaction rates between Reactions 2.1 and 2.11 is on the order of 100:1 for waters with moderate to high bromide concentrations, demonstrating that HOBr formation is not negligible under these conditions. Once formed, HOBr can participate in several reactions. It can react with ammonia to form monobromamine (Reaction 2.16) and with monochloramine to form chlorobromamine (Reaction 2.17). The rate constant for Reaction 2.16 is $7.5 \times 10^7\ M^{-1}s^{-1}$ (Wajon and Morris 1980), while the rate constant for Reaction 2.17 is $2.86 \times 10^5\ M^{-1}s^{-1}$ (Gadza and Margerum 1994). Although these rate constants are quite different, under typical conditions the relative concentrations of NH_3 and NH_2Cl may result in comparable reaction rates for Reactions 2.16 and 2.17.

Table 2.7
Bromide-chloramine reactions of importance in drinking water treatment

No.	Reaction
2.1	$HOCl + NH_3 \rightarrow NH_2Cl + H_2O$
2.11	$HOCl + Br^- \rightarrow HOBr + Cl^-$
2.12	$NH_2Cl + H^+ \leftrightarrow NH_3Cl^+$
2.13	$NH_3Cl^+ + Br^- \rightarrow NH_3Br^+ + Cl^-$
2.14	$NH_3Br^+ \leftrightarrow NH_2Br + H^+$
2.15	$NH_3Br^+ + NH_2Cl \rightarrow NHClBr + NH_4^+$
2.16	$HOBr + NH_3 \rightarrow NH_2Br + H_2O$
2.17	$HOBr + NH_2Cl \rightarrow NHClBr + H_2O$
2.18	$NHClBr \leftrightarrow NClBr^- + H^+$
2.19	$NClBr^- + Br^- + H^+ \leftrightarrow NHBr_2 + Cl^-$
2.20	$NHBr_2 \leftrightarrow NBr_2^- + H^+$
2.21	$NHClBr + NHCl_2 \rightarrow N_2 + Br^- + 2Cl^- + 3H^+$

Monochlorammonium ion, formed by the ionization of monochloramine (Reaction 2.12), also reacts with bromide to form monobromammonium ion (Reaction 2.13). NH_3Br^+ is a strong acid and dissociates to monobromamine (Reaction 2.14); NH_3Br^+ also reacts with monochloramine to form chlorobromamine (Reaction 2.15). Monobromamine is rapidly consumed as a result of Reactions 2.14 and 2.15, as well as disproportionation reactions not shown, so its concentration should be very low (Bousher et al., 1989). Chlorobromamine also can be formed by the reaction of HOBr with monochloramine (Reaction 2.17); however, Reaction 2.15 dominates under drinking water conditions (Vikesland et al. 2001). The pK_a of NHClBr is 4.9; therefore, $NClBr^-$ is the dominant species over the typical pH range of drinking water (Reaction 2.18) (Bousher et al., 1989). $NClBr^-$ equilibrates with dibromamine (Reaction 2.19), which has a pK_a of 6.9, resulting in the formation of NBr_2^- under drinking water conditions (Reaction 2.20). Chlorobromamine also reacts with monochloramine leading to haloamine decomposition to nitrogen gas, chloride, and bromide (Reaction 2.21).

Under typical drinking water treatment conditions, bromide oxidation by monochlorammonium usually dominates (Reaction 2.13) in comparison to the reaction between bromide and the free chlorine produced by monochloramine hydrolysis (Reaction 2.11) (Vikesland et al. 2001). Reaction 2.13 produces monobromammonium, which undergoes rapid reactions to produce chlorobromamines, primarily NHClBr (Valentine 1986 and Vikesland et al. 2001). $NClBr^-$, therefore, is expected to be the bromine-containing species present at the highest concentration (Bousher et al., 1989; Vikesland et al., 2001), while smaller concentrations of bromamines and HOBr are expected. Because of their relative insignificance, all the reactions involving NH_2Br and NBr_3 are not included in Table 2.7, but can be found in Bousher et al. (1989).

The reactions of the bromamines and chlorobromamines are generally much faster than those of the chloramines. For example, the half-life of dibromamine at pH 8 is 30 minutes and that of monobromamine is 19 hours (Wajon and Morris, 1980), whereas a half-life of 50 to 100 hours is typical for monochloramine (Jafvert and Valentine, 1992). The more rapid kinetics of bromine-containing haloamines means that these compounds do not accumulate to an appreciable extent and are present at substantially lower concentrations that monochloramine (Vikesland et al., 2001). From the viewpoint of monochloramine stability, bromide can be viewed primarily as a catalyst that increases the rate of monochloramine decomposition in drinking water (*e.g.*, Reaction 2.21) (Vikesland et al., 2001; Diehl et al., 2000). From the viewpoint of DBP formation, bromide can be viewed as substantially increasing the range of possible halogenating species, with chlorobromamine, free bromine, and dibromamine being the main chemicals of interest.

Haloamine Reactions with NOM

Two types of reactions can be identified between monochloramine and NOM. In the first, NOM catalyzes auto-decomposition of monochloramine through general acid catalysis, which results in the oxidation of ammonia. The second involves reduction of active chlorine through oxidation and substitution reactions with NOM. The presence of NOM clearly accelerates the rate of monochloramine decay, and either or both of the reaction types identified could account for the observed acceleration. Valentine et al. (1998) and Vikesland et al. (1998) investigated the relative significance of the two reaction types in controlled experiments in which chlorine and nitrogen mass balances were examined. These experiments demonstrated that NOM accelerates monochloramine decay primarily because NOM is readily oxidized; the role of NOM as a catalyst is negligible. The total production of the typical monochloramine oxidized decay products, nitrogen gas and nitrate, decreases in the presence of NOM, and the production of the isoelectronic decay product, ammonia, increases. This observation reflects the fact that the oxidizing power of monochloramine is used to oxidize NOM to various oxidized organic species at the expense of oxidizing nitrogen. The production of chloride is largely unaffected by the presence of NOM, indicating that the same extent of oxidation occurs with and without NOM. This result further supports the conclusion that NOM is oxidized by monochloramine. Nitrate formation increases in the presence of NOM, indicating that NOM causes a shift in product speciation away from nitrogen gas and toward nitrate.

These overall observations of NOM reactivity with monochloramine are likewise in agreement with more specific analyses that have identified formation of various DBPs as a result of monochloramine/NOM reactions (Speitel, 1999). DBP formation might result from reactions with monochloramine or through general acid catalysis reactions with monochlorammonium ion. In particular, Synder and Margerum (1982) and Issac and Morris (1983) showed that general acid catalysis reactions could transfer chlorine to nitrogenous organic chemicals at significant reaction rates through pH values as high as 8 to 8.5. Vikesland et al. (1998) suggest that NOM may contain two types of reactive sites: those that readily react with monochloramine and those that react only with the HOCl produced by monochloramine hydrolysis (Reaction 2.2). They further suggest that chloramination DBPs that are likewise formed during chlorination may result from reactions between NOM and the HOCl formed by monochloramine hydrolysis. Conversely, DBPs found primarily during chloramination may result from reactions between NOM and monochloramine (or monochlorammonium ion).

The formation of DBPs through reactions between bromine-containing haloamines and NOM has not been studied in detail, although the observed formation of bromine-substituted HAAs in the presence of bromide clearly indicates that such reactions are of practical importance. As noted above, chlorobromamine is the species present at highest concentration under typical drinking water conditions; thus, it is the most likely brominating agent. Valentine (1986) found that chlorobromamine reacts rapidly with *N,N*-diethyl-*p*-phenylenediamine (DPD), indicating that the bromine atom of chlorobromamine is very labile and reactive. This rapid reactivity led Valentine (1986) to suggest that the products of chlorobromamine/NOM reactions should be characteristic of those produced by free bromine. Little formation of THMs was found with dibromamine (Sugam and Helz, 1980), suggesting similar low reactivity for both chloramines and bromamines with respect to THM formation. The suggestion by Vikesland et al. (1998) that different types of reactive sites exist in NOM presumably can be extended to suggest that some DBP formation also may result from reactions between NOM and HOBr. Duirk et al. (2002b) studied the effect of bromide ion on monochloramine loss in the presence of natural organic matter in Iowa River water. Bromide ion catalyzes monochloramine autodecomposition by the formation of active bromine intermediates. However, in the presence of NOM, the active bromine may either react with NOM or monochloramine. From model development and experimental data collected, they hypothesized that most of the active bromine reacts with the NOM via oxidative and substitution reactions and not with monochloramine. A significant quantity of bromide appears to be incorporated into the NOM resulting in the formation of more brominated species. A better understanding of the reaction pathways of the bromine-substituted oxidant will help to better predict the formation of bromine-substituted HAAs.

SUMMARY

The literature indicates that many factors may influence HAA formation during chloramination. Unfortunately, the relative importance of these factors remains unclear. Also, the available data on precursors from various water sources and their concentrations are inadequate for the purposes of understanding the reactivity of NOM and various key fractions of NOM, as well as the effects of treatment on those fractions, both from the viewpoint of their removal and their resulting reactivity after treatment. Additionally, kinetic data are very limited for DXAA formation during chloramination, and many plants have incorporated a significant period of chlorination prior to ammonia addition for purposes of meeting CT requirements. Therefore, a more detailed assessment of HAA formation kinetics during chloramination and the impact of prechlorination on these kinetics is needed because of the practical importance of formation kinetics in meeting regulatory limits. The research undertaken on this project was designed to address these issues.

CHAPTER 3
MATERIALS AND METHODS

WATER COLLECTION, TREATMENT, FRACTIONATION, AND STORAGE

Source, Collection, and Storage

Water was collected from five different sources: Lake Austin, Metedeconk River, Charleston, Biscayne Aquifer, and St. Paul. These source waters provided a range of naturally occurring organic matter, bromide, alkalinity, and SUVA, which allowed the study of the significance of source water variability both spatially and temporally on HAA formation.
To ensure consistency of results, large volumes of water were obtained and stored at 4°C until used. Periodic measurements were made of pH, alkalinity, TOC, and UV_{254} to provide information on the stability of water during storage.

Lake Austin

Water was collected from Lake Austin, the drinking water source for the city of Austin, Texas. The water was collected immediately following the raw water intake pumps prior to chlorination at the Ullrich Water Treatment Plant, a chloramination plant that practices softening. 100-L samples were collected in February of 2001 and September of 2001 to provide a typical source and a source at summer algal levels, respectively.

Metedeconk River

Brick Utilities primarily draws its water from the Metedeconk River in New Jersey, which is treated by alum coagulation. Source water samples (30 L) were collected for characterization, treatment, fractionation, and storage. This source water, however, was not stable during long storage periods, and therefore was only stored for up to 1 week at 4 °C.

Charleston

The Hanahan Water Treatment Plant in Charleston, South Carolina primarily draws its water from a surface impoundment. This plant is a conventional coagulation plant with disinfection occurring after filtration in a disinfection contact basin where chlorine is added as the water enters and ammonia as it exits. The plant's treatment train allowed the collection of 30-L samples of both source and filtered water prior to any disinfectant addition.

Biscayne Aquifer

Boca Raton, Florida draws its water from the Biscayne aquifer. The water is lime softened and settled; chlorine is added before the multimedia filters, allowing for 10 – 20 minutes of contact time before ammonia addition at a transfer pump station prior to the clearwells. This treatment train allowed the collection of 30-L samples of both source and settled water without any disinfectant added.

St. Paul

The St. Paul Water Utility draws water from a series of impoundments fed by diverted Mississippi River water and water shed runoff. This plant is a combined lime softening and alum coagulation plant in which chlorine is added three minutes before ammonia after the water is softened/coagulated, flocculated and settled. 30-L samples of source and settled water were collected, again prior to disinfectant addition.

Treatment and Fractionation

Softening

The treatment plants that serve the city of Austin currently add chlorine in the transmission lines from their source water intakes; therefore, the full-scale plants cannot be used as sources of treated water. A pilot plant owned and operated by the City of Austin was run to generate treated water. The pilot plant, located at the City of Austin's Robert C. Ullrich Water Treatment Plant, was supplied continuously with Lake Austin water from the plant's raw water intake. The water underwent softening, sedimentation, and filtration under conditions that simulated full-scale plant performance.

Coagulation

As with Lake Austin, Brick Utilities cannot be used as a source of treated Metedeconk River water because disinfectant is added prior to coagulation. Therefore, source water underwent batch alum coagulation, sedimentation, and filtration at the University of New Hampshire and the University of Texas. Alum coagulation of Metedeconk River source water, performed 8 to 10 L at a time, simulated the conditions and coagulation procedure at Brick Utilities. Sodium bicarbonate was added to the source water before the treatment to allow for a residual alkalinity between 40 and 50 mg/L as $CaCO_3$. The water was rapid mixed at 300 rpm for 30 seconds in a jar test apparatus, then slowly mixed at 50 rpm for 15 minutes, and finally settled for 75 minutes. Coagulant, alum ($Al_2(SO_4)_3$ x $14H_2O$), and sodium bicarbonate were added simultaneously during rapid mixing. The supernatant was filtered through 47-mm diameter, 0.7-µm nominal pore size GF/F glass-fiber filters, (Whatman® International Ltd, Maidstone, England) immediately after settling and stored at 4°C for a maximum of one week before chloramination analyses. Both the alum coagulated and source waters were monitored for alkalinity, pH, UV, turbidity, and DOC.

Biodegradation

Natural organic matter in Metedeconk River source water was biodegraded using an adaptation of the "shaker-batch" method originally developed by Wang and Summers (1995) and Allgeier et al. (1996). The biodegraded water was prepared by allowing source water a 5-day contact time with natural bacteria attached to a sand media. The biofilter, which consisted of PVC columns packed with sand, was kept biologically active by continually recirculating Durham, NH source water in a constant temperature room at 20 °C. The Durham source served as the initial seed as well as the nutrient source. A fresh two-liter sample of Durham source water was obtained weekly to ensure an adequate supply of nutrients to the sand's biofilm. The sand columns were acclimated for three days with Metedeconk source water prior to analysis. The acclimated sand was washed immediately before use with chlorine-free water until the UV absorbance of the rinse water was between 0.02 and 0.05 cm^{-1}. Then, multiple 500 mL-aliquots of the source water sample was combined with 150 g of acclimated washed sand in 1-L amber bottles, placed on an orbital shaker table at 120 rpm, and left in the dark at a constant temperature of 20ºC for five days. At the end of the incubation time, all the aliquots of water sample were combined to generate a biodegraded water fraction of 8 to 10 L. At time zero, two, and five days, small aliquots of water were analyzed for dissolved organic carbon (DOC). The biodegraded DOC was determined by the difference between the initial and the remaining DOC after the 5-day incubation period.

Apparent Molecular Weight and Distribution

Waters were fractionated using an ultrafiltration membrane (Amicon Corp., Danvers, MA) with a molecular weight (MW) cut off of 3000 daltons (Collins et al., 1986). YM/YC type membranes (62 mm diameter) were prepared as recommended by the manufacturer and verified to meet required flux ratings using a nitrogen pressurized, stirred container. Apparent molecular weight (AMW) quantification was made by measuring DOC and UV absorbance at 254 nm from the ultrafilter permeate. In addition to collection of the permeate, DOC and UV absorbance grab samples were collected and monitored with permeate volume to correct for concentration polarization at the membrane surface (Logan and Jiang, 1990). Batch filtration was performed to generate a water fraction of 7 to 9 L.

NOM Hydrophibicity and Fractionation

Adsorption chromatography using XAD-8 and XAD-4 resins (Rohm and Haas, Co., Philadelphia, PA) was used to fractionate source water samples into hydrophobic, transphilic, and hydrophilic fractions (Malcolm 1991, Leenheer 1981, and Croué et al., 1993). Water samples were first fractionated into hydrophobic and trans/hydrophilic fractions using XAD-8 resin. The water sources were first filtered (0.7 µm GF/F, Whatman® International Ltd, Maidstone, England), acidified to pH 2, and then applied to the XAD-8 resin. The hydrophobic fraction of the TOC adsorbs to the XAD-8, while the transphilic and hydrophilic fractions pass through. Subsequent application of trans/hydrophilic fractions to the XAD-4 resin retained the transphilic fraction, while the hydrophilic fraction passed through. The hydrophobic and transphilic fractions were removed from the XAD-8 and XAD-4 resins, respectively, by eluting with 0.1 N NaOH. Periodic mass balances were performed on each resin to account for the entire sample TOC. Batch adsorption chromatography was performed using only XAD-8 resin to generate a hydrophilic and hydrophobic water fraction of 7 to 9 L.

Characterization

The waters studied were characterized by measuring the pH, alkalinity, UV_{254}, turbidity, and bromide concentration. The NOM in these waters was typically characterized by TOC, DOC, hydrophilic DOC, transphilic DOC, hydrophobic DOC, < 3K AMW, and > 3K AMW fractions. The DBPs were characterized by the uniform formation conditions (UFC) HAA_9 and softening uniform formation conditions (SUFC) HAA_9 tests. In addition, specific UV absorbance (SUVA), the ratio of UV_{254} and DOC, and Br^-/DOC ratios were calculated.

BATCH SCREENING EXPERIMENTAL PROCEDURE

Initial Batch Screening Experiments (Batch I)

Two water sources of differing water quality were studied in the initial batch screening experiments. These experiments were conducted on Lake Austin, Texas water and Metedeconk River, New Jersey water. The selected waters were subjected to various treatments and NOM fractionations. The Lake Austin waters tested included "typical" source, softened, the hydrophilic fraction of the source water, and source water at summer algal levels. The Metedeconk River waters tested consist of source, coagulated, < 3K AMW, biodegraded, and the hydrophilic fraction of the source water. The hydrophobic fractions of both waters were not tested in the laboratory, but were determined by the difference between the source and hydrophilic fractions. Similarly, the > 3K AMW fraction of the Metedeconk River water was determined by the difference between the source and < 3K AMW fraction. The Lake Austin and Metedeconk River source, treated, and NOM fractions resulted in the study of 12 different waters. These 12 waters were selected to both simulate the effect of treatment and to study the reactivity of specific NOM fractions.

The key objective of the batch screening experiments was to identify the water quality and chloramination variables that are most important in DXAA formation, which was achieved by testing a series of source waters, treated waters, and NOM fractions as outlined above. The screening approach used a fractional experimental design to systematically evaluate the many factors that may influence DBP formation during chloramination. Each water was exposed to sixteen separate experimental conditions that were used to evaluate the influence of Cl_2/N ratio, pH, temperature, chloramine residual concentration, and bromide concentration on DXAA formation. Sixteen trials for these five factors resulted in an experimental resolution level of three meaning that all the main factors and two-factor interactions can be estimated by analysis of variance (ANOVA) without confounding. A subsequent section of this chapter describes the use of ANOVA in more detail. Two levels of each of the five variables were selected for these experiments (Table 3.1). The levels of the experimental variables were chosen to be representative of current water treatment practices and to provide an adequate range between the two variables for the purposes of observing effects. The resulting experimental design (Table 3.2) allowed the identification of the major water quality and chloramination variables that contribute to DXAA formation.

Table 3.1
Chloramination and water quality variables to be evaluated in the initial batch screening experiments

Factor Levels	Cl_2/N Ratio	pH	Temperature	Chloramine Residual	Bromide Concentration
1	3/1	7	Ambient (~70°F)	2 mg/L	Ambient
2	5/1	9	85°F	4 mg/L	Ambient + 0.5 mg/L

Table 3.2
Summary of experimental variables for fractional factorial batch tests

	Experimental Variables				
Trial No.	pH	Cl_2/N Mass Ratio	Temperature (°F)	Bromide Concentration (mg/L)	Chloramine Residual (mg/L)
1	7	3/1	70	Ambient	2
2	7	3/1	70	Ambient + 0.5	4
3	7	3/1	85	Ambient	4
4	7	3/1	85	Ambient + 0.5	2
5	7	5/1	70	Ambient	4
6	7	5/1	70	Ambient + 0.5	2
7	7	5/1	85	Ambient	2
8	7	5/1	85	Ambient + 0.5	4
9	9	5/1	70	Ambient	4
10	9	5/1	70	Ambient + 0.5	2
11	9	5/1	85	Ambient	2
12	9	5/1	85	Ambient + 0.5	4
13	9	3/1	70	Ambient	2
14	9	3/1	70	Ambient + 0.5	4
15	9	3/1	85	Ambient	4
16	9	3/1	85	Ambient + 0.5	2

Samples were incubated in 250-mL brown, glass bottles capped with a Teflon-lined septa to simulate the formation of DBPs through a treatment plant and distribution system. Prior to dosing with chloramines, the water to be tested was brought to the appropriate temperature and the pH was adjusted with nitric acid or sodium hydroxide. Strong acid or base was used, rather than a buffer, to eliminate any competing reactions. A sequential filling procedure was used in all the batch screening experiments to achieve mixing. First, the bottles were partially filled with sample water, then the appropriate dosing solutions (e.g., chloramines or bromide) were added to the bottles, which were then completely filled and capped head-space-free with Teflon-lined septa. The bottles were then incubated for 48 hours. A standard, 48-hour incubation period was used to match the conditions in the work of Symons et al. (1998), to allow the DXAA formation reactions to proceed further toward completion, and to account for the fact that maximum rather than average distribution system residence times may be of regulatory interest (Diehl et al. 2000). In many experiments, two or three bottles were set up for each experimental condition and dosed with different initial concentrations of chloramines to ensure that at least one bottle was within 10% of the target chloramine residual concentration at the end of the incubation period. Bromide was added using a potassium bromide solution, and chloramines were added in the form of preformed chloramines. At the end of the incubation period, the chloramine residual was measured, and replicate samples were taken for HAA analyses from the bottle meeting the target residual concentration.

Bottles used for sample treatment in the batch screening experiments were prepared by soaking overnight in detergent, rinsing four times with hot tap water, and twice with distilled water. The glassware was then soaked for 24 hours in a 10 - 20 mg/L chlorine solution. Finally, the glassware was rinsed four times with distilled water, then twice with distilled, deionized water produced on a Milli-Q filter apparatus, and baked at 140 °C overnight.

Reagent grade chemicals were used for all solutions, which were mixed with distilled, deionized water produced on a Milli-Q filter apparatus. Before use, this water was determined to be free of chlorine demand.

Follow-up Batch Screening Experiments (Batch II)

Additional batch experiments were undertaken to further study the most important variables that were identified in the initial batch screening experiments. The two factors studied were Br^-/DOC ratio and pH. The Br^-/DOC ratio was selected instead of bromide alone to normalize the DOC concentrations among the waters. Bromide was added, when necessary, using a potassium bromide solution to achieve the desired Br^-/DOC ratio. The experiments were conducted on source and softened waters from Biscayne Aquifer, Florida and St. Paul, Minnesota, as well as Metedeconk River source water. Two experimental designs were created for the follow-up batch screening experiments: one for the source waters and one for the treated waters. The three source waters were exposed to a nine-trial orthogonal array (OA L9). Three levels of each of the two variables (pH and Br^-/DOC ratio) were selected to evaluate non-linear effects of variables (Table 3.3). A summary of the variables held constant in the OA L9 is provided in Table 3.4. Two levels of each of the two variables were selected for the experiments with the treated waters, resulting in the four-trial orthogonal array (OA L4) shown in Table 3.5.

Table 3.3
Summary of experimental variables for the follow-up batch experiments nine trial fractional factorial batch tests (OA L9)

Trial No.	pH	Br^-/DOC
1	7	<15
2	7	30-50
3	7	75-100
4	8	<15
5	8	30-50
6	8	75-100
7	9	<15
8	9	30-50
9	9	75-100

Table 3.4
Summary of constant variables in the follow-up batch experiments

Factor	Level
Cl_2/N (Mass ratio)	4/1
Chloramine Residual (mg/L)	2
Temperature (°F)	70
Incubation Time (hr.)	48

Table 3.5
Summary of experimental variables for the follow-up batch experiments four trial fractional factorial batch tests (OA L4)

Trial No.	pH	Br^-/DOC
1	7	<15
2	7	75-100
3	9	<15
4	9	75-100

The four experimental conditions of the OA L4 for the treated waters were also included in the OA L9 for the source waters, which allowed all water sources, including selected NOM characteristics, to be evaluated simultaneously using the ANOVA. The five NOM characteristics studied were DOC, UV_{254}, SUVA, and both the hydrophobic and hydrophilic fractions. To evaluate the effect the different NOM characteristics had on DXAA formation, the ANOVA was recalculated with the selected NOM characteristics of interest.

A separate set of 32 experiments with Charleston, SC source water was used to assist in modeling efforts. A summary of factors and levels used for the Charleston experiments is listed in Table 3.6 and Table 3.7. Four levels were selected to evaluate the influence of pH, DOC, and bromide concentration, and two levels were selected to evaluate the influence of alkalinity. The levels of the experimental variables were chosen to be representative of current water treatment practices. The ambient bromide concentration of Charleston source water was too high (183 μg/L) to achieve some of the desired Br^-/DOC ratios. Therefore, the organic carbon in the source water was concentrated by ultrafiltration with a YM1 membrane (Amicon Corp., Danvers, MA) characterized by a nominal molecular weight cutoff of 1000 daltons. The membrane was prepared as recommended by the manufacturer and verified to meet required flux ratings using a nitrogen pressurized, stirred container. Three organic carbon-concentrating steps were required to achieve the necessary organic carbon concentration. First, 300 mL of Charleston source water was filtered until approximately 150 mL of organic carbon-concentrated source water remained in the retentate. Second, 150 mL of source water was combined with the retentate and filtered down to 150 mL, and third, the second step was repeated and the DOC concentration was determined. This concentration procedure, which retained about 90% of the organic carbon while allowing the bromide to pass through the membrane, was performed until 1.3 L of carbon-concentrated Charleston source water was obtained. The 16 different Br^-/DOC ratios were prepared by mixing different ratios of Charleston source, organic carbon-concentrated Charleston source, and distilled water. Alkalinity was added using a concentrated sodium bicarbonate solution.

Table 3.6
Summary of experimental variables for the 32 Charleston batch experiments

Factor	Factors			
Level	pH	DOC (mg/L)	Bromide(μg/L)	Alkalinity (mg/L $CaCO_3$)
1	6	3.0	46	27 (Ambient)
2	7	4.98 (Ambient)	383	200
3	8	7.0	183 (Ambient)	N/A
4	9	1.0	583	N/A

N/A = Not Applicable

Table 3.7
Summary of constant variables in the Charleston batch experiments

Factor	Level
Cl_2/N (Mass Ratio)	4/1
Chloramine Residual (mg/L)	2
Temperature (°C)	70

The follow-up batch screening experiments were performed with the same incubation and dosing methodologies used in the initial batch screening experiments.

Kinetic Experiments without Prechorination

The experiments conducted to characterize the DXAA formation kinetics during traditional chloramination (simultaneous ammonia and chlorine addition) are summarized in Table 3.8. Samples were incubated in 250-mL brown glass bottles and capped with Teflon-lined septa to simulate the formation of DBPs through a treatment plant and distribution system. Prior to dosing with chloramines, the water to be tested was brought to the appropriate temperature and the pH was adjusted with nitric acid or sodium hydroxide. The appropriate dosing solutions were added to the bottles, which were capped head-space-free with Teflon-lined septa. All bottles in each experiment received the same initial dose of chloramines. A Cl_2/N mass ratio of 4/1 was used in all experiments, and the dose was selected to provide a target residual concentration of 2 mg/L at 48 hours. At the end of the incubation period, the chloramine residual was measured, and replicate samples were taken for HAA analyses. DXAA measurements were made after incubation times of 0.5, 1, 4, 8, 24, 48, and 72 hours. The effect of pH, bromide concentration, and treatment on DXAA formation kinetics was examined. The focuses of the kinetic experiments for each water are described in Table 3.8.

Table 3.8
Kinetic experiments without prechlorination

Water	pH	Bromide	Focus
Lake Austin source	7	Ambient	pH and
Lake Austin source	9	Ambient	Treatment
Lake Austin softened	9	Ambient	
Metedeconk River source	8	Ambient	Br^- and
Metedeconk River coagulated	8	Ambient	Treatment
Metedeconk River source	8	+0.5 mg/L	
Lake Austin softened	9	Ambient	High
Charleston filtered	8	Ambient	Temperature
Charleston source	8	Ambient	Treatment
Charleston filtered	8	Ambient	
Biscayne Aquifer source	9	Ambient	Treatment
Biscayne Aquifer settled	9	Ambient	

Kinetic Experiments with Prechlorination

Experiments were also conducted to determine the effects of prechlorination on HAA formation kinetics (Table 3.9). The temperature and pH of the waters were adjusted by the same methodology used for the kinetic experiments without prechlorination. A 4.3-L brown glass bottle was filled with sample water, dosed with the appropriate concentration of chlorine dosing solution, and capped head space-free with Teflon-lined septa. The water was incubated under these conditions for a portion of the prechlorination period prior to being transferred to a stainless steel tank with a floating cover and a sampling valve. Replicate samples were taken in 125-mL brown glass bottles prior to the transfer of exactly 4-L to the stainless steel tank. These samples were incubated for the remainder of the specified lag time (5 or 20 minutes) and subsequently analyzed for total chlorine residual, HAA_9 and THMs. The appropriate amount of an ammonium sulfate solution was then added to the tank after the specified lag time to achieve a 4/1 Cl_2/N mass ratio. Samples were collected from the tank in 250-mL brown glass bottles. This experimental setup was used to minimize THM volatilization losses. Because free chlorine was used, both THMs and HAA_9 were measured. THMs were measured immediately after the period of free chlorination and again at 72 hours, and HAA9 was measured after the period of prechlorination, as well as after 0.5, 1, 4, 8, 24, 48, and 72 hours. The initial chlorine dose was selected such that only ammonia had to be added after the period of free chlorination to achieve the target residual combined chlorine concentration of 2 mg/L at 48 hours.

Table 3.9
Kinetic experiments with prechlorination

Water	pH	Bromide	Lag Time
Lake Austin source	9	Ambient	5 min
Lake Austin source	9	Ambient	20 min
Metedeconk River coagulated	8	Ambient	5 min
Metedeconk River coagulated	8	Ambient	20 min
Biscayne Aquifer settled	9	Ambient	5 min
Biscayne Aquifer settled	9	Ambient	20 min

Preformed Chloramines

Preformed chloramines were created by mixing aqueous ammonium sulfate ($(NH_4)_2SO_4$) and sodium hypochlorite (NaOCl) solutions. These solutions were formulated so that approximately equal volumes of the two, when combined, produced the desired Cl_2/N ratio. Both solutions were adjusted to pH 9 with nitric acid and/or sodium hydroxide. The concentration of the chlorine solution was measured according to Standard Method 4500-Cl B (APHA, 1998) prior to creating preformed chloramines, and the volume of ammonium solution added was adjusted to ensure the correct Cl_2/N ratio. The chlorine solution was added slowly to the ammonium solution with constant mixing in an ice bath at 1 °C. After 15 minutes of mixing, the concentration of the chloramine solution was measured, again using Standard Method 4500-Cl B, prior to dosing the samples. Two measurements were made. If these measurements differed by more than 0.1 mg/L a second pair was made. If the second pair of measurements differed by more than 0.1 mg/L the solution was discarded and remixed. An average of the two appropriate measurements was used in the calculations. All solutions were mixed with distilled, deionized water produced on a Milli-Q filter apparatus and were mixed immediately before use and discarded after use. This procedure was used successfully in previous chloramination research (Symons et al., 1998).

Chlorine Dosing Solution

Hypochlorite stock solution was prepared with 4% minimum available chlorine Aldrich reagent grade sodium hypochlorite. To determine the concentration of the NaOCl solution, 2 mL of NaOCl was diluted to 25 mL with distilled, deionized water, and the concentration was measured according to Standard Method 4500-Cl B. The 4% minimum available chlorine reagent grade NaOCl was then diluted to the desired concentration of 5 mg/mL, and its pH was adjusted to pH 9 with nitric acid and/or sodium hydroxide. The chlorine concentration was determined prior to use using Standard Method 4500-Cl B, and the solution was stored at 4 °C until its concentration dropped below 4 mg/mL.

UFC and SUFC

Both the UFC and SUFC tests were used to characterize and compare the HAA_9 formation of the source and treated waters. The conditions of the UFC test (24 ± 1 hr incubation at 20 ± 1 °C and pH 8 ± 0.2 with a chlorine residual of 1 ± 0.4 mg/L) represent the average conditions in US distribution systems. The UFC tests were performed using the method described by Summers et al. (1996). The SUFC test was performed identically except that the pH of combined hypochlorite-buffer dosing solution and sample were adjusted to 9 instead of 8 (Ralls 1999).

ANALYTICAL METHODS

Total and Dissolved Organic Carbon

TOC was determined by UV-light promoted persulfate oxidation with infrared carbon dioxide detection using a Sievers® Model 800 TOC Analyzer with autosampler. The TOC concentrations were measured as NPOC (non-purgeable organic carbon), in which inorganic carbon was removed by acidification with ACS grade phosphoric acid and sparging with prepurified nitrogen (Merriam-Grave Industrial, Springvale, ME).

Dehydrated potassium biphthalate (KHP) was used to prepare a concentrated stock (1000 mg C/L) for preparation of calibration standards in distilled deionized water. Concentrated KHP stock was acidified with 1 mL of phosphoric acid and refrigerated for reuse up to one month. A calibration curve was generated by analyzing standards prepared as 0.2, 1, 5, 8, and 10 mg C/L. Calibration standards were generally prepared fresh for each run; otherwise, they were stored at 4 °C for up to two weeks. When reused, standards were checked and accepted only if within 10% of their original concentration. Samples were preserved with ACS grade phosphoric acid to pH 1-2, stored at 4 °C for up to two weeks, and equilibrated to room temperature prior to analyses.

DOC analysis was identical to TOC analysis except that the samples were filtered with GF/F glass-fiber filters, 47 mm diameter, 0.7 μm nominal pore size (Whatman® International Ltd, Maidstone, England). To minimize leaching of organic carbon into the sample, filters were first ignited at 500 °C for one hour, then placed on a glass vacuum filtration apparatus (Kontes) and rinsed with 750 mL of distilled deionized water. One, 50 mL-aliquot of sample was filtered and discarded prior to filtrate collection.

Glassware for organic carbon analysis was washed with Alconox, rinsed three times with tap water and three times with distilled water and soaked in a 10% nitric acid bath for 2 hours. The glassware was then rinsed 3 times with distilled deionized water. Sample vials were baked at 550 °C for two hours, brown glass bottles were baked overnight at 400 °C, and volumetric glassware was rinsed with reagent grade acetone and allowed to air-dry.

Bromide

Bromide analyses were performed by the Brick Township Municipal Utilities Authority (Brick, NJ) by ion chromatography with conductivity detection in accordance with EPA Method 300.1. Samples were collected, ice-packed, and shipped overnight for analysis.

Residual Analysis

Chloramine residual concentrations were measured by spectrophotometry as total chlorine using Hach DPD (N,N-diethyl-p-phenylenediamine) Total Chlorine Reagent Powder

Pillows with Hach DPD Method 8021 adapted form Standard Method 4500-Cl. Initially, free chlorine was also measured with Hach DPD Free Chlorine Reagent Powder Pillows according to Hach DPD Method 8167, but under the typical drinking treatment conditions (neutral pH and Cl_2/N ratios of 3-5/1) virtually all of the chlorine is present as combined chlorine. Subsequently, free chlorine was not measured.

Standards of approximately 0, 1, 2, 3, 4, and 5 mg/L Cl_2 were analyzed in triplicate prior to sample analysis. The standards were created from the chlorine dosing solution described above. First, the exact concentration of the chlorine dosing solution was determined in accordance with Standard Method 4500-Cl B. Then, a primary dilution standard (PDS) was created by diluting the dosing solution to a concentration of 10 mg/L Cl_2 with distilled, deionized water. Finally, the standards were prepared by diluting the PDS to the desired concentration.

Glassware was prepared as described above in the batch screening experiments, except that glass was allowed to air-dry instead of being baked. All solutions were mixed with distilled, deionized water produced on a Milli-Q filter apparatus.

Disinfection By-Product Analysis

Haloacetic Acids

The nine haloacetic acids (HAA_9) were measured at the University of Texas by liquid-liquid extraction in accordance with USEPA Method 552.2 with modifications (Ralls 1999). Samples were preserved with 4.5 mg of ammonium chloride, producing a concentration of 100 mg/L, and stored at 4 °C for up to two weeks. Prior to analysis, samples were equilibrated to room temperature and weighed. Fifteen mL of sample was removed, and the vial was reweighed, resulting in approximately a 30-mL aliquot with the exact mass determined by the difference in the vial weights. Three grams of copper sulfate, 12 g of sodium sulfate, 1.5 mL of concentrated sulfuric acid, and 3.0 mL of MTBE were then added to the vial. The vial was immediately capped and shaken briefly by hand, then stored on its side to prevent the salts from solidifying. Samples were then placed on a mechanical shaker for 30 minutes. The samples were then allowed to stand quiescently for several minutes to allow separation of the aqueous and organic layers. To convert the HAAs that partitioned into the organic phase to their methyl esters, exactly 2 mL of the ether layer was transferred to a vial containing 2 mL of acidic methanol (10% H_2SO_4/methanol solution) and placed in a 50 °C bath for 3 hours. The vials were then allowed to cool to room temperature and the acidic extract was neutralized by a back extraction with 4 mL of a saturated sodium bicarbonate solution. After the phases separated, a pasteur pipette was used to transfer the extract from the top of the vial to a GC autosampler vial. Extracts were stored at -50 °C for up to 7 days before analysis.

Glassware was washed with Alconox, rinsed three times with tap water and three times with distilled water, and soaked in a 10 % nitric acid bath for 2 hours. The glassware was then rinsed 3 times with distilled deionized water. Sample vials were baked at 550 °C for two hours, brown glass bottles were baked overnight at 400 °C, and volumetric glassware was rinsed with reagent grade acetone and allowed to air-dry.

Gas chromatographic analysis was performed on a Hewlett Packard 6890A Gas Chromatograph with a micro electron capture device (μECD) equipped with an autosampler. A 30 meter J&W DB1701 capillary column was used with hydrogen carrier gas, constant velocity, and splitless injection. Three μL of sample was injected, three injections per standard and two per sample. The initial oven temperature was 40 °C for 2 minutes, followed by a 2.5 °C/minute temperature ramp to 65 °C, followed by a 10 °C/minute temperature ramp to 85 °C, followed by a 20 °C/minute temperature ramp to 205 °C. An injector temperature of 210 °C, and a detector temperature of 290 °C were used.

Aqueous standards were extracted and analyzed in the same manner as the samples to compensate for extraction efficiency. 1,2-DBP was used as the internal standard. The ratio of the area of the peak of interest to the peak of the internal standard was used to quantify concentration.

Two, five-point standard curves were run with every GC run to eliminate the effect of possible equipment operating condition variations. The nominal concentrations of the two standard curves ranged from 0-3 μg/L and 0-20 μg/L. In addition, calibration check standards, duplicate samples, and blanks were analyzed at a frequency of 10% of the analytical load. The calibration check standards were analyzed to ensure that the instrument was still in calibration. Recoveries of the check standards fell between 70% and 130% for all the HAAs, ensuring that the preceding samples were valid.

HAA analyses were performed both at the University Texas and at the Brick Township Municipal Utilities Authority (Brick, NJ). The Brick Utilities laboratory, which has been certified by the New Jersey Department of Environmental Protection for a variety of drinking water and wastewater analyses since 1975, measured HAAs in accordance with USEPA Method 552.2. Split sampling was performed periodically to confirm that both labs were achieving comparable results.

Trihalomethanes

THMs were analyzed according to USEPA Method 551.1. THM samples were dosed with 0.75 g of sample preservative (5 g Na_2HPO_4, 195 g KH_2PO_4, and 1.2 g NH_4Cl) and stored at 4 °C for up to 14 days. Glassware was prepared in accordance with the procedure developed for HAA analysis.

Gas chromatographic analysis was performed on a Hewlett Packard 5890A Gas Chromatograph with an electron capture device (ECD) equipped with an autosampler. A J&W DB-5 capillary column was used with a helium carrier gas, constant velocity, and splitless injection. One μL of sample was injected, three injections per standard and two per sample. The initial oven temperature was 32 °C for 3.5 minutes, followed by a 20 °C/minute temperature ramp to 72 °C, which was held for 2 minutes.

Aqueous standards were extracted and analyzed in the same manner as the samples to compensate for extraction efficiency. TCE was used as the internal standard. The ratio of the area of the peak of interest to the peak of the internal standard was used to quantify concentration.

A five point standard curve was developed for every GC run. The standard curve nominal concentrations ranged from 0-100 μg/L. In addition, calibration check standards, duplicate samples, and blanks were analyzed at a frequency of 10% of the analytical load. The calibration check standards were analyzed to ensure that the instrument was still in calibration. According to USEPA Method 551.1, recoveries of the check standards must fall between 75% and 125% for all the THMs. Recoveries between 95% and 105% were obtained in this research.

Other Analytical Methods

Alkalinity was measured following Standard Method 2320 (APHA, 1998), turbidity was measured using Standard Method 2130 (APHA, 1998), the pH of various solutions and samples was determined using Standard Method 4500 H^+ (APHA, 1998), and the ultra-violet light absorbance was measured at 253.7 nm in accordance with Standard Method 5910 (APHA, 1998).

STATSTICAL ANALYSIS

Analysis of Variance (ANOVA)

Analysis of variance (ANOVA) was used to identify the water quality and chloramination variables that were most important in DXAA formation. This screening approach used a fractional factorial experimental design to systematically evaluate the many factors that may influence DXAA formation during chloramination (*i.e.*, Cl_2/N ratio, pH, temperature, chloramine residual concentration, and bromide concentration). The sixteen trials for these five factors on each water provided an experimental resolution level of three meaning that all main effects and two-factor interactions could be statistically estimated by ANOVA without confounding. The ANOVA highlights the relative significance and contribution of each evaluated factor to DXAA formation.

As mentioned above (Table 3.1), two levels of each of the five variables were selected for these experiments. A summary of the levels of each of the five experimental variables is provided in Table 3.2. The fractional factorial experimental design facilitated statistical assessment and comparison of the numerous factors that influence DXAA formation. The key statistical results were found in the F-statistic and the percent contribution. The former determined the significance of a given parameter (at various levels of confidence), while the latter ranked the parameters according to their contribution to the variation in the DXAA formation data. All main effects (Cl_2/N ratio, pH, temperature, Br^- concentration, and chloramine residual) and two-factor interactions for each water type could be statistically evaluated. Moreover, each water type could be blocked so any desired combination of water type(s) could be statistically compared against other water type(s).

In the follow-up batch screening experiments, two experimental designs were created: one for the source waters and one for the treated waters. Three levels of each of the two variables were selected for the experiments with the raw waters (Table 3.3). All main effects (pH and Br^-/DOC concentration) and the one two-factor interaction (*i.e.*, pH x Br^-/DOC concentration) for each source water type could be statistically evaluated. Two levels of each of the two variables were selected for the experiments with the treated waters (Table 3.5). All main effects (pH and Br^-/DOC concentration) and the one two-factor interaction for each treated water could also be statistically evaluated. As in the initial batch screening experiments, each water type could be blocked so any desired combination of water type(s) could be statistically compared against other water type(s), which translates into comparing the orthogonal arrays of each water type between each other. When doing so, one additional factor becomes incorporated. In the follow-up batch screening experiments, the additional factor could be any one of the five water quality variables measured: DOC, UV, SUVA, hydrophobic fraction, or hydrophilic fraction. The ANOVA was calculated using one of the five water quality variables as the additional factor and then recalculated with the other water quality variables one at a time.

Modeling

The objective of this task was to predict the DXAA concentration from a number of selected water quality characteristics. Numerous analytical measurements of water quality characteristics and DXAA data were gathered throughout the research in order to create an extensive database, which was used for the modeling. The water quality variables were identified as the predictors and the DXAA data were identified as the response. Various regression analyses were created using the program JMP® IN (SAS Institute Inc., 2001). Three models exhibiting the best fit to the complete data set were selected.

SAMPLING OF FULL SCALE PLANTS AND DISTRIBUTION SYSTEMS

Sampling of three treatment plants and distribution systems was undertaken to establish the level of correlation that exists between the relatively simple laboratory tests performed in this research and full-scale system performance. Two of the utilities participating in this evaluation were chloramination-only plants, which limited the geographic diversity for plant selection. In order to meet CT without any free prechlorination period, the options were restricted to utilities located in the southern United States where the climate is warmer. The selected utilities include:

- Manatee
  - The Manatee County water treatment plant in Florida draws source water from the Floridan Aquifer and the Lake Manatee Reservoir. Water is treated in three parallel treatment trains. Two trains treat surface water by alum coagulation, and the third train treats groundwater by lime softening. At the time of sampling, the two surface water trains were treating a combined flow rate of 40 MGD, while the groundwater train was treating 16 MGD. The settled water in the surface water trains was prechlorinated prior to filtration by adding a low dose (approximately 1.5 mg/L) of free chlorine for about 10 minutes; the groundwater was not prechlorinated. In the blend chamber following filtration, ammonia and chlorine are added to achieve a 4/1 Cl_2/N mass ratio. This utility was included in the study because it was challenging to find utilities that were strictly chloraminating and because independent DBP studies conducted by the utility indicated prechlorination contributed less than 10% to the total DBP formation (MacLeod 2003).
- Port Orange, Florida
  - The Charles B. Harnoly Water Treatment Plant in Port Orange, Florida also treats groundwater with conventional lime softening and disinfects with chloramines in the recarbonation chamber prior to filtration by adding ammonia 9.6 seconds before chlorine to achieve approximately a 3/1 Cl_2/N ratio.
- Chalmette, Louisiana
  - The Water Treatment Division of Chalmette, Louisiana primarily draws its water from the Mississippi River. This treatment plant practices both alum coagulation and chloramination. Ammonia is added to the source water 10 feet before chlorine prior to the precipitation upflow basin to achieve a 4/1 Cl_2/N mass ratio.

Three distinct water samples were collected; source water, treated water at the point of chloramination just before disinfectant addition, and distribution system water. Duplicate samples were collected at two different locations of the distribution system and analyzed for DXAA concentration to both confirm and validate the DXAA formation model. Various additional water quality measurements were also made at key sampling points to assist in interpreting the HAA data. The treated water was analyzed for all the water quality parameters present in the model for predicting DXAA formation. These measurements included the chloramine dose, chloramine residual concentration, pH, UV_{254} , DOC, hydrophilic DOC, hydrophobic DOC, and bromide concentration. In addition, the chloramine demand was also determined from these measurements and the Cl_2/N mass ratio was obtained from the treatment plants. The source water was sampled simply to obtain selected basic water quality characteristics as background information; the measurements consisted of pH, UV_{254}, DOC, and alkalinity.

In addition to sampling chloramination-only plants, the Davis Drinking Water Treatment Plant and associated distribution system were also sampled to confirm the HAA formation kinetics measured in the laboratory. This plant chlorinates Lake Austin source water for approximately 6 minutes prior to ammonia addition and softening. Duplicate samples were taken immediately before ammonia addition, after rapid mixing (ammonia and lime addition), after settling, and after filtration. The distribution system was also sampled at three different points that provided approximate residence times of 1.1, 2.4, and 3.1 days. The pH and chloramine residual concentration were measured at the time of sample collection. In addition, source and treated water was collected and characterized by measuring the pH, UV_{254}, DOC concentration, hydrophilic and hydrophobic DOC, and bromide concentration.

CHAPTER 4
BATCH SCREENING EXPERIMENTS

WATER CHARACTERISTICS

General Water Quality Characteristics for Lake Austin and Metedeconk River Waters

The water quality characteristics for Lake Austin and Metedeconk River waters are shown in Table 4.1 and Table 4.2. Lake Austin can be characterized as a low SUVA, high bromide water, while Metedeconk River is characterized as a moderately high SUVA, low bromide water. Over the course of this study, the DOC values were fairly consistent for the Lake Austin source water at around 3.7 mg/L. However, the Metedeconk River source water DOC varied from 2.6 to 4.2 mg/L because of severe drought conditions over the fall of 2001. Metedeconk River source water is more turbid (2.6 NTU) than Lake Austin (0.66 NTU). Another significant difference between source waters is the alkalinity, with Lake Austin having a concentration roughly fifteen times that of Metedeconk River. The NOM characterizations suggested that Lake Austin had a higher apparent molecular weight (AMW) and hydrophilic (XAD-4) fraction compared to Metedeconk River. The reactivity of both source waters was measured by chlorination under uniform formation conditions (UFC tests). Chlorination under these conditions produced 220 µg/L HAA_9 in Metedeconk River source water, whereas Lake Austin source water formed 90 µg/L HAA_9, indicating that the NOM in Metedeconk River water is more reactive with chlorine.

In addition to the study of source waters, several treatments and fractions were also studied. The Lake Austin waters tested included softened water, the hydrophilic fraction of source water, and an additional source water at summer algal levels. Similar water types were tested for Metedeconk River: alum coagulated, biodegraded, hydrophilic fraction of the source, and < 3K AMW fraction of the source water. Treatment had a significant impact on the water quality characteristics. Alum coagulation reduced the DOC concentration of Metedeconk River water, removing about 37%, whereas softening was less effective for Lake Austin, removing only about 11%. Similarly, softening only caused an 18% SUVA reduction in Lake Austin, whereas alum coagulation of Metedeconk River water resulted in a SUVA decrease of 51%. The reactivity of the treated waters due to chloramination was measured using UFC or SUFC tests where appropriate, since Lake Austin water is softened the SUFC test was used, whereas the UFC test was used in the alum coagulated Metedeconk River source water. Treatment reduced the HAA_9 formations in both Lake Austin and Metedeconk River waters. Alum coagulating Metedeconk River source water dramatically decreased HAA_9 formation, forming 65.0 µg/L (30% of that in the source water), whereas the HAA_9 formation in softened Lake Austin source water was 58.1 µg/L (64% of that in the source water).

Table 4.1
Summary of Lake Austin water characteristics

Parameter	Source	Softened	Algal Source	Hydrophilic
pH	8.2	9.7	8.1	ND
TOC (mg/L)	3.87	3.3	ND	ND
DOC (mg/L)	3.74	3.32	3.7	1.67
Hydrophobic DOC (%)	50	49	57	0
Transphilic DOC (%)	17	16	18	ND
Hydrophilic DOC (%)	20	20	17	100
AMW < 3K (mg/L DOC)	1.31	1.26	1.1	1.42
AMW > 3K (mg/L DOC)	2.67	2.06	2.6	0.25
UV_{254}	0.093	0.068	0.095	0.051
Alkalinity (mg $CaCO_3$/L)	148	61	158	ND
UFC HAA_9 (μg/L)	90.2	ND	ND	ND
SUFC HAA_9 (μg/L)	ND	58.1	ND	ND
Turbidity (NTU)	0.66	0.09	0.67	ND
Bromide (μg/L)	168	164	170	168*
SUVA (L/mg-m)	2.49	2.05	2.57	3.05
Ambient Br^-/DOC (μg/mg)	44.9	49.4	45.9	101
Spiked Br^-/DOC (μg/mg)	179	200	181	400

ND = Not determined; * = assumed to be equal to source water

Table 4.2
Summary of Metedeconk River water characteristics

Parameter	Source	Alum Coagulated	Biodegraded	Hydrophilic	AMW < 3K
pH	6.79	6.71	ND	ND	ND
TOC (mg/L)	3.14	ND	ND	ND	ND
DOC (mg/L)	3.05	1.92	2.31	1.27	1.42
Hydrophobic DOC (%)	51	51	51	0	48
Transphilic DOC (%)	15	24	20	ND	21
Hydrophilic DOC (%)	15	20	16	ND	27
AMW < 3K (mg/L DOC)	1.42	1.7	0.71	1.02	ND
AMW > 3K (mg/L DOC)	1.78	0.47	1.6	0.08	ND
BDOC (mg/L)	0.59	0.28	0	ND	ND
UV_{254}	0.107	0.033	0.078	0.044	0.0121
Alkalinity (mg $CaCO_3$/L)	10	46	ND	ND	ND
UFC HAA_9 (μg/L)	220	65.0	ND	ND	ND
Turbidity (NTU)	2.6	0.6	ND	ND	ND
Bromide (μg/L)	27	49	38	27*	44
SUVA (L/mg-m)	3.48	1.72	3.37	3.46	1.69
Ambient Br^-/DOC (μg/mg)	8.85	25.5	16.5	21.3	31
Spiked Br^-/DOC (μg/mg)	173	286	233	415	383

ND = Not determined; * = assumed to be equal to source water

RELATIVE IMPORTANCE OF FACTORS IN DXAA FORMATION

The key objective of the batch screening experiments was the identification of the water quality and chloramination variables that are most important in DXAA formation, which was achieved by testing a series of source waters, treated waters, and NOM fractions. The basis of this screening approach was the use of a fractional factorial experimental design to systematically evaluate the many factors that may influence HAA formation (*i.e.*, Cl_2/N mass ratio, pH, temperature, chloramine residual, and bromide concentration). This approach resulted in sixteen trials for these five factors per water, providing an experimental resolution level of three. Therefore, all the main factors and two factor interactions can be statistically estimated by analysis of variance (ANOVA) without confounding. The ANOVA highlights the relative significance and contribution of each evaluated factor to DXAA formation. Chapter 3 explains the use of ANOVA in more detail.

A summary of the Lake Austin water batch screening experiments is provided in Figure 4.1 and Figure 4.2, which represent the ANOVA results on a mass and molar basis, respectively. The waters screened include source, softened, and source water collected during a period of algal bloom, as well as both hydrophilic and hydrophobic fractions. The hydrophobic fraction was determined by difference in DXAA formation between the source and hydrophilic fraction.

In all the Lake Austin waters studied, pH and bromide concentration were the two most significant contributors to DXAA formation. The contribution of bromide was somewhat less when ANOVA calculations were done on a molar basis instead of a mass basis. This decreased contribution of bromide is to be expected since the bromine-substituted DXAAs are essentially weighted more heavily when the ANOVA calculations are done on a mass basis, whereas the molar calculations weight all three DXAAs equally. The trends and conclusions, however, remain the same for both ways of examining the data. After pH and bromide, Cl_2/N mass ratio was the next most significant contributor to DXAA formation, accounting for approximately 9 (mass concentration based ANOVA) or 12 (molar concentration based ANOVA) percent of the formation. Temperature and chloramine residual also impacted DXAA formation, but were much less influential than the aforementioned factors, generally accounting for 5 percent or less of the DXAA formation. The possibility of interactions between Cl_2/N mass ratio and pH was also investigated; however, in general this parameter contributed minimal significance to DXAA formation.

The percent contribution of the error or unexplained variation of the ANOVA analysis was typically less than 15%, meaning that no significant variables contributing to DXAA formation were omitted, and the experiment was under controlled conditions. Chapter 3 explains the significance of the unexplained variation in ANOVA analyses in more detail. Two of the five Lake Austin waters exhibited errors greater than 15%. Lake Austin source water and hydrophobic fractions had unexplained variations of about 25 and 50 percent, respectively. Since three of the four Lake Austin waters tested maintained errors less than 15%, the higher unexplained variation in the source water (25%) can be attributed to experimental error. The hydrophobic fraction was determined by the difference between Lake Austin source water and the hydrophilic fraction. The relatively large error in this fraction is likely attributed to error accumulated from differencing the source and hydrophilic factions, each of which had errors of approximately 25 and 5 percent, respectively.

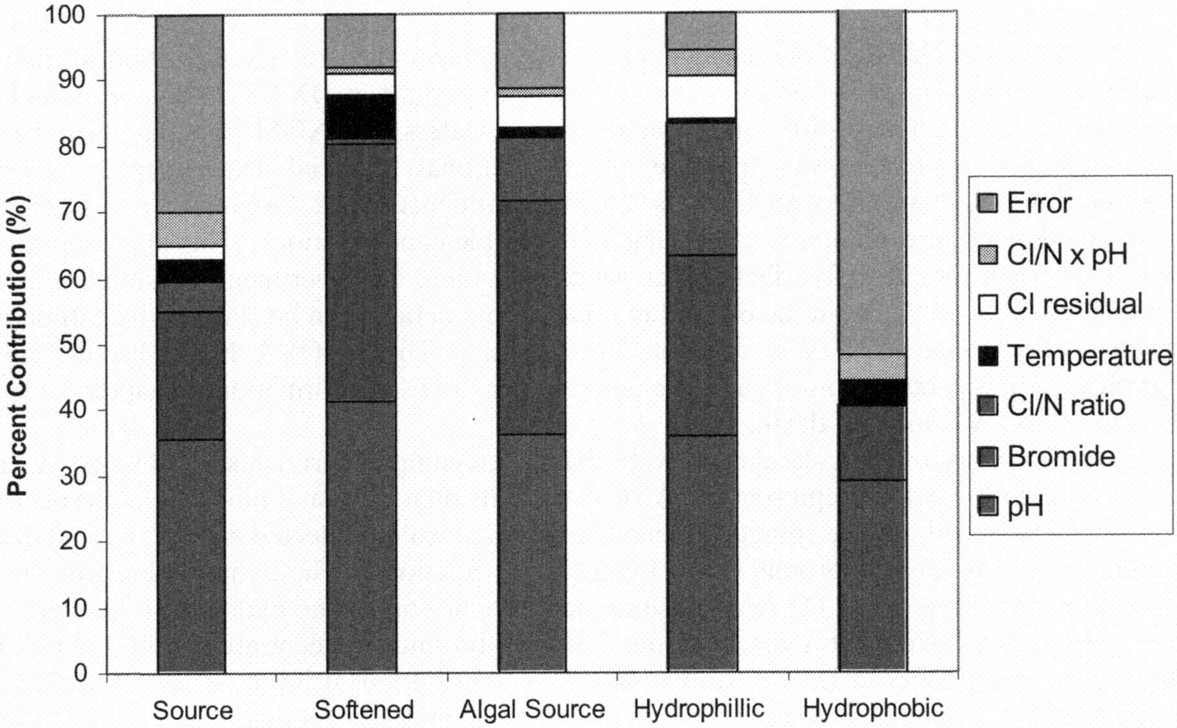

Figure 4.1 ANOVA results for Lake Austin water on a mass basis (µg/L)

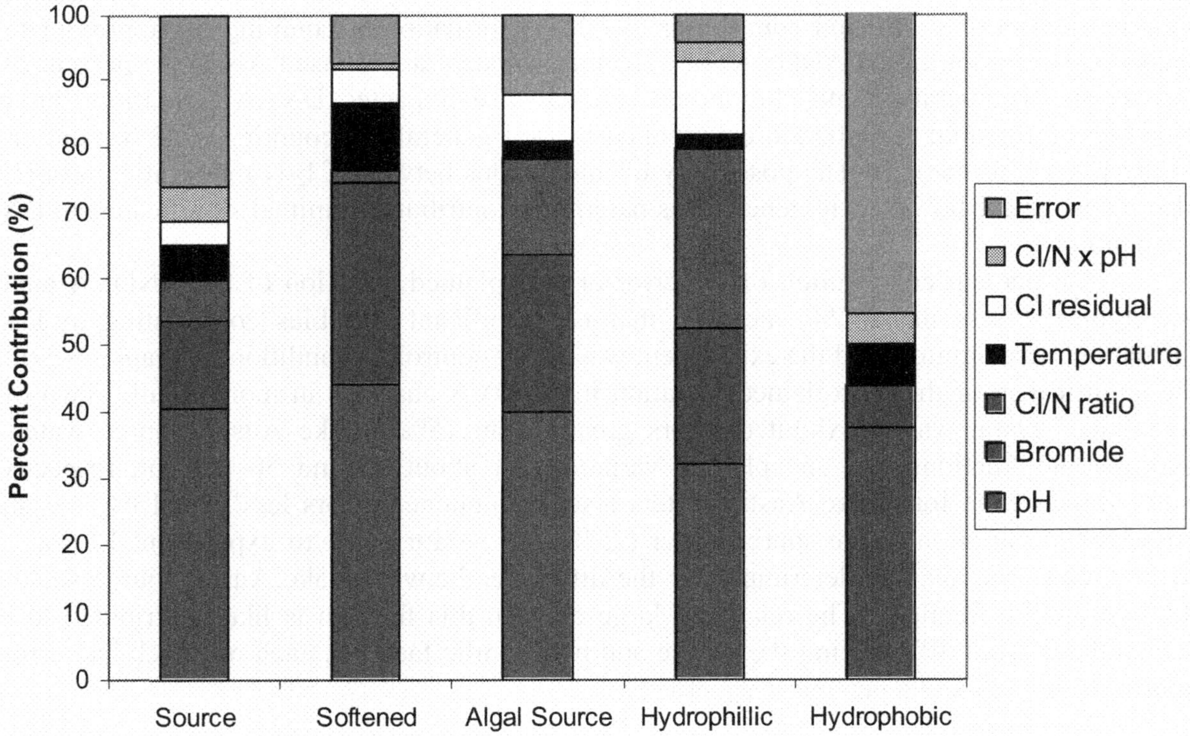

Figure 4.2 ANOVA results for Lake Austin water on a molar basis (µmol/L)

A summary of the Metedeconk River water batch screening experiments is provided in Figure 4.3 and Figure 4.4, which represent the ANOVA results on a mass and molar basis for all the waters screened, including source, alum coagulated, biodegraded, both < 3K AMU and > 3K AMU fractions, as well as both hydrophilic and hydrophobic fractions. The hydrophobic and > 3K AMU fractions were determined by difference in DXAA formation between the source and hydrophilic and source and > 3K AMU fractions, respectively.

In general, the two most significant factors in DXAA formation were again pH and bromide concentration, although temperature was an important factor in the water that underwent biodegradation. As with Lake Austin, the contribution of bromide concentration was somewhat less when the ANOVA was calculated on a molar rather than a mass basis, which in two of the seven waters caused the contribution of Cl_2/N ratio to be greater than that of the bromide concentration. Analogous to Lake Austin water, Cl_2/N mass ratio typically was the most significant contributor to DXAA formation after pH and bromide, on average accounting for about 10 or 13 percent of the DXAA formation on a mass or molar basis, respectively. Unlike Lake Austin water, however, temperature played a larger role in DXAA formation, contributing on average 7 or 10 percent for mass or molar ANOVA based calculations, respectively, and accounting for greater than 18% of the formation in the biodegraded water. Again, as with Lake Austin water, chloramine residual and the interaction between Cl_2/N mass ratio and pH were minimal, generally accounting for 6 percent or less of the DXAA formation.

The unexplained variation of the ANOVA analyses was again controlled satisfactorily, with only two of the experimental data sets having errors greater than 15%. The water that underwent biodegradation had about 25% error and the alum coagulated source water had errors of 27 and 17 percent on a mass and molar basis, respectively. Again, like Lake Austin water, the fractions obtained by difference (>3K AMU and hydrophobic) both had errors greater than 15%; however, these errors were less than those seen in the Lake Austin fractions obtained by difference. Nevertheless, the large errors observed in the fractions obtained by differencing two data sets indicates caution must be exercised in drawing conclusions from them.

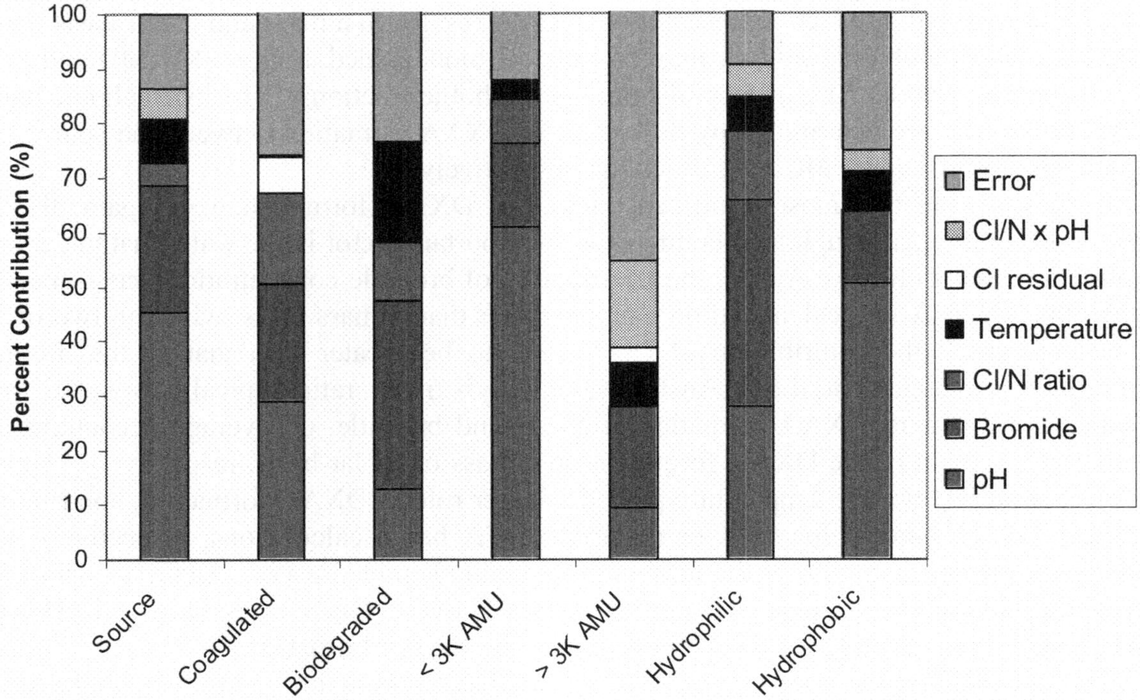

Figure 4.3 ANOVA results for Metedeconk River water on a mass basis (μg/L)

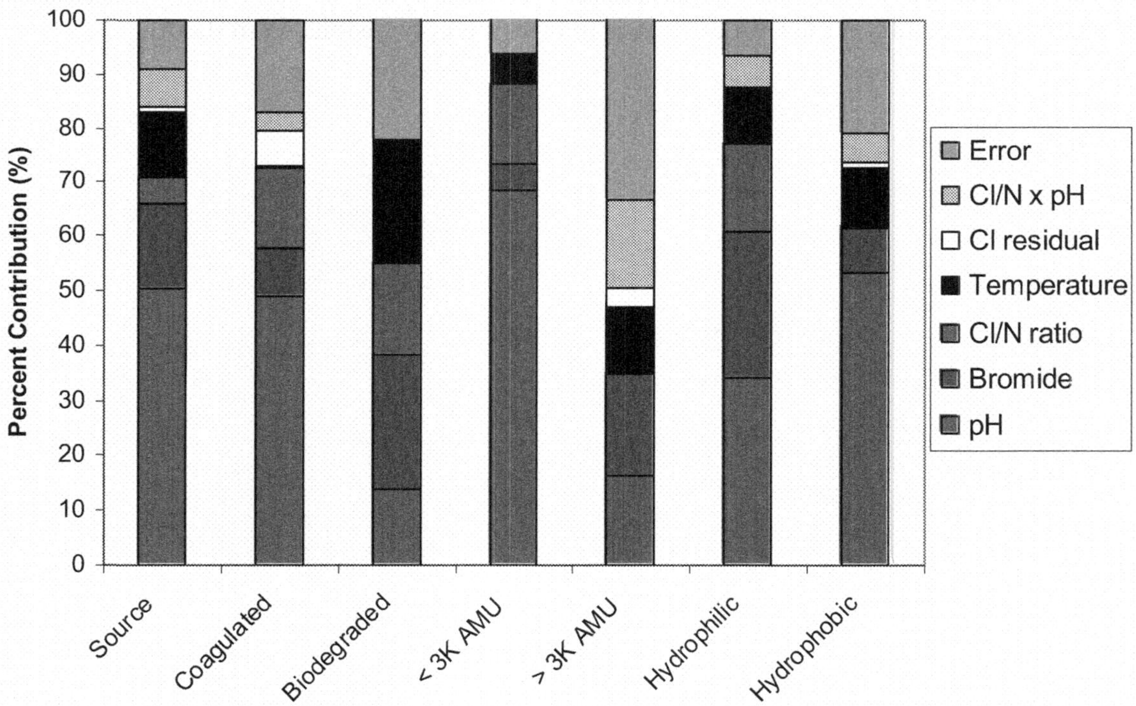

Figure 4.4 ANOVA results for Metedeconk River water on a molar basis (μmol/L)

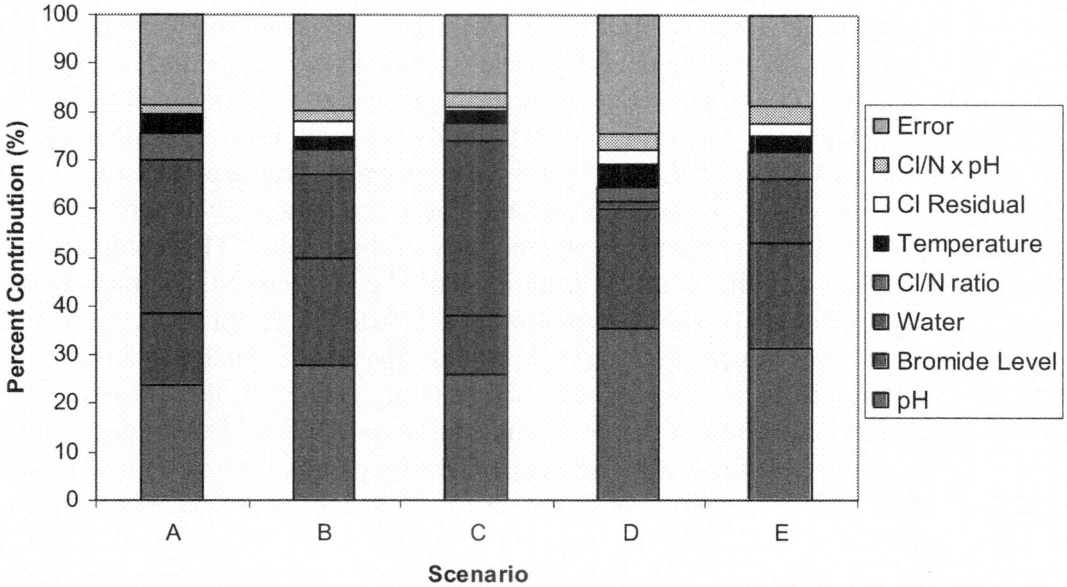

A: Metedeconk River, NJ source, alum coagulated, biodegraded, <3K amu, hydrophilic fraction
B: Lake Austin, TX source, algal source, softened, hydrophilic fraction
C: Metedeconk River, NJ source and alum coagulated
D: Lake Austin, TX source and softened
E: Metedeconk River, NJ source; Lake Austin, TX source and algal source

Figure 4.5 ANOVA results for various combinations of waters on a mass basis

The differences in source waters, treatments, and fractions also have a significant influence on DXAA formation. ANOVA analyses of several scenarios combining the trials of the individual experiments can be considered, as shown in Figure 4.5. Scenario A, showing all the Metedeconk River waters, clearly demonstrates the role of treatment and the different contributions of the various NOM fractions to DXAA formation based on the large contribution from the "waters" factor of the ANOVA analysis. In contrast, the "waters" factor had a small contribution in Lake Austin water (Scenario B), because the water was not fractionated as extensively as Metedeconk River and treatment had a lesser effect. To better characterize the impact of treatment, the various fractions were excluded from the analysis and only the source and treated waters were considered in scenarios C and D. These scenarios illustrate the effect of treatment of Metedeconk River and Lake Austin source waters, respectively. Alum coagulation of Metedeconk River water had a significant impact on DXAA formation, contributing greater than 35%, making it the most important factor. Softening Lake Austin has a minimal influence, however, contributing less than 2%. This relative difference is possibly related to TOC removal, minimal TOC removal (15%) was obtained from softening Lake Austin water, whereas coagulation of Metedeconk River water resulted in a rather large TOC removal (36%). Another aspect of HAA formation is the significance of source water variability (*e.g.*, NOM and SUVA), either spatially or temporally. Scenario E depicts the influence of the differing source water quality characteristics, comparing the Metedeconk river source with both the Lake Austin source waters. The pH and bromide are the two most important variables, with the water characteristics third in importance.

HAA formation is strongly influenced by DOC concentration, which varied from experiment to experiment depending upon the water treatment or fraction tested. To determine if observed differences in DXAA formation were truly related to differing reactivity of water or just a reflection of DOC concentration, DXAA yields were studied. Examination of DXAA yields (μg or μmol DXAA/mg DOC) in essence is a normalization approach to account for the different DOC concentrations. Figure 4.6 presents the same data as in Figure 4.5, except on a mass yield basis, allowing the comparison of reactivity of the waters. This comparison made the effect of treatment and the different contributions of the various NOM fractions to DXAA formation much less influential. The contribution of the "waters" factor of the ANOVA analysis for the Metedeconk River waters (Scenario A) is less than 16% when the influence of DOC concentration is normalized in Figure 4.6, whereas in Figure 5 the "waters" component contributed 32%. Normalization of the DOC concentration (DXAA yield) reduced the "waters" component in the Lake Austin waters (Scenario B) to essentially zero; in Figure 4.5 this component contributed 17%. Scenarios C and D illustrate the impact of treatment on the two waters. The ANOVA results, analyzed on a mass DXAA yield basis, revealed that softening Lake Austin water had essentially no effect on DXAA yield, and similarly, when analyzed on a mass basis alone, the "waters" component contributed approximately 2%. Coagulating Metedeconk River water, however, had a more significant impact. Figure 4.6 (ANOVA calculations analyzed on a mass yield basis) indicates that coagulating Metedeconk River water contributed less than 6%, whereas in Figure 4.5 (ANOVA calculations analyzed on a mass basis) it had a significant impact, accounting for 36% of the ANOVA analysis of DXAA formation. This reduction in significance of the "waters" component between Figure 4.5 and Figure 4.6 can be attributed to DOC removal. As mentioned above, analyzing the results on a mass yield basis normalizes the effect of DOC concentration, allowing the reactivity of the waters to be compared. Treatment of Metedeconk River water, therefore, decreased the reactivity, whereas softening Lake Austin water had essentially no effect. The difference of reactivity of the source waters (Figure 4.6), illustrated in scenario E, is also insignificant (approximately 2%). When analyzed on a mass basis (Figure 4.5), however, the difference in water sources contributed 13%. Again, the difference in significance of the "waters" factor between the two figures can be attributed to normalizing the DOC concentration. This normalization approach indicated that the reactivity of the source waters studied was similar.

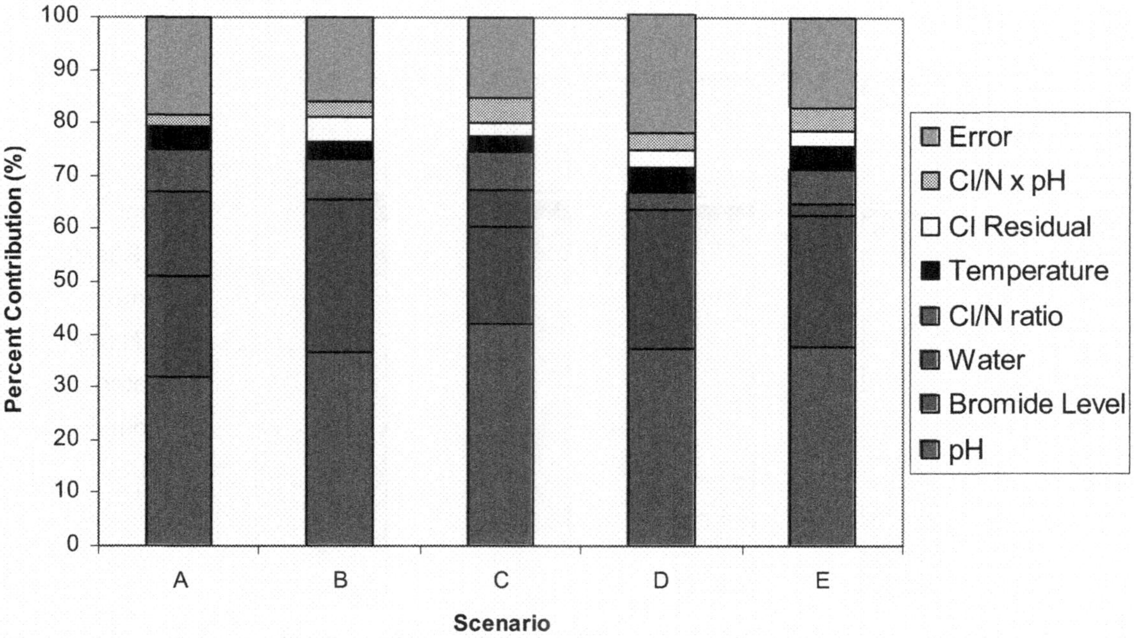

A: Metedeconk River, NJ source, alum coagulated, biodegraded, <3K amu, hydrophilic fraction
B: Lake Austin, TX source, algal source, softened, hydrophilic fraction
C: Metedeconk River, NJ source and alum coagulated
D: Lake Austin, TX source and softened
E: Metedeconk River, NJ source; Lake Austin, TX source and algal source

Figure 4.6 ANOVA results for various combinations of waters on a mass yield basis (μg DXAA/mg DOC)

Lastly, all nine waters (sources, treated, and NOM fractions), from Lake Austin and Metedeconk River were compared using a combined ANOVA (Figure 4.7). The mass and molar concentration basis showed that "water" was the most important parameter in DXAA variation, followed, in order, by pH, bromide concentration, and Cl_2/N ratio. It is important to note, however, that the second level of bromide concentration used in the initial batch screening experiments (ambient plus 0.5 mg/L) was greater than the concentrations typically encountered in practice, which may have artificially increased the influence of bromide concentration. As noted above, the influence of the bromide concentration was higher when DXAA formation was analyzed on a mass concentration basis and lower when analyzed on a molar concentration basis, because of the molecular weight difference between chlorine and bromine. The ANOVA was also performed on a mass and molar yield basis to normalize the effect of the varying DOC concentrations of the different waters, allowing a comparison of the reactivity of the waters. Again, the influence of the "water" parameter was diminished when analyzed on a yield basis indicating the large influence of DOC concentration encompassed within the "water" parameter. The "water" parameter became third in importance on a yield basis but remained more influential than the Cl_2/N ratio, showing that the relative reactivity of NOM in a water source plays a significant role in DXAA formation.

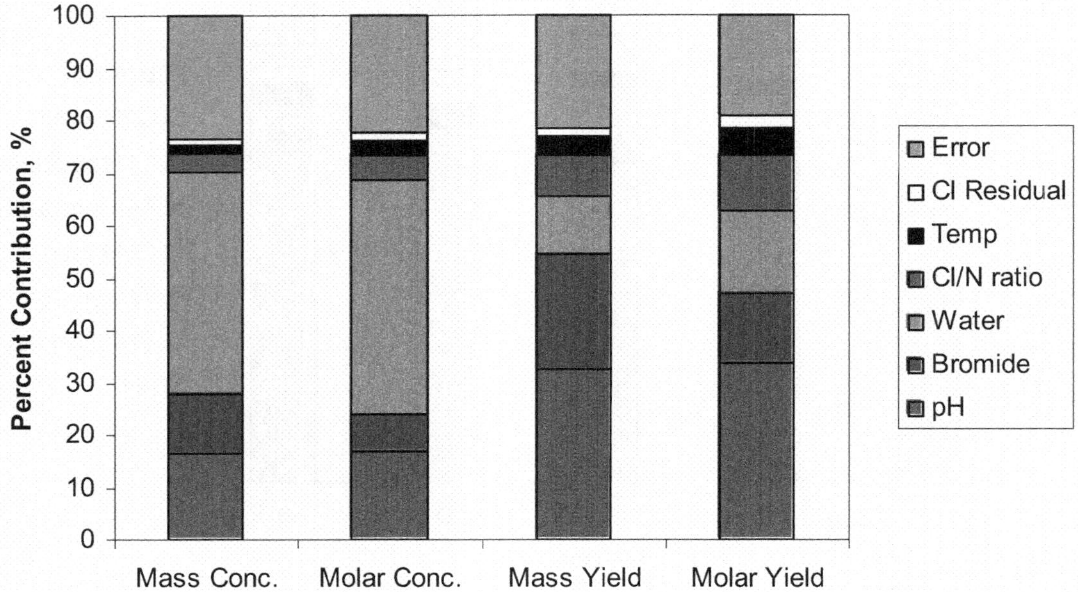

Figure 4.7 ANOVA results comparing the nine waters (sources, treated, and NOM fractions) from Lake Austin and Metedeconk River studied in the initial batch screening experiments using mass concentration, molar concentration, mass yield and molar yield unit basis

ROLE OF pH and BROMIDE CONCENTRATION IN DXAA FORMATION

The ANOVA analysis of the batch screening experiments indicated that pH and bromide were the two most significant variables contributing to DXAA formation. DXAA concentrations can be averaged across all trials of an experiment to show the impact of a selected factor, even as other factors vary. Each bar in Figure 4.8 represents the average of eight data points. This figure shows the effect of pH, demonstrating diminished DXAA formation in all batch screening experiments as the pH increased. These experiments encompass a broad range of different NOM fractions; therefore, the impact of pH appears to be independent of NOM characteristics. The figure also shows the impact of treatment on the two waters tested. The DXAA formation in coagulated Metedeconk River water was 47% of that in the source water, whereas softened Lake Austin water formed 87% of the DXAA formation of the source. As noted in the discussion of the ANOVA results, this relative difference is linked to the degree of DOC removal.

The general trend of increased HAA formation at lower pH indicates that acid catalyzed monochloramine reactions may be responsible. NOM reactions with increased concentrations of one or more of three reactive species (dichloramine, free chlorine, and monochlorammonium ion) are the likely cause of increased DXAA formation as the pH decreased. The overall rate of chloramine decomposition is primarily limited by the rate of formation of dichloramine, which then rapidly decomposes (Jafvert and Valentine 1992). Dichloramine formation occurs in two ways, through monochloramine hydrolysis (Reaction 2.2) and the subsequent monochloramine reaction with free chlorine (Reaction 2.3), and through monochlormine disproportionation (Reaction 2.5). Reaction 2.5 is general acid catalyzed, and a variety of proton donating species (general acid catalysts such as bicarbonate, carbonic acid, phosphate, sulfate, and acetic acid) can act to accelerate monochloramine decay (Vikesland *et al.* 2001).

$NH_2Cl + H_2O \rightarrow NH_3 + HOCl$ (Reaction 2.2)
$NH_2Cl + HOCl \rightarrow NHCl_2 + H_2O$ (Reaction 2.3)
$NH_2Cl + NH_2Cl \rightarrow NHCl_2 + NH_3$ (Reaction 2.5)

Therefore, when the solution pH is decreased the rate of monochloramine loss is increased. This accelerated decay is due to the enhanced rate of dichloramine formation at lower pH. In experiments with Lake Austin and Lake Houston source water, Diehl (2001) observed that the formation of HAAs increased under conditions favorable for the presence of dichloramine. Therefore, dichloramine or its resultant decomposition products (e.g., free chlorine) may be the active halogenating agent. Monochlorammonium ion (Reaction 2.12) is another possible halogenating species.

$NH_2Cl + H^+ \leftrightarrow NH_3Cl^+$ (Reaction 2.12)

This reaction has a pK_a of 1.5, indicating that any reaction involving NH_3Cl^+ in natural waters can be expressed as an acid catalyzed reaction of NH_2Cl (Jafvert and Valentine 1992). Symons et al. (1998) postulated that NH_3Cl^+ may be an active halogenating agent that plays in increasingly important role as the pH decreases. In addition to haloamine speciation, pH can also effect the 3-D structure of NOM molecules. At low pH, the carboxylic and phenolic functional groups of the NOM are neutralized, decreasing the electrostatic repulsion between adjacent functional groups and causing the NOM molecule to coil (Ghosh and Schnitzer 1980; Murphy et al. 1994). This change in structure may influence the reactivity of the NOM.

The presence of bromide increases the number of possible halogenating species. Bousher *et al.* (1989) and Vikesland *et al.* (2001) determined that two rate limiting reactions are important when bromide is oxidized in the presence of monochloramine. Free chlorine produced by monochloramine hydrolysis reacts with bromide to form hypobromous acid (HOBr) (Reaction 2.11), and the oxidation of bromide by the monochloramonium (NH_3Cl^+) ion produces monobromamine (NH_2Br) (Reactions 2.13 and 2.14). During chloramination, bromide oxidation by monochloramonium ion usually dominates (Vikesland *et al.* 2001); therefore, conditions (e.g., lower pH) that promote increased monochlorammonium ion and free chlorine concentrations are likely to also promote the formation of bromine-containing halomamines. Once HOBr and NH_2Br are formed they undergo further rapid reactions (Reactions 2.15 and 2.17) to produce chlorobromamines, primarily NHBrCl, which is a very reactive brominating agent (Valentine 1986; Vikesland 2001).

$HOCl + Br^- \rightarrow HOBr + Cl^-$ (Reaction 2.11)
$NH_3Cl^+ + Br^- \rightarrow NH_3Br^+ + Cl^-$ (Reaction 2.13)
$NH_3Br^+ \leftrightarrow NH_2Br + H^+$ (Reaction 2.14)
$NH_3Br^+ + NH_2Cl \rightarrow NHClBr + NH_4^+$ (Reaction 2.15)
$HOBr + NH_2Cl \rightarrow NHClBr + H_2O$ (Reaction 2.17)

Thus, the presence of bromide substantially increases the range of possible halogenating species that can react with NOM and be incorporated into DBPs.

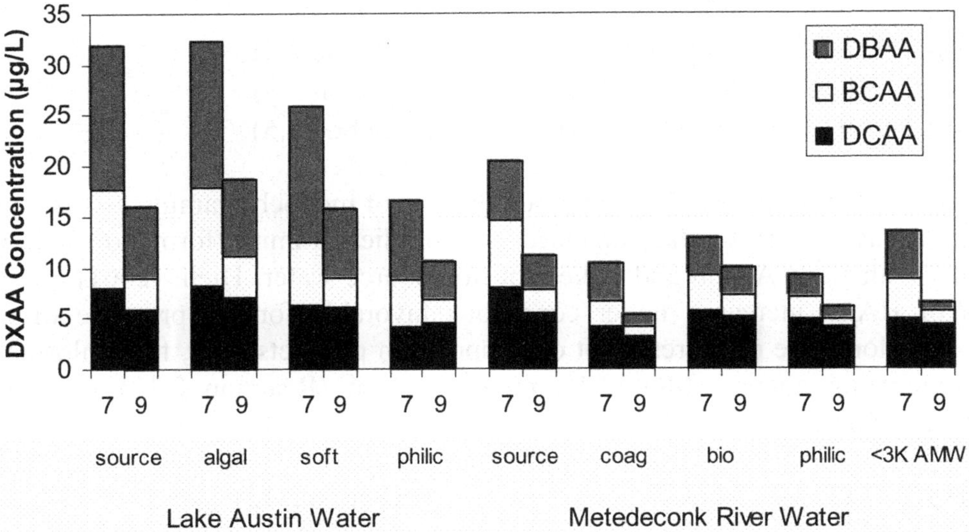

Figure 4.8 Effect of pH on DXAA formation for batch screening experiments

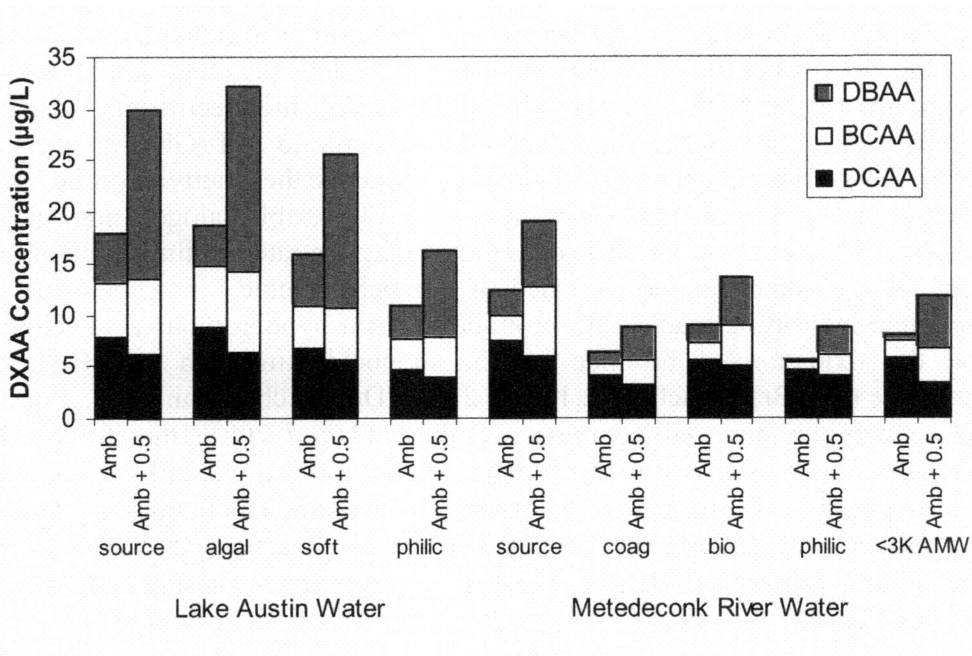

Figure 4.9 Effect of bromide concentration on DXAA formation for batch screening experiments (Amb = ambient bromide concentration; Amb + 0.5 = ambient bromide concentration + 0.5 mg/L Br$^-$)

DXAA concentrations were averaged across all trials of each experiment to discern the influence of bromide, which is shown in Figure 4.9. As in Figure 4.8, each bar represents 8 data points. The addition of bromide caused both an increase in DXAA formation and a shift in speciation to the bromine-substituted species; all waters spiked with bromide formed less DCAA and more BCAA and DBAA than those incubated at ambient levels. Bromide is incorporated into haloamines, forming bromamines and chlorobromamines, and also is present as HOBr. Increases in bromide concentration thus increase the concentration of bromine-substituted reactive species that are available to react with NOM, and thereby increase the concentration of bromine-substituted DXAAs.

Duirk *et al.* (2002b) proposed that NOM has two types of reactive sites, fast and slow. Investigating the formation kinetics of each species (presented in Chapter 6) can explain the decreased DCAA formation at higher bromide concentrations. Briefly, these experiments indicated a fast period of DXAA formation followed by a slow period. During the fast period of DXAA formation, both the bromine-substituted and the chlorine-substituted reactive species may be in competition with a finite number of fast reactive NOM sites. In addition, the bromine-substituted HAAs exhibit faster reaction kinetics than the chlorine-substituted species. The competition for fast NOM reactive sites and the relative kinetics of each species leads to more formation of bromine-substituted HAA species and less formation of chlorine-substituted species.

To better characterize the contribution of pH and bromide on DXAA formation, the data was analyzed on both a mass and molar yield basis. In each experiment, two different pH values and two different bromide concentrations were used, such that the 16 trials of each experiment fell into four groups of four trials each based on the four combinations of these factors in each experiment. DXAA yield was examined across all 16 trials and within the four groups to discern the impact of the major factors on DXAA yield. The character of the natural organic matter (NOM) plays a major role in disinfection by-product (DBP) formation. One of the simplest characteristic of NOM is SUVA, which was constant for all 16 trials in each experiment, but varied from experiment to experiment. Furthermore, the impact of bromide on DBP formation is most commonly characterized not by the bromide concentration, but rather by the bromide to DOC (Br^-/DOC) ratio (μg/mg). The DOC concentration for each experiment was constant, so the two bromide concentrations per experiment resulted in two bromide to DOC ratios. These values are reported in Table 4.1 and Table 4.2.

The mass and molar DXAA yields for the experiments with Lake Austin water are shown in Figure 4.10 and Figure 4.11, respectively. The data in all four experiments show the largest DXAA yield with bromide addition at pH 7 and the smallest DXAA yield with the ambient bromide concentration at pH 9. The trends are the same with both the mass and molar DXAA yields. The DXAA yields only changed to a small extent as a result of softening, indicating that softening did not preferentially remove DXAA precursors.

The mass and molar DXAA yields for the experiments with Metedeconk River water are shown in Figure 4.12 and Figure 4.13, respectively. As with Lake Austin water, bromide addition increased the HAA yield on both a mass and molar basis and the yields were always less at pH 9 than pH 7 for comparable bromide concentrations. Also, alum coagulation was beneficial in decreasing both the mass and molar yields, indicating preferential removal of DXAA precursors, whereas softening of Lake Austin water had little effect on yields. The < 3K AMW fraction had the greatest yield of the Metedeconk waters tested, followed by the hydrophilic fraction, indicating that these fractions of NOM contained more reactive DXAA precursors.

In comparing the two water sources, the main observation is that the DXAA yield on a mass basis was greater with Lake Austin than with Metedeconk River water, even for the source waters spiked with bromide. For the sources spiked with bromide (0.5 mg/L), the bromide to DOC ratios of both were essentially equal (179 μg/mg for Lake Austin versus 173 μg/mg for Metedeconk River). Therefore, the greater DXAA yield in Lake Austin water is unexpected considering the higher SUVA for Metedeconk River. However, the DXAA molar yields are more comparable for the two waters. The discrepancy between the mass and molar DXAA yields is caused by the degree of bromine substitution of the DXAAs. More bromine substitution was observed in Lake Austin water, even when the bromide to DOC ratios were the same for both waters. These results indicate that something in addition to SUVA and bromide to DOC ratio is an important factor in determining DXAA speciation.

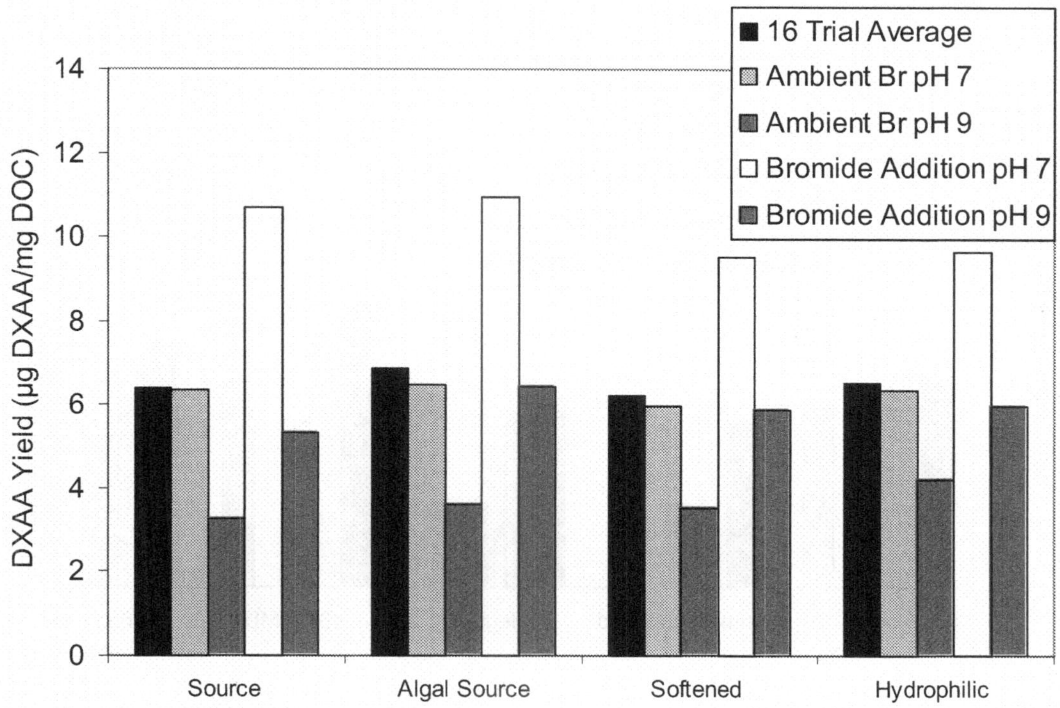

Figure 4.10 Mass DXAA yield in Lake Austin water

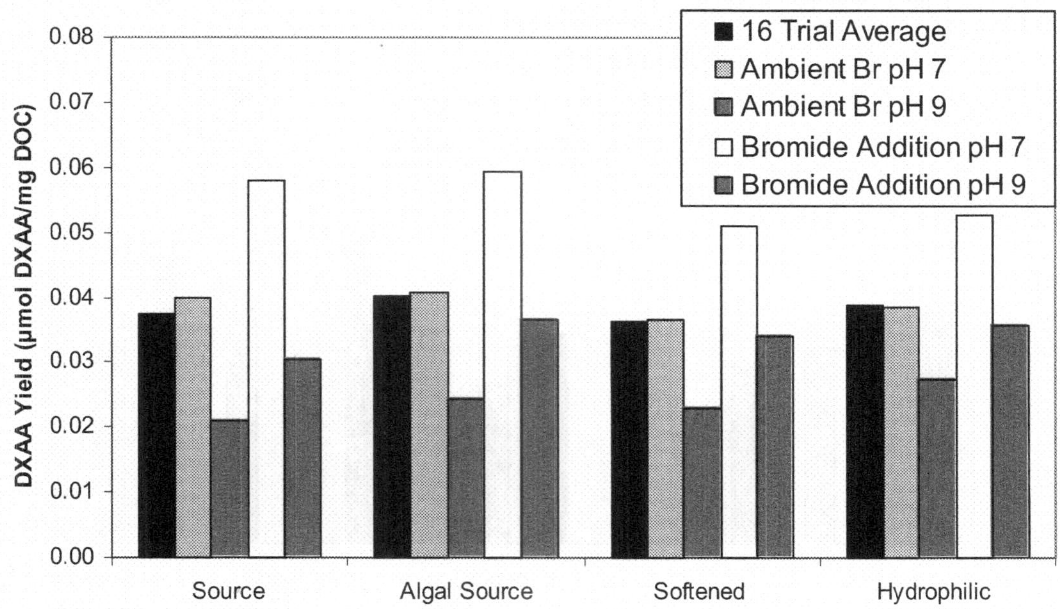

Figure 4.11 Molar DXAA yield in Lake Austin water

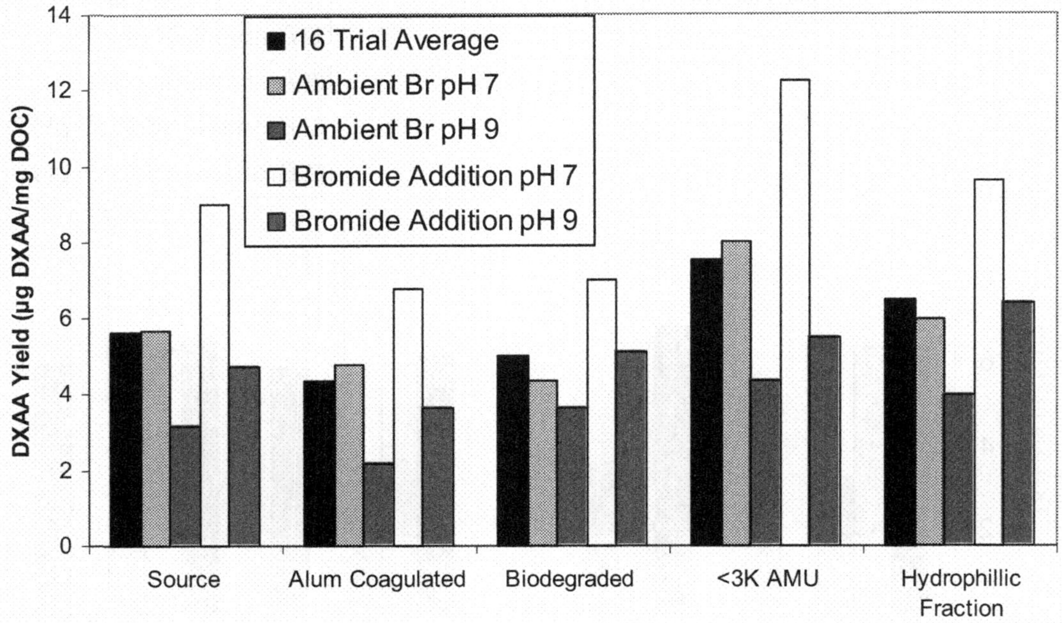

Figure 4.12 Mass DXAA yield in Metedeconk River water

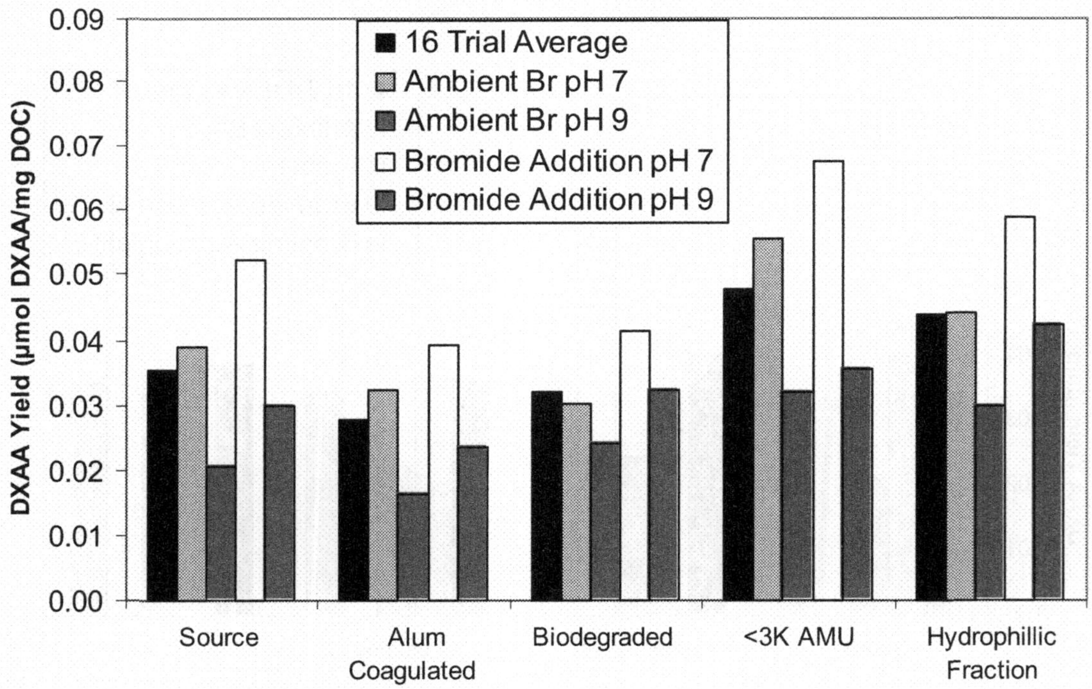

Figure 4.13 Molar DXAA yield in Metedeconk River water

BROMINE INCORPORATION

DXAA speciation also is of interest as the health effects of bromine-substituted HAAs may be of more concern. Previous research (Diehl *et al.* 2000) introduced the factor, n', to describe the degree of bromine substitution of HAAs, which is analogous to the factor, n, used to characterize the degree of bromine substitution of trihalomethanes. In this research, the factor, n_D', is used to characterize the bromine substitution of DXAAs. The value of n_D' can range from zero (all DCAA) to two (all DBAA). Figure 4.14 shows the values of n_D' for the experiments with Lake Austin water. As expected, n_D' was larger for the trials with bromide addition. The larger bromide to DOC ratio promoted bromine substitution since the presence of bromide substantially increases the range of possible halogenating species that can react with NOM and be incorporated into DBPs. As discussed above, increasing bromide concentrations lead to more bromine-substituted reactive species that are available to react with NOM, leading to increased formation of bromine-substituted DXAAs. Cowman and Singer (1996) showed that the formation of bromine-substituted HAAs during chlorination, most notably DBAA, BrClAA, $BrCl_2AA$, and Br_2ClAA, was enhanced by lowering the pH from 8 to 6. Similar trends occur during chloramination. Higher relative concentrations of bromine-substituted species tend to form at lower pH, as n_D' was somewhat larger at pH 7 than at pH 9. Also, softening had very little effect on DXAA speciation relative to that seen in the source water.

Figure 4.15 shows the values of n_D' for the experiments with Metedeconk River water. As with Lake Austin water, the addition of bromide caused greater formation of the bromine-substituted DXAAs. The degree of bromine substitution at pH 9 was less than at pH 7 for all the Metedeconk waters studied except for two instances, the source water and biodegraded water at the ambient bromide concentration. However, when these two waters were spiked with 0.5 mg/L of bromide the n_D' at pH 9 was less than the n_D' at pH 7. As noted previously, n_D' for Metedeconk River water was less than that observed for Lake Austin water, even when the bromide to DOC ratio in Metedeconk River equaled or exceeded the value in the comparable Lake Austin water. For example, the bromide-spiked Lake Austin source water had a bromide to DOC ratio of 179 μg/mg and an n_D' of about 1.2 at pH 7 and 1.1 at pH 9, while the Metedeconk River source water had a bromide to DOC ratio of 173 μg/mg and an n_D' of about 0.9 at pH 7 and 0.7 at pH 9. Similarly, the bromide-spiked coagulated Metedeconk River water had a larger bromide to DOC ratio than bromide-spiked softened Lake Austin water, but the n_D' values were again smaller than the Lake Austin values.

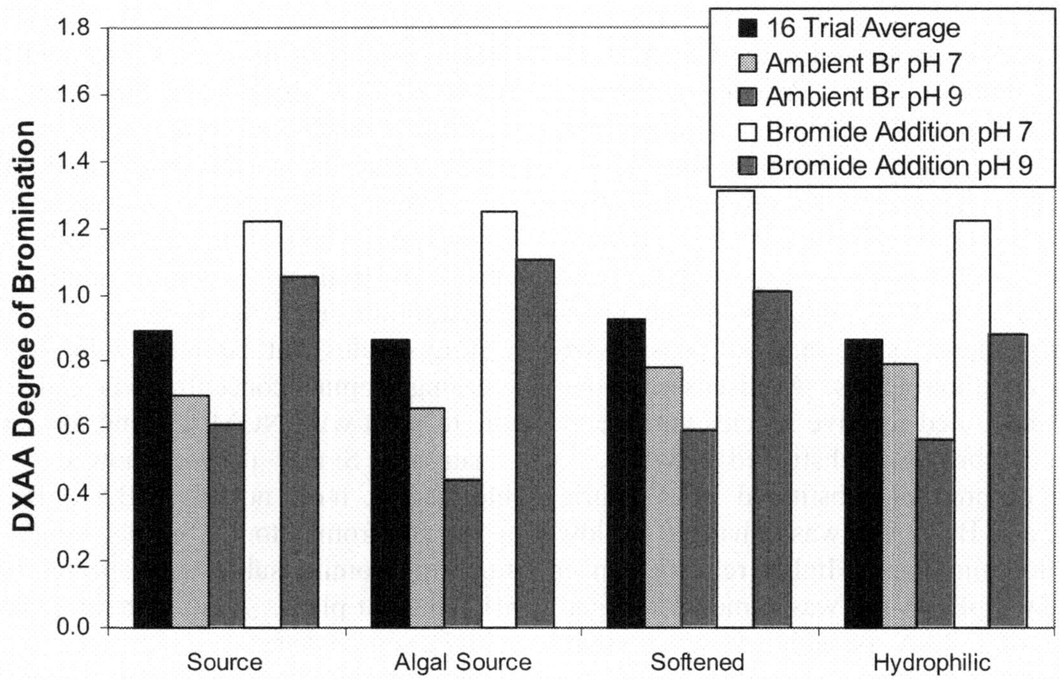

Figure 4.14 Degree of bromine substitution formed in Lake Austin water

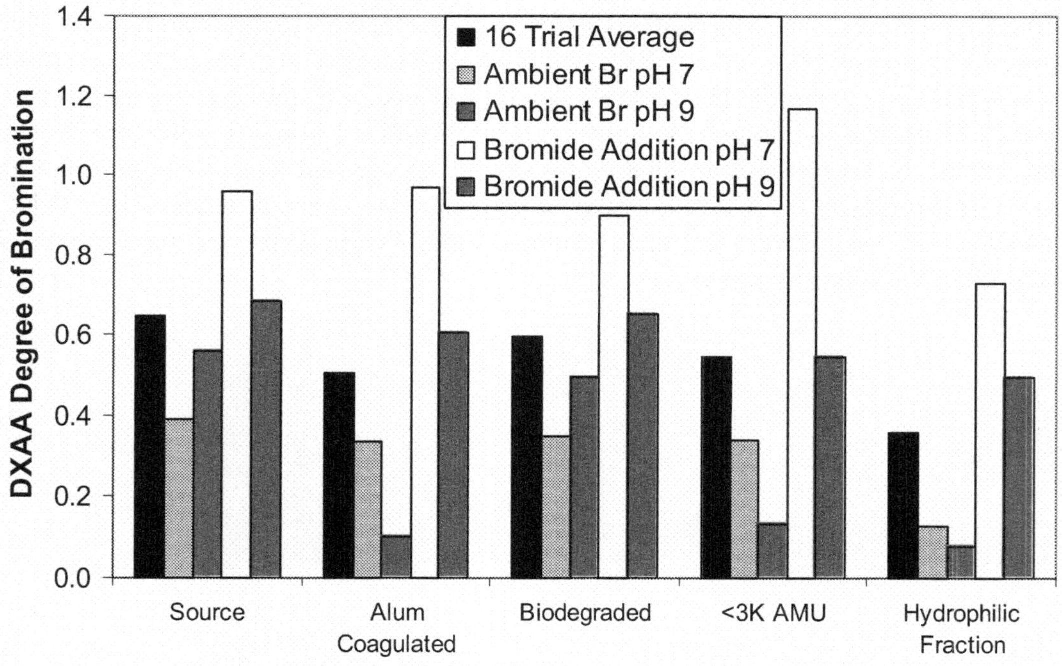

Figure 4.15 Degree of bromine substitution formed in Metedeconk River water

As mentioned above, bromide may be incorporated into haloamines such as bromamines and chlorobromamines, and may also be present as HOBr. Increasing bromide concentrations increase the bromine-substituted reactive species that are available to react with NOM. The bromine-substituted and the chlorine-substituted reactive species may compete for NOM reactive sites. These species may react with NOM and form both bromine-substituted and chlorine-substituted HAAs. To further evaluate bromine-substituted DXAAs, the effect of both Br^-/average Cl^+ molar ratios and Br^-/DOC ratios were investigated for the Task 1 data. Symons et al. (1993) investigated the influence of these two ratios on bromine substitution during chlorination; higher Br^-/DOC and Br^-/average Cl^+ ratios led to greater bromine incorporation into DBPs. Figure 4.16 illustrates the influence of both the Br^-/DOC ratios and Br^-/average Cl^+ molar ratios, as well as pH, on the degree of bromine incorporation into DXAAs. At lower Br^-/average Cl^+ molar ratios, increases in Br^-/average Cl^+ caused an increase in DXAA bromine incorporation. However, at higher Br^-/average Cl^+ molar ratios, there was no discernable impact. The Br^-/DOC ratio had a similar effect on bromine incorporation, and as before, higher concentrations of bromine-substituted species formed at lower pH. Also, at similar Br^-/DOC ratios and Br^-/average Cl^+ molar ratios Lake Austin waters tended to exhibit more bromine incorporation into DXAAs than Metedeconk River waters. The effect of chloramine residual on n_D' is also noted. Higher chloramine residuals, and therefore lower Br^-/average Cl^+ molar ratios exhibited similar n_D's for each water tested. Therefore, Br^-/DOC ratio may be a more accurate way to predict the extent of bromine-substitution of HAAs, especially at lower Br^-/DOC ratios.

During chloramination, bromide present may be oxidized by the monochloramonium ion to produce monobromamine, which undergoes further rapid reactions to produce chlorobromamines, primarily chlorobromamine (Valentine 1986; Vikesland et al. 2001). The reaction that produces the monochloramonuim ion (Reaction 2.12) is general acid catalyzed and has a pK_a of 1.5, indicating that at low pH and in the presence of proton donating species (such as bicarbonate, carbonic acid, phosphate, sulfate, and acetic acid) more monochloramonium ion would be available to drive the aforementioned bromine-substitution reactions. In addition, conditions that favor chloramine decomposition, and therefore the largest concentrations of free chlorine (high Cl_2/N ratios and low pH) should promote the largest formation of free and combined bromine (Symons et al. (1998). These pH dependencies may explain the increased bromine incorporation at lower pH.

Jafvert and Valentine (1992) observed that the rate of dichloramine decomposition accelerated in the presence of carbonate. Metedeconk River source water has a lower alkalinity (10 mg $CaCO_3$/L) than Lake Austin source water (148 mg $CaCO_3$/L). Symons et al. (1998) postulated that low alkalinity may account for greater stability of dichloramines which would lead to less decomposition and less free chlorine production, a decomposition product. In turn, less free chlorine would be available to drive bromination reactions through the conversion of bromide to hypobromous acid. In addition, the lower alkalinity would lead to less production of the monochloramonium ion, which drives the bromine-substitution reactions that lead to the formation of chlorobromamine.

Chlorine-substituted and bromine-substituted halogenating agents are in competition for NOM reactive sites. Valentine (1986) concluded that the bromine atom of chlorobromamine is very labile and reactive and in reactions with organic chemicals it may be an active halogenating agent. The bromine-substituted reactive species, therefore, have a relatively higher reactivity than their chlorine-substituted counterparts. Consequently, it is expected that bromine-substituted reactive species will out compete the chlorine-substituted reactive species for the available NOM reactive sites. Increasing Br^-/DOC ratios resulted in increased bromine incorporation; however, each water studied achieved a maximum amount of bromine incorporation. After this maximum was achieved, increased Br^-/DOC ratios did not result in an increased n_D' (Figure 4.16). Investigation of HAA formation kinetics (Chapter 6) reveals that NOM has fast and slow reactive sites. Therefore, these results imply that beyond a certain Br^-/DOC ratio the bromine-substituted reactive species have achieved the maximum possible kinetic advantage in reacting with fast reactive NOM sites.

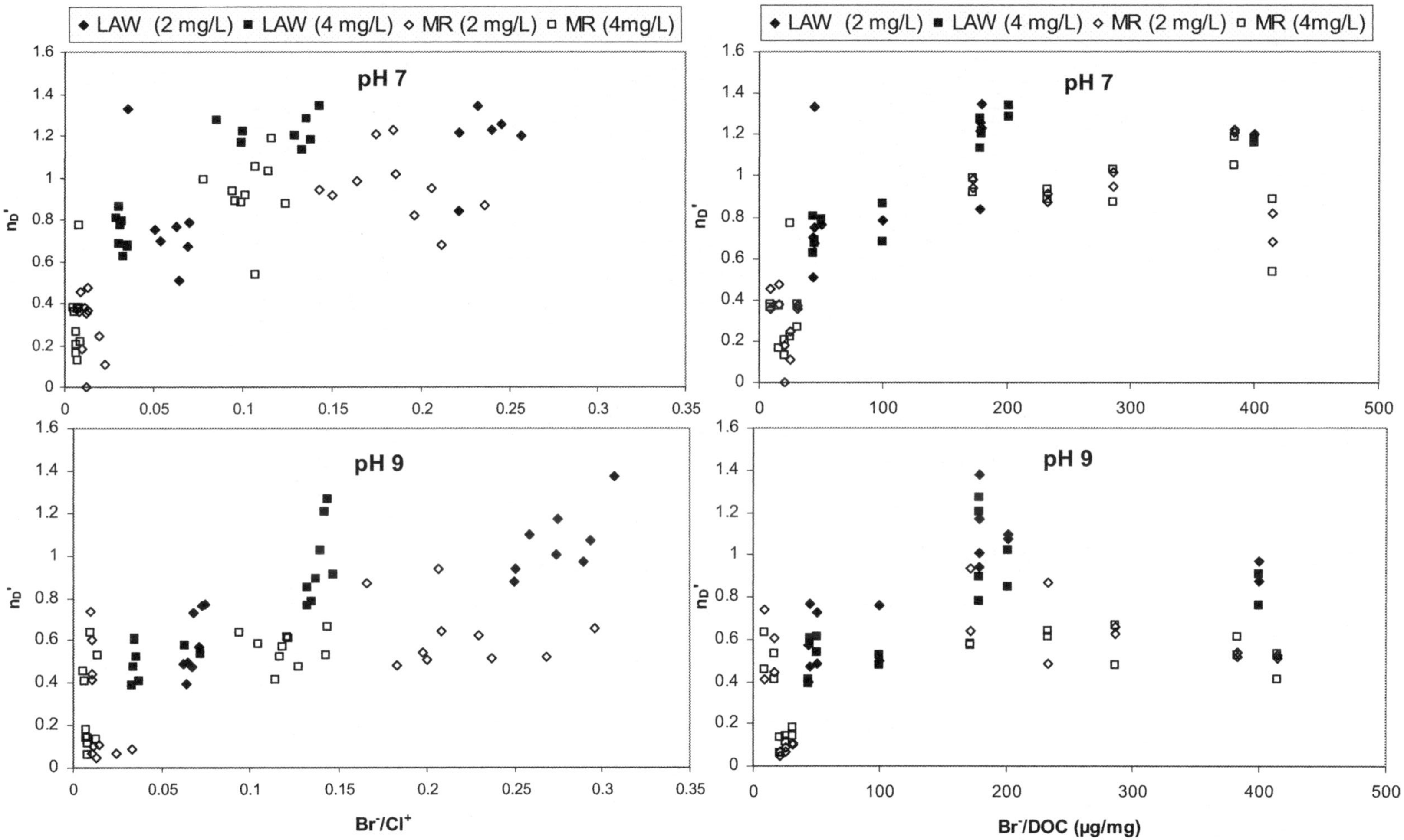

Figure 4.16 Effect of bromine to average chloramine concentration ratio and bromide to DOC ratio on DXAA degree of bromine-substitution for all batch screening experiments (parentheses show chloramine residual at 48 hours)

OTHER FACTORS

The ANOVA analysis of the batch screening experiments indicated that pH and bromide were the two most significant variables contributing to DXAA formation. Although Cl_2/N mass ratio, temperature, and chloramine residual are less influential, they still have noticeable impacts.

Chlorine to Ammonia Mass Ratio

ANOVA results indicate that the chlorine to ammonia mass ratio influences DXAA formation; however, it is less influential than pH and bromide concentrations. DXAA concentrations were averaged across all 16 trials of each batch screening experiment to show the impact of a selected factor, even as other factors may vary. Two Cl_2/N mass ratios (3/1 and 5/1) were studied in the batch screening experiments, and Figure 4.17 illustrated that the DXAA concentration decreased as the Cl_2/N mass ratio decreased. Symons et al. (1998) and Diehl et al. (2000) have also observed this trend in HAA formation. Chloramines autodecompose through a complex set of reactions (Chapter 2). Under drinking water treatment conditions, larger Cl_2/N ratios promote more rapid decomposition, resulting in the formation of dichloramine or its resultant decomposition products (i.e. free chlorine), which may be active halogenating agents that promote HAA formation.

Temperature

Most of the previous HAA formation studies during chloramination have been done at constant temperature. However, it is reasonable to expect HAA concentrations to increase with increasing temperature because of faster reaction kinetics. The batch screening experiments incorporated temperatures of 70 and 85 °F to study the impact of elevated summer water temperatures on HAA formation. As above, DXAA concentrations were averaged across all trials of each experiment to show the impact of temperature (Figure 4.18), and as expected, waters incubated at 85 °F exhibited higher DXAA concentrations than those at 70 °F.

Chloramine Residual Concentration

The ANOVA analysis of the batch screening experiments indicated that the chloramine residual was of minimal significance to DXAA formation. Chloramine residuals of 2 and 4 mg/L were studied, and again the DXAA concentrations were averaged across all trials of each experiment. Figure 4.19 indicates that increasing the chloramine residual concentration caused an increase in DXAA formation in Lake Austin water, but had a minimal impact on DXAA concentrations in Metedeconk River.

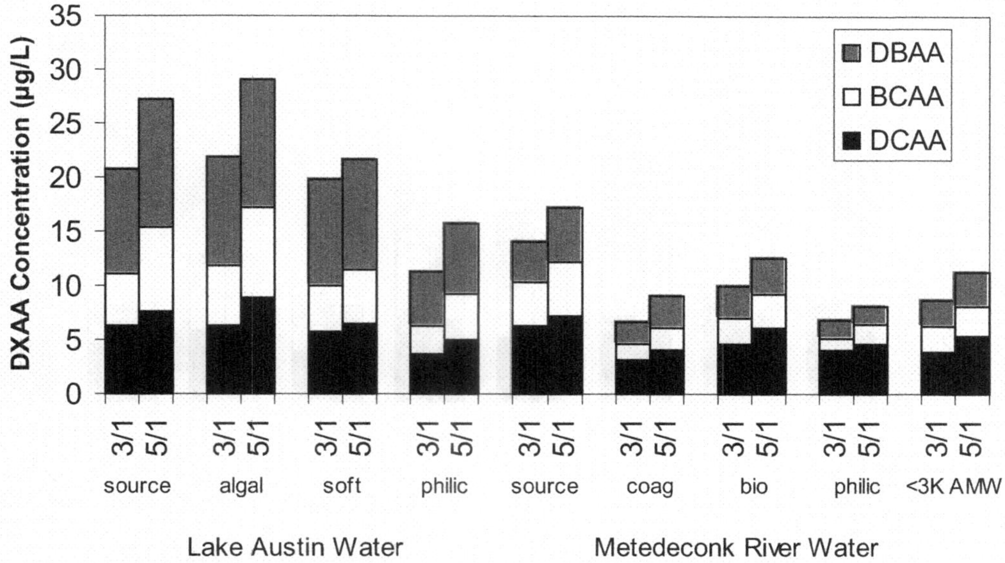

Figure 4.17 Effect of chlorine to ammonia mass ratio on DXAA formation for batch screening experiments

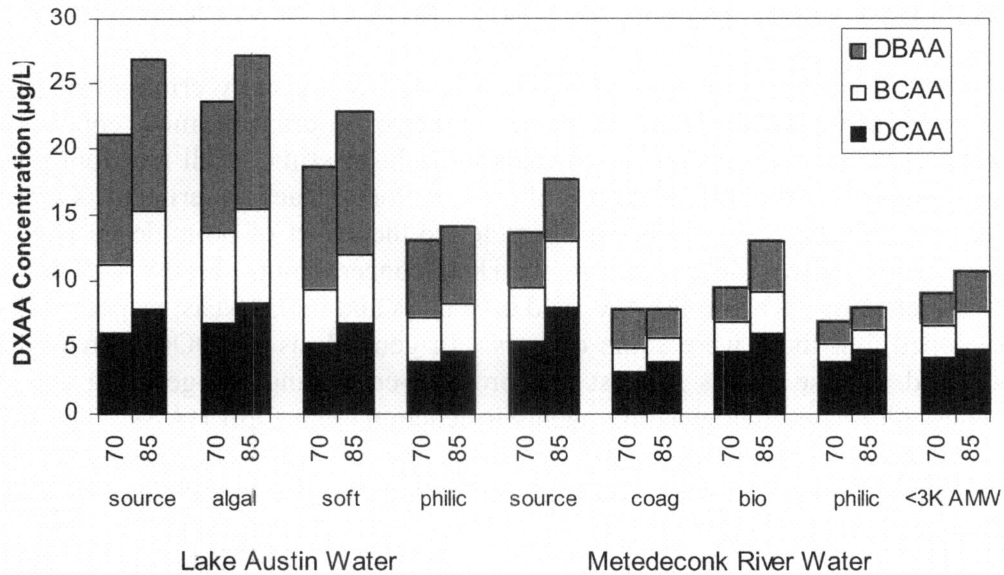

Figure 4.18 Effect of temperature (°F) on DXAA formation for batch screening experiments

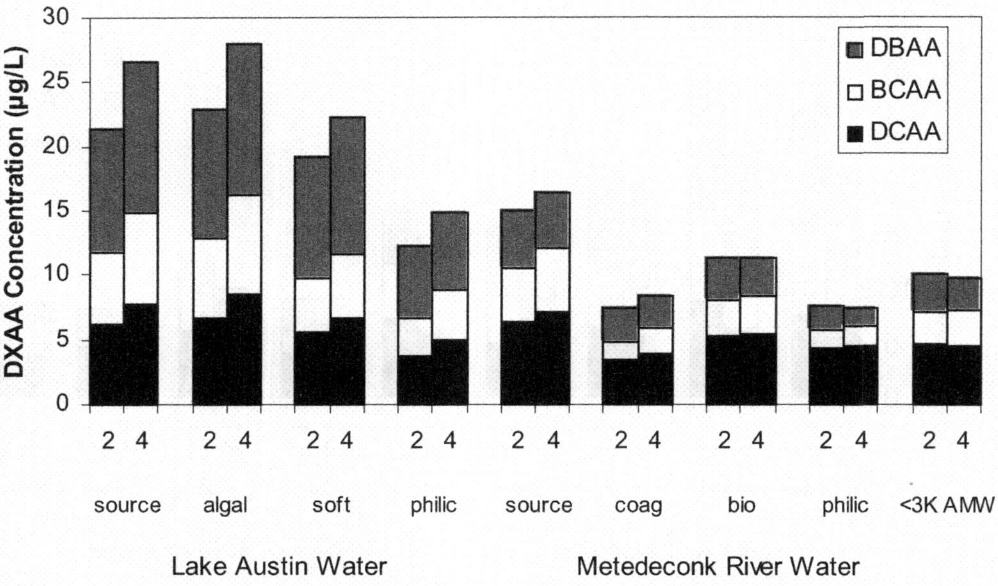

Figure 4.19 Effect of chloramine residual (mg Cl_2/L) on DXAA formation for batch screening experiments

TRENDS IN DXAA FORMATION AS A FUNCTION OF WATER QUALITY

Predicting DXAA formation based upon basic water quality criteria can be useful. As discussed previously, two different pH values and two different bromide concentrations were used in each experiment such that the 16 trials in each experiment fell into four groups of four trials each based on the four combinations of these factors in each experiment. General trends in these data were investigated. To better characterize the effect of bromide on DXAA yield the bromide concentration was normalized by the DOC concentration. Figure 4.20 and Figure 4.21 depict the effect of Br^-/DOC on DXAA yield on a mass and molar basis, respectively. A general trend developed, but there were some outliers. In general, as Br^-/DOC increased the DXAA yield increased. These results suggest that bromine-containing halogenating agents are more reactive than chlorinated halogenating agents in reactions that form HAAs. The pH effects on DXAA yield are also apparent, as pH decreased the DXAA yield increased. To be more chemically rigorous, the data were also analyzed on a molar basis; however, the trend remained about the same.

Because Br^-/DOC ratio does not take into account the reactivity of the NOM, the influence of Br^- x SUVA on DXAA yield was investigated. This approach normalized the product of UV_{254} and bromide with DOC. Because a correlation exists between the UV absorbance and the aromatic carbon content of a sample, the UV absorbance at a wavelength of 254 nm (UV_{254}) can be used to characterize NOM reactivity. As discussed above, Figure 4.20 presents a trend between the bromide concentration normalized with the DOC concentration and DXAA yield. These two water quality characteristics were combined (the product of Br^- and SUVA) to determine their effect on DXAA formation. Figure 4.22 shows the influence of Br^- x SUVA on mass DXAA yield. This approach resulted in less scatter of the data, but still some outliers existed. An increasing Br^- x SUVA value correlated with increasing DXAA formation. The dramatic impacts of pH can also be seen, and the effect of Br^- x SUVA on molar DXAA yield was essentially the same (data not shown).

The impact of Br^-/DOC on each of the three DXAA species (DCAA, BCAA, and DBAA) at pH 7 and 9 is illustrated in Figure 4.23 and Figure 4.24, respectively. This analysis provides a rough estimate of DXAA formation as well as showing the expected effect of increasing Br^- concentration on speciation. In general, increasing the Br^- concentration resulted in greater DBAA formation and less DCAA formation, thus shifting the speciation to the more bromine-substituted DXAAs.

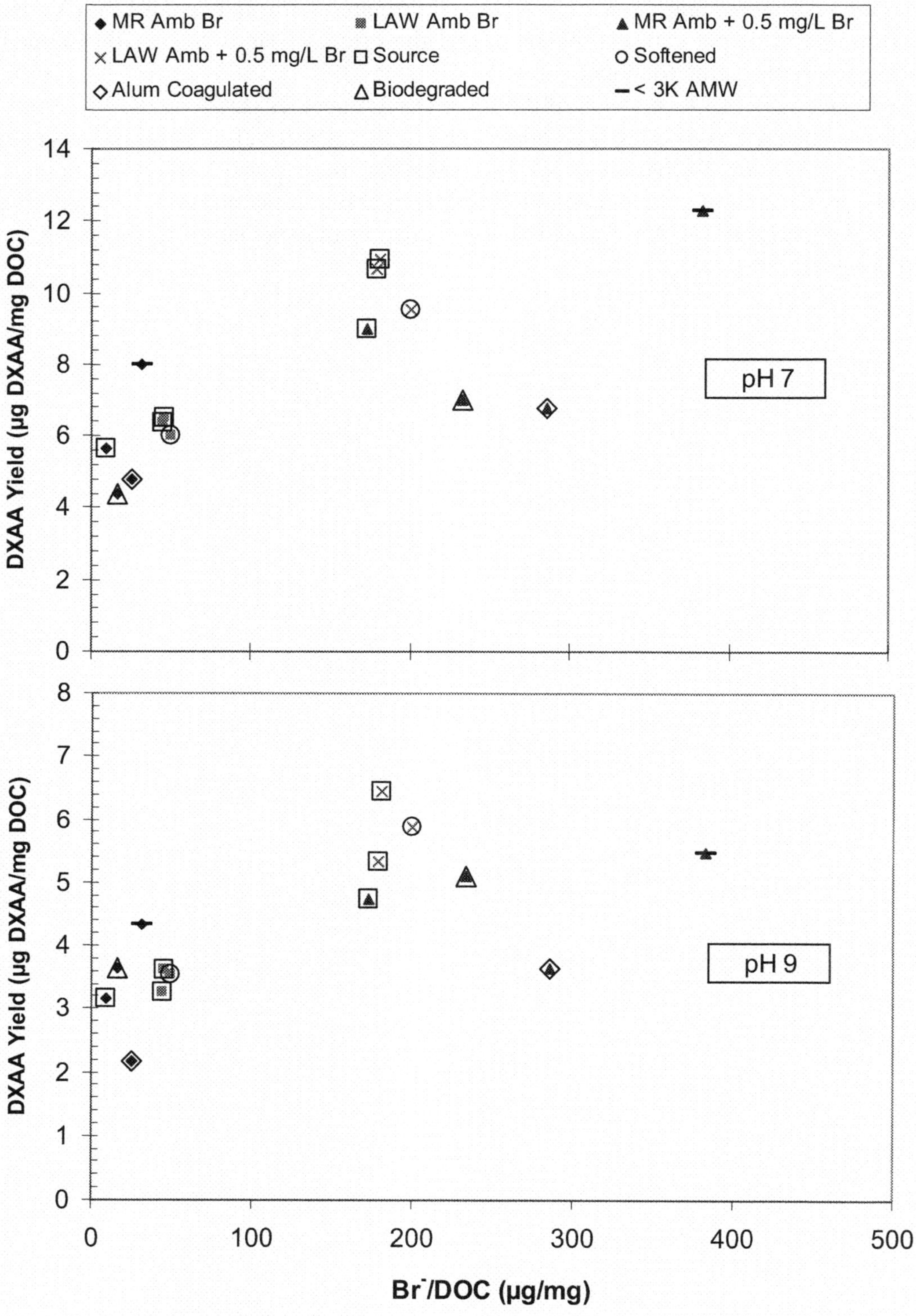

Figure 4.20 Influence of Br⁻/DOC ratio on mass DXAA yield in Lake Austin and Metedeconk River waters

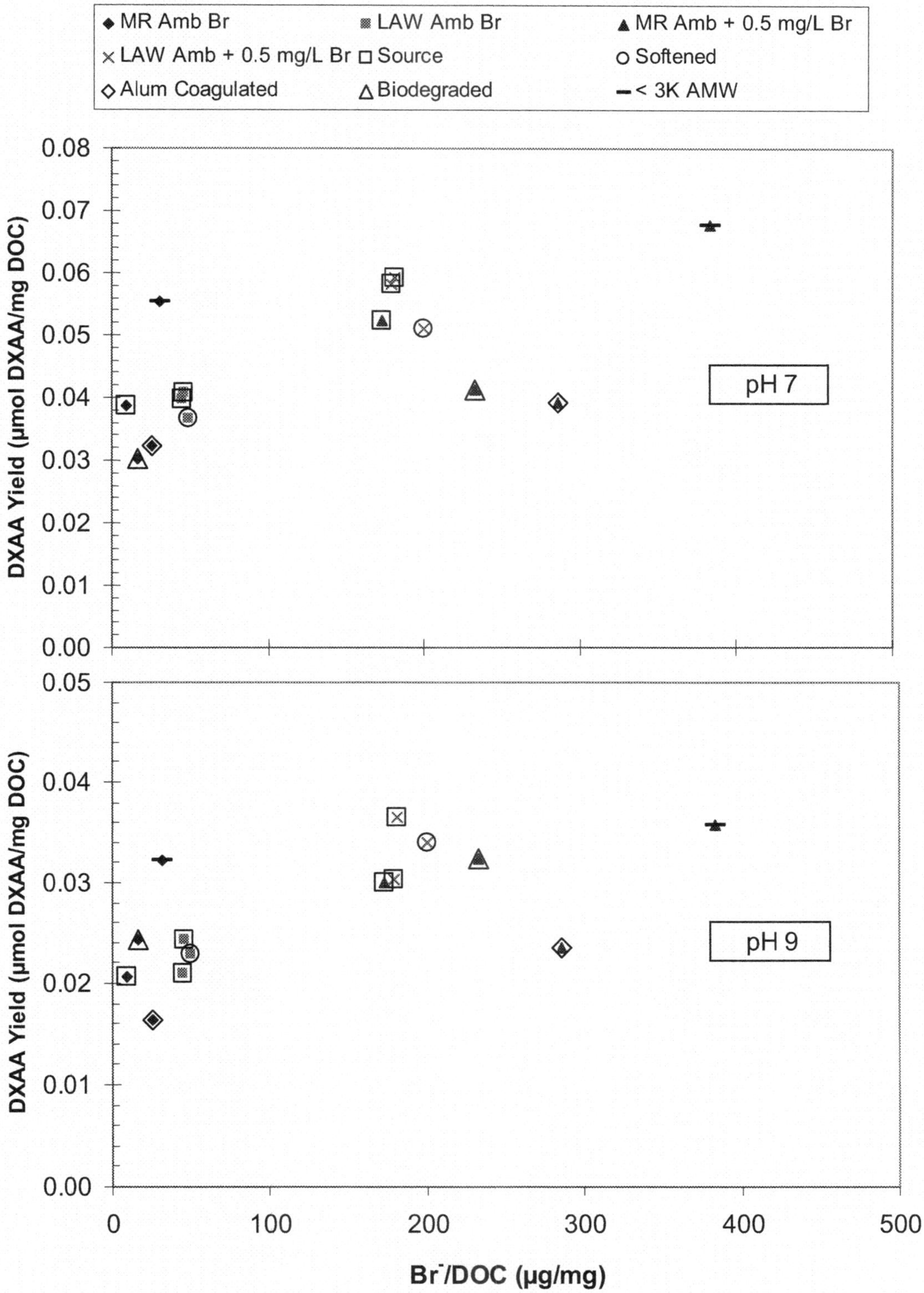

Figure 4.21 Influence of Br⁻/DOC ratio on molar DXAA yield in Lake Austin and Metedeconk River waters

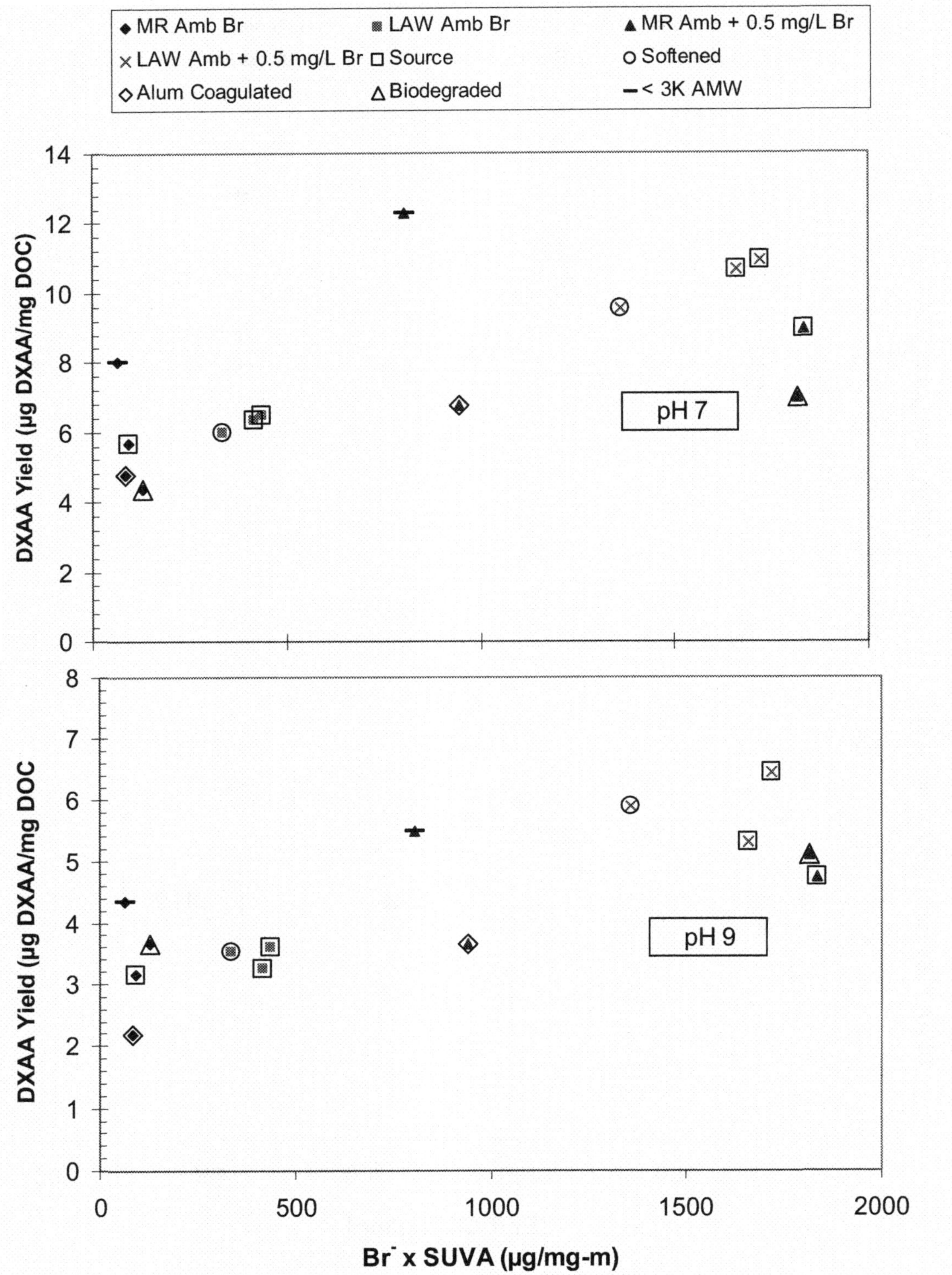

Figure 4.22 Influence of Br⁻ x SUVA on mass DXAA yield in Lake Austin and Metedeconk River waters

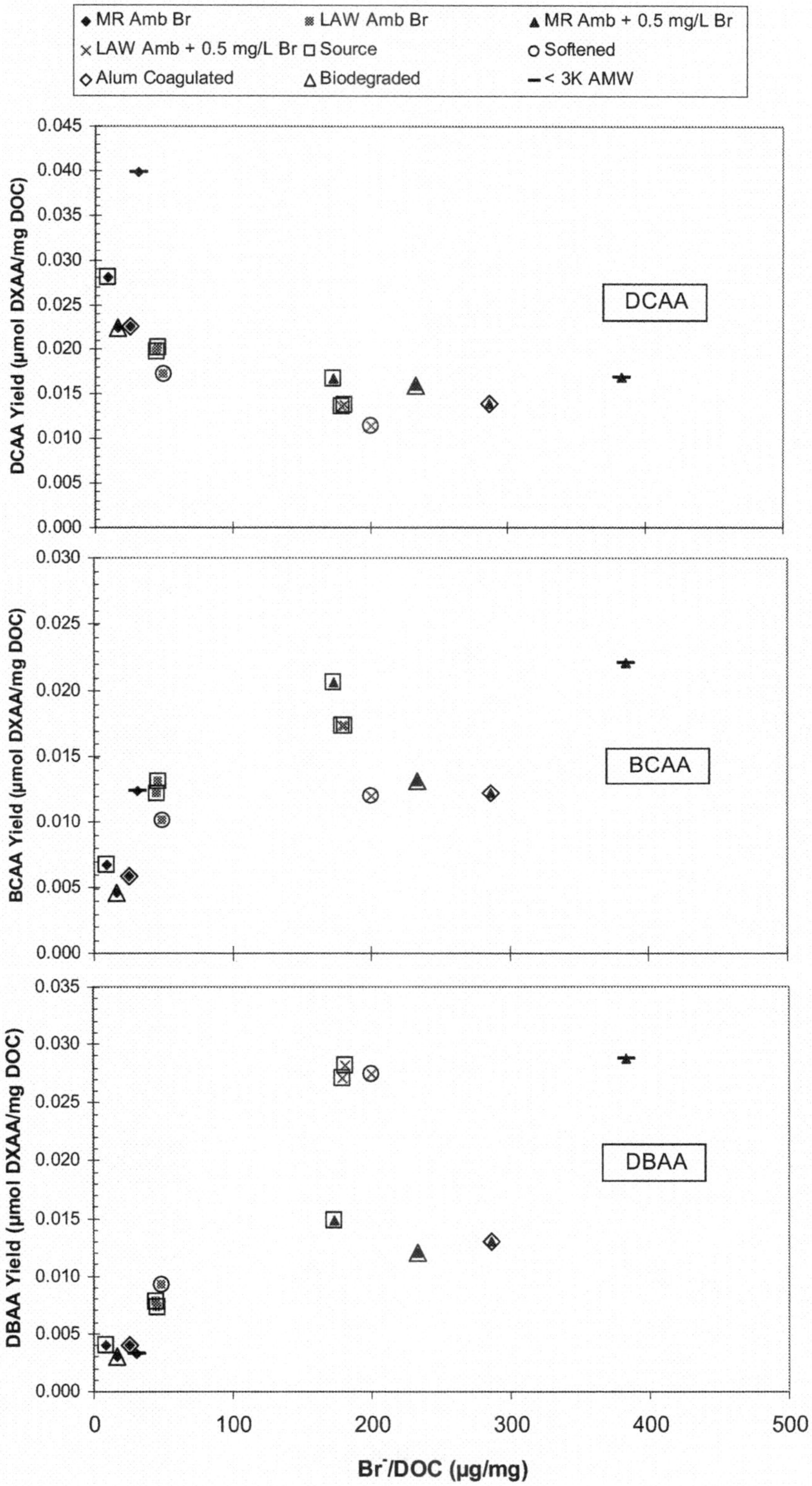

Figure 4.23 Influence on Br^-/DOC ratio on mass DXAA yields at pH 7 in Lake Austin and Metedeconk River waters

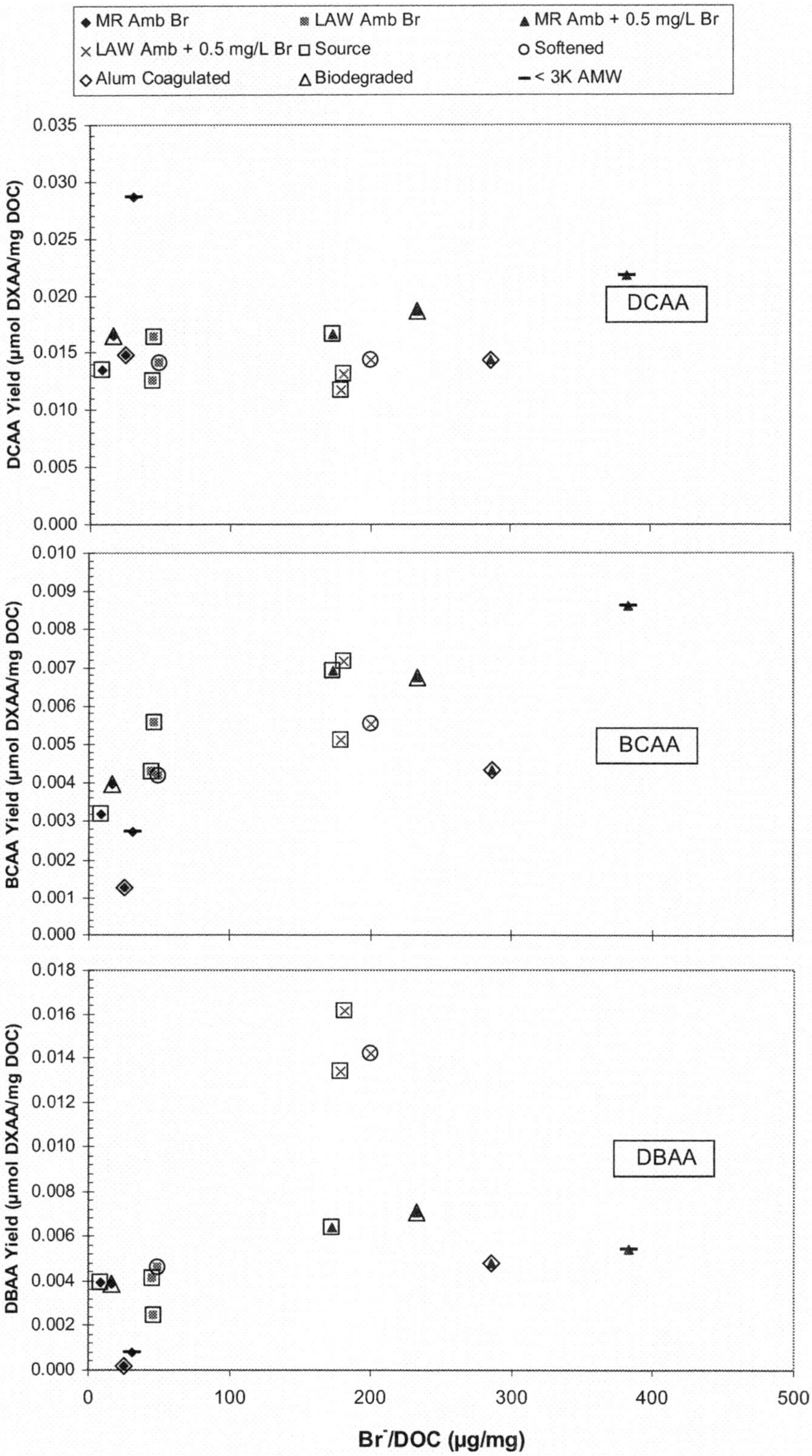

Figure 4.24 Influence of Br^-/DOC ratio on molar DXAA yields at pH 9 in Lake Austin and Metedeconk River waters

SUMMARY

Batch Screening experiments were conducted on two diverse water sources: Lake Austin water, which is a low SUVA, hard, high alkalinity source treated by lime softening, and Metedeconk River water, which is a high-SUVA, soft, low alkalinity source treated by alum coagulation. The key objective of the experiments was the identification of the water quality and chloramination variables that are most important in DXAA formation, which was achieved by testing source and treated waters as well as NOM fractions.

Several factors that influence DXAA formation during chloramination were evaluated: pH, bromide, Cl_2/N ratio, temperature, and chloramine residual. A fractional factorial experimental design allowed the significance and contribution of each factor to DXAA formation to be evaluated using ANOVA. The pH and bromide concentration were determined to be the most significant variables. The Cl_2/N mass ratio as the next most significant contributor to DXAA formation; temperature and chloramine residual also impacted DXAA formation, but were much less influential than the aforementioned factors. The effect of each factor on DXAA formation was also determined:

- pH: DXAA formation decreased as the pH increased.
- Bromide concentration: DXAA formation increased and a shift in speciation to the bromine-substituted species occurred as the bromide concentration increased.
- Cl_2/N mass ratio: DXAA formation decreased as the Cl_2/N ratio decreased.
- Temperature: DXAA formation increased as the temperature increased.
- Chloramine residual: Increasing the chloramine residual concentration caused an increase in the DXAA formation in Lake Austin water, but had a minimal impact on DXAA formation in Metedeconk River.

Treatment can also have a significant effect on DXAA formation. The effectiveness of treatment on DXAA formation is related to DOC removal; softening Lake Austin water resulted in minimal DOC removal, and therefore only a slight decrease in DXAA formation. However, alum coagulation of Metedeconk River water removed a larger fraction of the DOC, resulting in a more significant decrease in DXAA formation. To determine if observed differences in DXAA formation were related to differing reactivity of water or just a reflection of DOC concentration, DXAA yields were studied. Alum coagulation was beneficial in decreasing DXAA yield, indicating preferential removal of DXAA precursors, whereas softening Lake Austin water had a minimal effect on yield.

Predicting DXAA formation based upon basic water quality criteria can be useful. The influence of Br^- x SUVA on DXAA yield was investigated. This approach normalized the product of UV_{254} and bromide with DOC, accounting for both the concentration and reactivity of the NOM. A general trend developed, but there were some outliers. In general, increasing Br^- x SUVA values correlated with increasing DXAA formations.

DXAA speciation is also important because the health effects of bromine-substituted HAAs may be of greater concern. In general, higher Br^-/DOC ratios and lower pH led to greater bromine incorporation into DXAAs. At Br^-/DOC ratios of less than 200 μg/mg, increases in the Br^-/DOC ratio caused an increase in DXAA bromine incorporation. At Br^-/DOC ratios above 200 μg/mg, further increases caused no discernable impact on bromine incorporation into DXAAs. The impact of Br^-/DOC on each of the three DXAA species (DCAA, BCAA, and

DBAA) was investigated as well. In general, increasing Br^-/DOC ratio resulted in greater DBAA formation and less DCAA formation, thus shifting the speciation to the bromine-substituted DXAAs.

CHAPTER 5
FOLLOW-UP BATCH SCREENING EXPERIMENTS (BATCH II)

WATER SOURCE SELECTION

The initial batch screening experiments (Chapter 4) revealed that pH, bromide concentration and water source were influential factors in DXAA formation. Total organic carbon and SUVA are often used as surrogate parameters for characterizing or estimating the reactivity of a water source. Thus, to complement the waters used in the initial batch experiments, TOC, SUVA, and bromide concentration were the criteria chosen as the basis for selecting the additional waters for the second batch of screening experiments. Waters with low bromide concentrations were preferred because bromide was easier to add than to remove during laboratory analyses. As the pH does not significantly vary between source waters, it was not used as a criterion for water source selection, but as one of the controlled factors in the experimental design of the follow-up batch screening experiments. The source waters were also selected to achieve geographic diversity among the utility locations.

High SUVA, Low Bromide Water

Charleston, SC source water was originally selected for this category based on the information collected for the Information Collection Rule (ICR, EPA 2000), but was found to contain high bromide levels. For this reason, a separate set of experiments was developed for the Charleston water source, which is described in Chapter 3. The data collected were used in the modeling effort, and therefore are in Chapter 7. Metedeconk River, NJ, was ultimately selected as the high SUVA, low bromide water since it had been extensively characterized in the first batch screening experiments and satisfied all the criteria of this category.

High SUVA, High TOC Water

Biscayne Aquifer, FL, was selected because it had both a higher SUVA and TOC concentration than the waters used in the batch screening experiments. Biscayne Aquifer softened water was also studied, and can be classified as having an average to high SUVA, high TOC concentration, and average bromide concentration.

Low SUVA, High TOC Water

St. Paul, MN, was the low SUVA, high TOC water source chosen. In addition, St. Paul softened water was also studied because it too had a low SUVA and high TOC concentration. St. Paul source and treated water also had very low bromide concentrations, which allowed for better control of the bromide concentration adjustments.

WATER CHARACTERISTICS

The water quality characteristics of the water sources used in the follow-up batch screening experiments are shown in Table 5.1. Biscayne Aquifer and St. Paul source waters had very high DOC concentrations, while Metedeconk River had a much lower DOC concentration. Both Biscayne Aquifer and St. Paul source waters were softened, which resulted in a reduction of the DOC concentration (21% and 40%, respectively), UV_{254} absorbance (33% and 62%), turbidity (34% and 82%), and the AMW, especially the lower molecular weight fraction (46% and 54%). Therefore, softening had a more significant impact on St. Paul source water. The NOM characterizations of the source and treated waters suggested that all of the waters selected had a high hydrophobic content (69-78%), except for Metedeconk River and St. Paul source waters (51% and 59%, respectively), compared to the water sources selected in the initial batch screening experiments (near 50%). In addition, softening did not preferentially remove either the hydrophobic or hydrophilic fractions as their percentages did not vary significantly between the source and softened waters. In general, St. Paul water had a lower AMW, aromaticity (low SUVA) and Br^-/DOC ratio than observed in both Biscayne Aquifer and Metedeconk River waters. Conversely, Biscayne Aquifer source water had the highest hydrophobic fraction, molecular weight, UV_{254} absorbance, turbidity, as well as alkalinity.

Table 5.1
Summary of source water characteristics for follow-up batch screening experiments

Parameter	Metedeconk River Source	Biscayne Aquifer Source	Biscayne Aquifer Softened	St. Paul Source	St. Paul Softened
pH	7.02	7.53	9.61	8.00	9.62
TOC (mg/L)	N/A	12.24	9.36	11.41	6.42
DOC (mg/L)	4.32	11.95	9.45	11.07	6.6
Hydrophobic DOC (%)	51	78	74	59	69
Hydrophilic DOC (%)	34	26	29	41	40
AMW < 3K (mg/L DOC)	1.42*	5.1	2.97	6.74	3.09
AMW > 3K (mg/L DOC)	1.78*	6.85	6.48	4.33	3.51
UV_{254} (cm^{-1})	0. 171	0.412	0.274	0.271	0.102
Alkalinity (mg $CaCO_3$/L)	10	216	43.6	128	40
Turbidity (NTU)	ND	1	0.66	0.53	0.095
Bromide (μg/L)	33	100	135	6.9	12
SUVA (L/mg-m)	3.96	3.45	2.90	2.45	1.55
Ambient Br^-/DOC (μg/mg)	7.64	8.4	14.3	0.6	1.8
Spiked Br^-/DOC (μg/mg)	87.5	87.5	87.5	87.5	87.5

* = determined in initial batch screening experiments

INFLUENCE OF NOM CHARACTERISTICS AND CHLORAMINE DEMAND ON DXAA FORMATION

Various NOM characteristics, water quality parameters, and chloramination conditions were monitored during this study, which permitted the assessment of DXAA formation as a function of a diversity of variables. The hydrophobic and hydrophilic content of the waters studied in the follow-up batch screening experiments showed interesting tendencies with DXAA formation. It is useful to note that DOC can be subdivided in hydrophilic, transphilic, and hydrophobic fractions. In most of the literature, XAD-8 non-ionic resins are used to separate the hydrophobic and hydrophilic fractions. As the transphilic is included in the hydrophilic fraction, a further separation step is sometimes undertaken to isolate the transphilic fraction from the hydrophilic fraction. In this chapter, fractionation is discussed based on XAD-8 separation only; therefore, the transphilic fraction is included in the hydrophilic fraction.

A correlation between the DXAA mass yield and the hydrophobic fraction was observed. In general, DXAA mass yield decreased as the hydrophobic/DOC ratio increased (Figure 5.1). The R^2 for the correlations of the nine different experimental conditions ranged from 0.80 to 0.95 except for the condition at pH 7 and a 40 µg/mg Br^-/DOC ratio, which had a R^2 of 0.50. In addition, the hydrophobic fraction was more reactive as the pH decreased. For example the slope of the waters incubated at pH 7 (represented by the square symbols) is greater than the slope of the waters incubated at pH 9 (represented by the triangle symbols). The DXAA yield showed no apparent trend with Br^-/DOC in these experiments, which contrasts with the initial screening experiments (Chapter 4). The difference is most likely attributable to the narrower range of Br^-/DOC ratios used in these experiments, which were more typical of the range commonly found in drinking water sources.

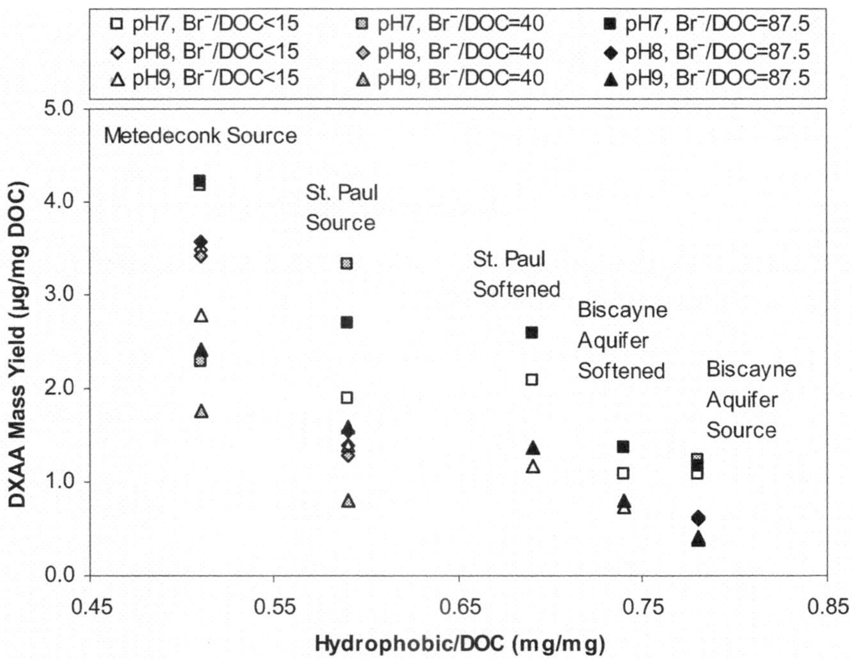

Figure 5.1 Influence of hydrophobic/DOC ratio on DXAA mass yield (µg DXAA/mg DOC) in the follow-up batch screening experiments

As would be expected when a water source is divided into only two fractions, a relationship was also found between the hydrophilic fraction and the DXAA mass yield (Figure 5.2). In general, DXAA mass yield increased as the hydrophilic/DOC ratio increased, suggesting that the hydrophilic fraction preferentially contributed to DXAA formation. Similarly, Hwang et al. (2000) determined that the hydrophilic fraction was significantly more reactive than the hydrophobic fraction during chloramination experiments with Colorado River water.

The influence of chloramine demand on DXAA formation was also evaluated. Although the data as a whole were somewhat scattered, for a given water the DXAA mass yield was approximately proportional to the chloramine demand (Figure 5.3). As the chloramine demand increased, the DXAA mass yield increased.

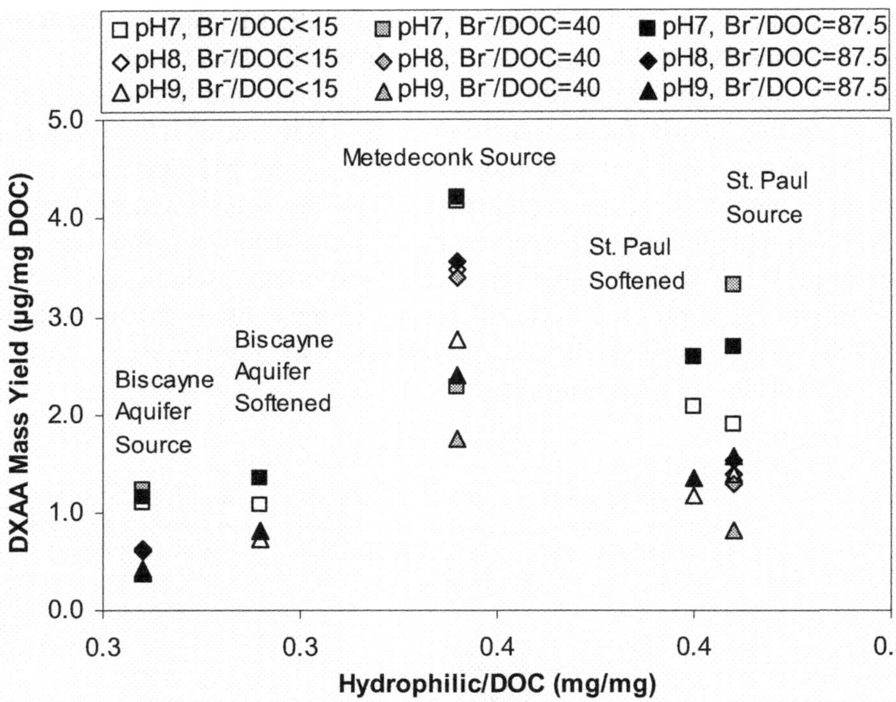

Figure 5.2 Influence of Hydrophilic/DOC ratio on DXAA mass yield (µg DXAA/mg DOC) in the follow-up batch screening experiments.

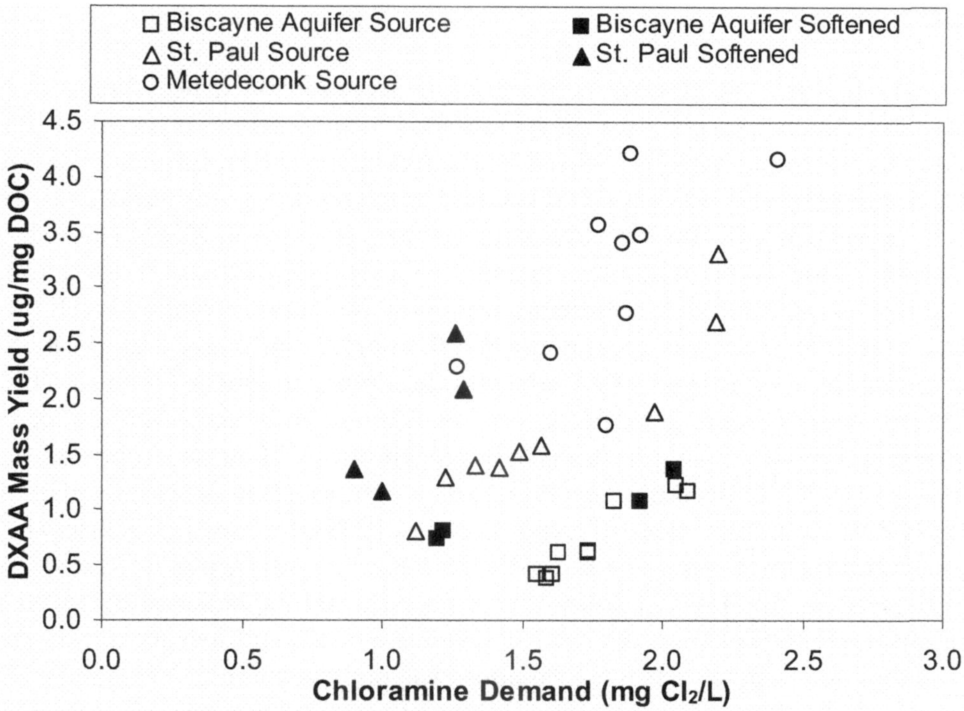

Figure 5.3 Influence of chloramine demand on DXAA mass yield (μg DXAA/mg DOC) in the follow-up batch screening experiments

Many other water quality characteristics such as DOC, UV, SUVA, SUVA x Br^-, hydrophilic DOC, and hydrophobic DOC were considered and plotted against DXAA formation to search for meaningful relationships. However, none of the aforementioned water quality characteristics yielded significant correlations with DXAA formation. Previous research (reported in Chapter 2) suggested that a relationship between DXAA yield and SUVA might exist (Figure 2.4); however, these experiments did not support the significance of such a relationship. Given the substantial reactivity of the hydrophilic fraction, as demonstrated recently by others (Hwang et al., 2000) and these experiments, the absence of a relationship between DXAA yield and SUVA is now to be expected when comparing waters from different sources.

RELATIVE CONTRIBUTION OF pH, Br^-/DOC RATIO, AND SELECTED NOM CHARACTERISTICS TO DXAA VARIATION

To investigate the potential non-linear effects of the most influential variables on DXAA formation three different levels of pH and Br^-/DOC ratio were used as a follow-up to the initial batch screening experiments. Therefore, three different pH values (7,8, and 9) and three different Br^-/DOC ratios (<15, 40, and 87.5 μg/mg) were studied (See Chapter 3 for more details).

Influence of pH

With one exception, the DXAA concentration decreased in a near-linear fashion with increasing pH. The influence of pH at Br^-/DOC ratios of <15, 40 and 87.5 µg/mg is illustrated in Figure 5.4. The DXAA molar yield decreased as the pH increased for all of the Br^-/DOC ratios studied. This general pH trend was expected because batch studies by Symons et al. (1998) and the initial batch experiments in this research showed that, in general, DBPs (including HAAs) decreased in concentration as the pH increased and the Cl_2/N ratio decreased. Metedeconk River source water, however, exhibited a non-linear trend for the middle range Br^-/DOC ratio (40 µg/mg); the DXAA molar yield was greater at pH 7 than pH 9, and highest at pH 8. Symons et al. (1998) also observed a significant DOX concentration peak around pH 8 for a Cl_2/N ratio of 5/1 for Lake Austin water (Figure 2.1). However, insufficient HAA data were collected in their study to determine if similar peaking in the HAA concentrations likewise occurred. In addition, Stumm and Morgan (1996) showed that reaction rates for chlorination of various phenols peak in the vicinity of pH 8 as well. The work of Symons et al. (1998), Stumm and Morgan (1996), and the DXAA concentration peak around pH 8 at a Br^-/DOC of 40 µg/mg in Metedeconk River source water observed in this present study may suggest a pH dependence of NOM reactivity (similar to that observed with phenols) or a pH dependence of haloamine speciation. However, it is most probable that not as much DXAA formation was observed at pH 7 for the Br^-/DOC ratio of 40 µg/mg as would be expected based upon the results observed at the other Br^-/DOC ratios, and because the Br^-/DOC ratio did not have a significant impact on DXAA molar yields (Figure 5.5) in most of the waters tested.

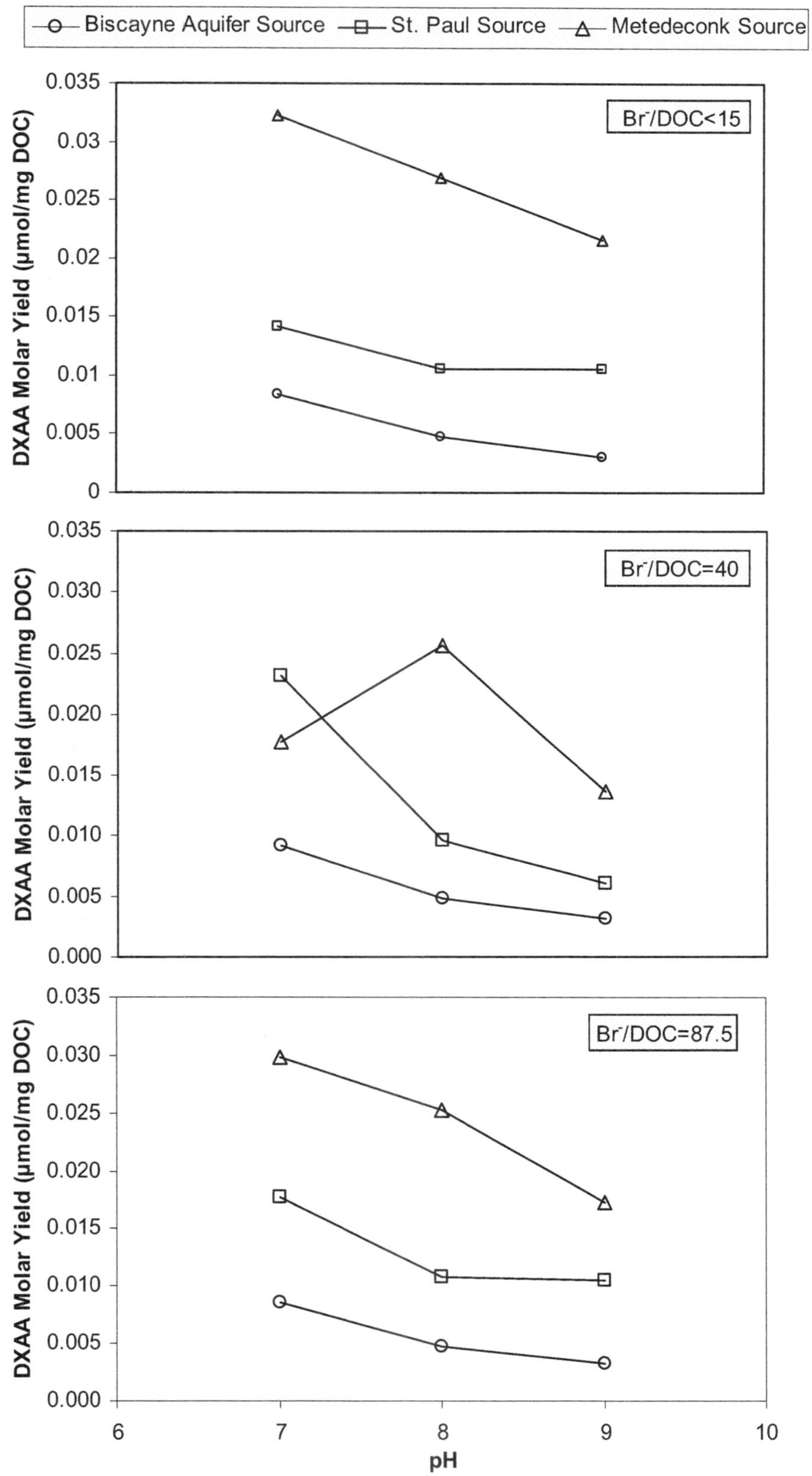

Figure 5.4 Influence of pH 7, 8 and 9 on DXAA molar yield (µmol DXAA/mg DOC) at Br^-/DOC ratios of <15 µg/mg, 40 µg/mg, and 87.5 µg/mg

Influence of Br^-/DOC

The Br^-/DOC ratio is a way of normalizing the bromide concentration among various water sources, and more accurately reflects the potential for forming bromine-substituted DBPs than does the bromide concentration alone. Regardless of the pH, the Br^-/DOC ratio had a minor influence on DXAA molar yield for the waters tested (Figure 5.5). These results were somewhat unexpected as the ANOVA from the initial batch screening experiments revealed that bromide had a significant influence on DXAA concentration. However, the range Br^-/DOC ratios used in the initial batch screening experiments was greater than in the follow-up batch screening experiments, where the Br^-/DOC levels were lower and more typical of the range of typical drinking water sources. Thus, very high bromide or Br^-/DOC ratios will affect DXAA yield, but small effects should be observed over the range of typical practice as previous research also revealed (Cowman and Singer, 1996; Symons et al., 1998; Singer et al., 1999; Zhang et al., 1999; Diehl et al., 2000).

DXAA speciation is also of interest because the bromine-substituted HAAs may pose a greater carcinogenic risk. In this research, the factor n_D' was used to characterize the bromine substitution of DXAAs. The value of n_D' can range from zero (all DCAA) to two (all DBAA). The impact of pH and Br^-/DOC ratio on the formation of bromine-substituted DXAAs in Biscayne Aquifer and St. Paul source and softened water, as well as Metedeconk River source water, is shown in Figure 5.6. In general, higher Br^-/DOC ratios and lower pH resulted in a greater degree of bromine substitution. Similar trends were observed in Metedeconk River and Lake Austin waters in the initial batch screening experiments (Chapter 4). The larger Br^-/DOC ratio promoted more bromine incorporation into DXAAs because the presence of bromide substantially increases the range of possible halogenating species that can react with the NOM.

Overall, increasing Br^-/DOC ratios resulted in a shift toward the bromine-substituted species without significantly influencing the DXAA molar yield. However, in the source waters, treatments, and NOM fractions of both Metedeconk River and Lake Austin studied in the initial batch screening experiments (Figure 4.10 and Figure 4.12), increasing bromide concentrations caused a more significant increase in DXAA formation, quite probably because the maximum Br^-/DOC ratios were larger than in these follow-up experiments. Increased HAA formation due to the presence of bromide may be water specific. For example, Symons et al. (1998) observed that the addition of bromide to Lake Houston water did not have a large effect on the HAA_6 concentration, but significantly increased the HAA_6 concentration in California State Project water.

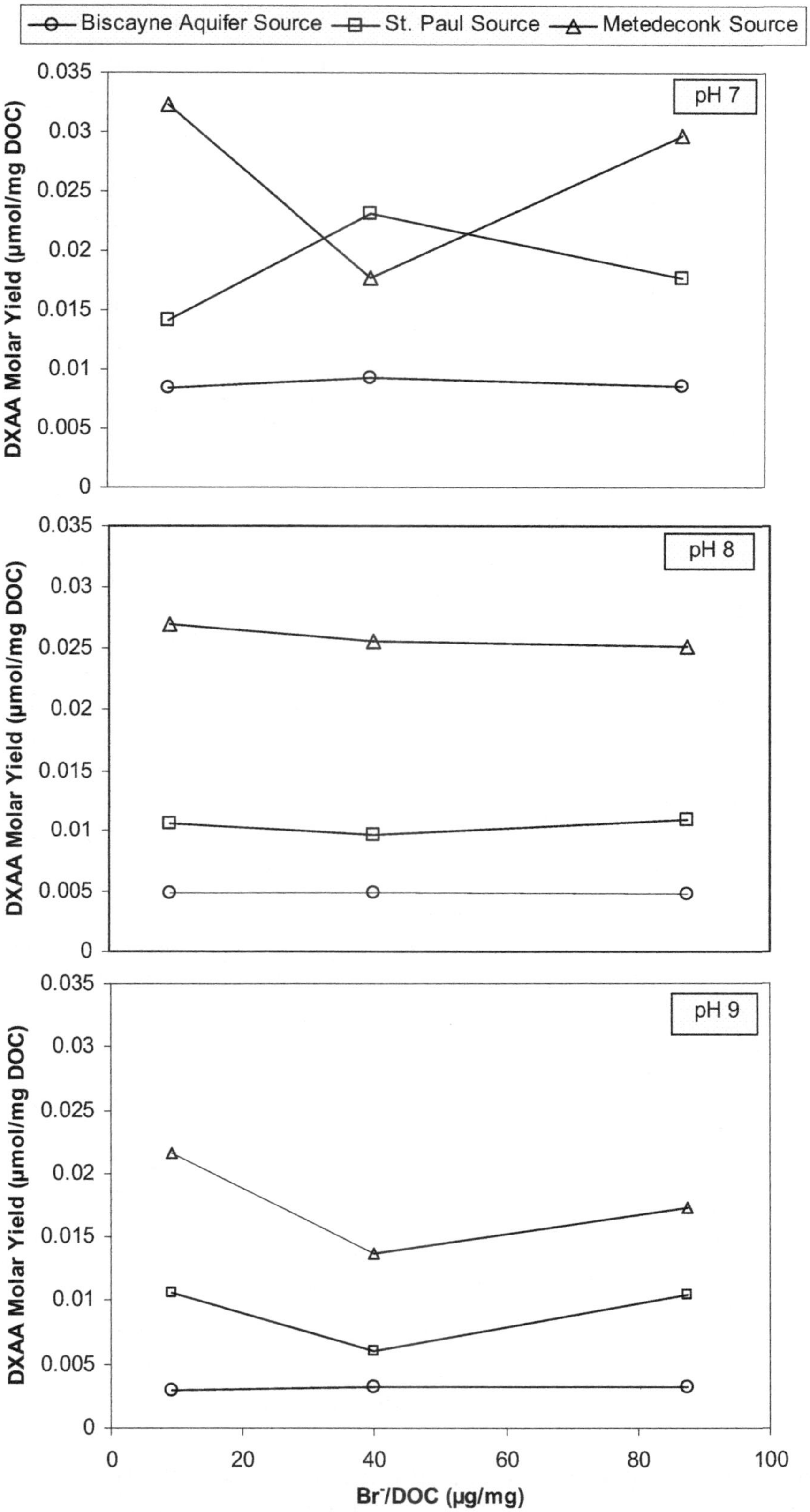

Figure 5.5 Influence of Br⁻/DOC ratios <15 µg/mg, 40 µg/mg, and 87.5 µg/mg on DXAA molar yield (µmol DXAA/mg DOC) at pH 7, 8, and 9

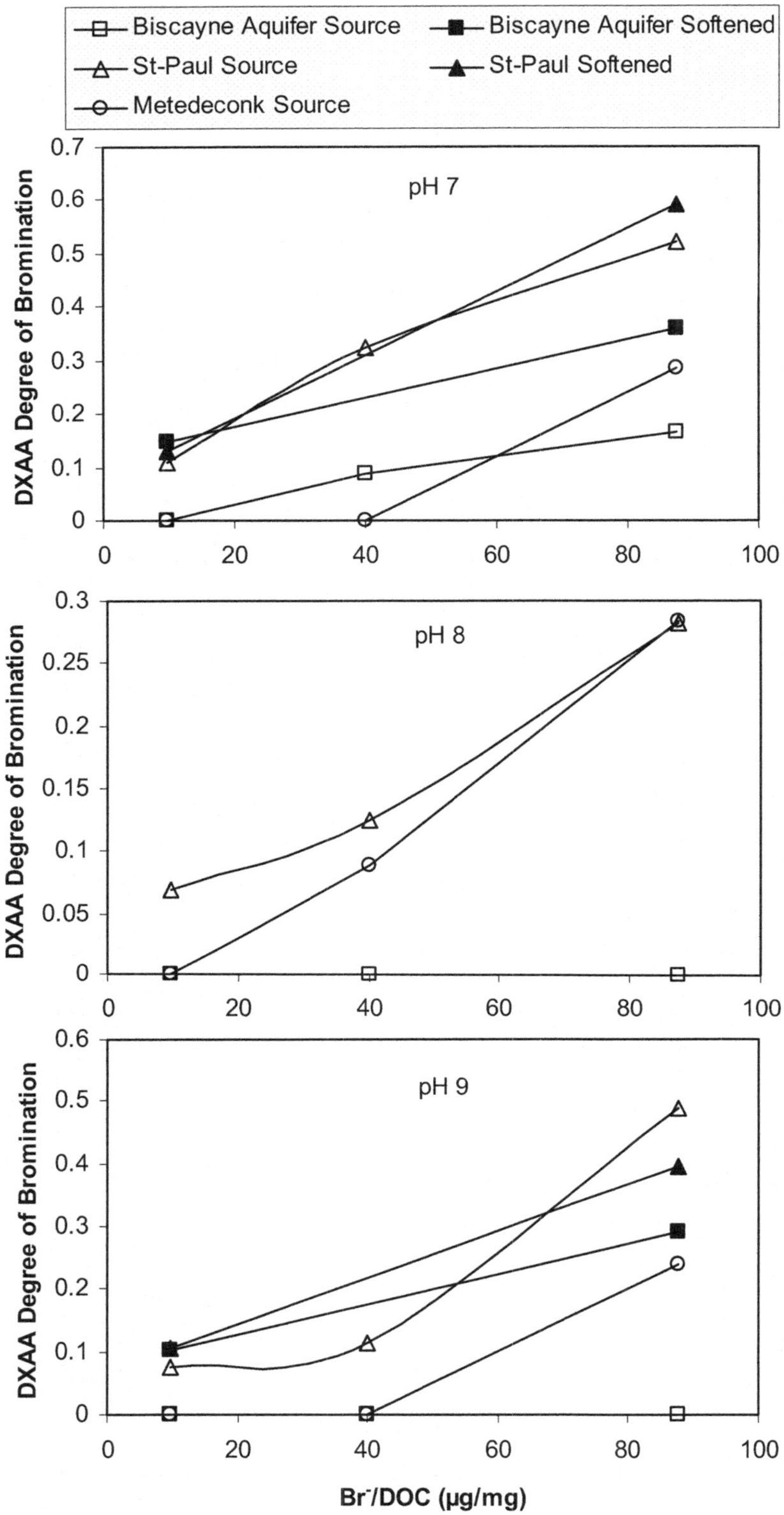

Figure 5.6 Influence of experimental conditions (pH 7, 8, and 9; Br^-/DOC ratio <15 µg/mg, 40 µg/mg, and 87.5 µg/mg) and water type on DXAA degree of bromination in the follow-up batch screening experiments

In the follow-up batch screening experiments, a fractional factorial experimental design was also used to systematically rank the influence of pH and Br^-/DOC ratio as well as the water source on DXAA formation. The contribution of each factor was statistically estimated by analysis of variance (ANOVA), which highlights the relative significance and percent contribution of each evaluated factor on DXAA formation.

Percent contributions of each factor obtained from the ANOVA analysis of the Biscayne Aquifer, St. Paul, and Metedeconk River source waters are shown in Figure 5.7, while the softened Biscayne Aquifer and St. Paul waters are shown in Figure 5.8. In each water, pH was a significant contributor to DXAA formation, while Br^-/DOC ratios were insignificant. The considerable influence of pH on DXAA formation during chloramination was likewise observed in other studies (Krasner et al., 1996, and Symons et al., 1998). Bromide has generally been found to be influential in DXAA formation during chloramination as well as chlorination. However, it is less influential at low concentrations (Cowman and Singer, 1996; Symons et al., 1998; Sinha et al., 1997; Singer et al., 1999; Zhang et al., 1999; Diehl et al., 2000; Duirk et al. 2002a; Krasner et al, 1996). The diminished role of Br^-/DOC was due to the use of lower Br^-/DOC ratios, more typical of drinking water sources, in the follow-up batch screening experiments compared to the higher Br^-/DOC ratios used in the initial batch screening experiments as explained above. In general, the error, or unexplained variation, was satisfactory (between 0-20%), indicating that the experimental variations in DXAA formation could be adequately explained by the changes in pH and all the other factors were adequately controlled. However, St. Paul and Metedeconk River source waters had a much larger error (48-77%), suggesting that one or more main factors were missing and/or the experimental conditions were not adequately controlled.

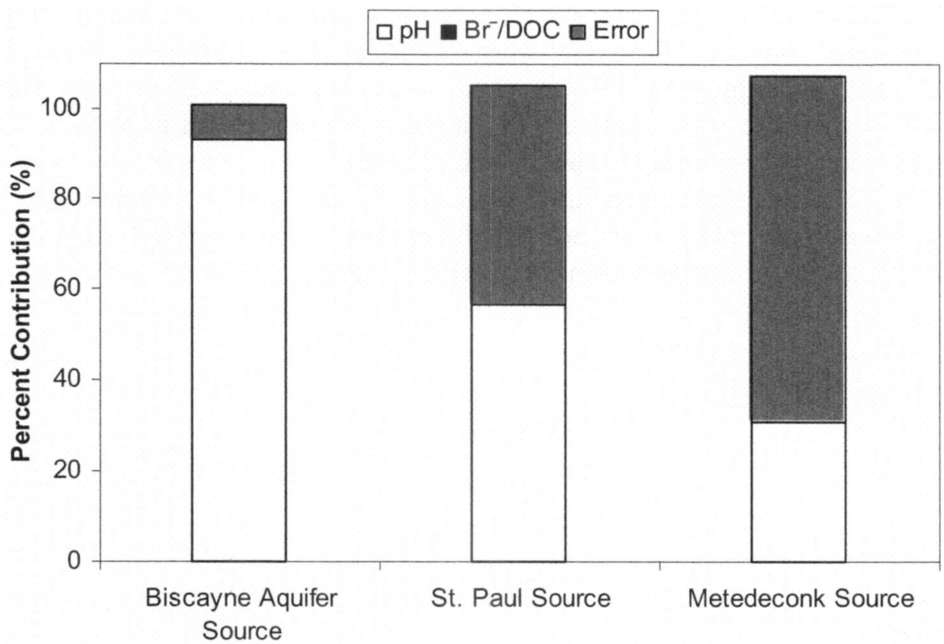

Figure 5.7 Influence of pH and Br^-/DOC ratio on DXAA formation in Biscayne Aquifer, St. Paul, and Metedeconk source waters (ANOVA calculations performed on a DXAA molar concentration basis)

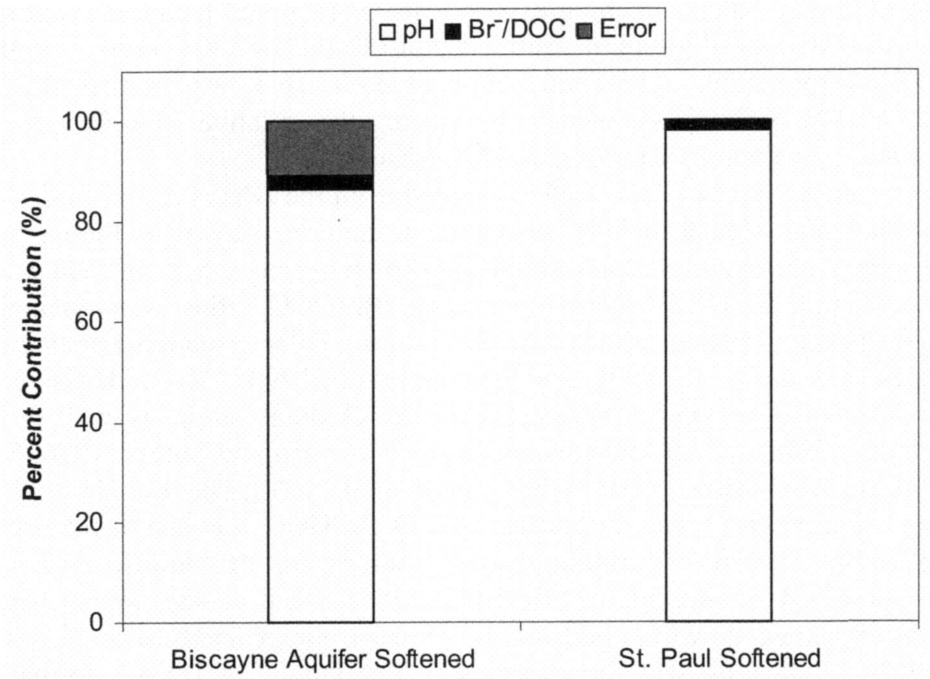

Figure 5.8 Influence of pH and Br^-/DOC ratio on DXAA formation in Biscayne Aquifer and St. Paul softened waters (ANOVA calculations performed on a DXAA molar concentration basis)

The influence of source water on DXAA formation was investigated. To determine if observed differences in DXAA formation were truly related to differing water reactivity or just a reflection of DOC concentration, DXAA yields were studied. Percent contributions from a combined ANOVA of the water sources are shown in Figure 5.9 on a molar yield basis. The water factor played a very significant role in DXAA molar yield, contributing as much as 71% to the total molar yield, while pH contributed 14%, and the Br^-/DOC ratio factor was insignificant. Therefore, the water and pH factors adequately explained the DXAA formation (molar yield based ANOVA), as the error was within the acceptable range of 0-20%.

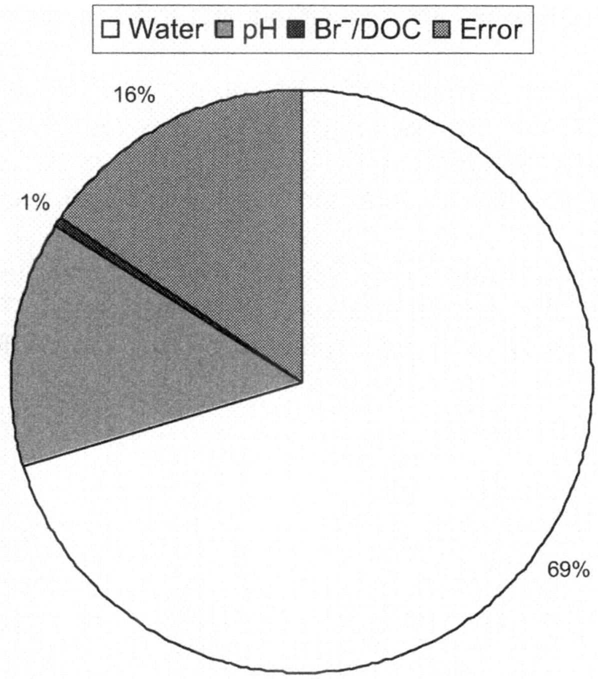

Figure 5.9 Influence of water source on DXAA molar yield (ANOVA calculations performed on a DXAA molar yield basis)

SUMMARY

A series of five waters with varying water quality and NOM characteristics was studied to compliment the waters used and the results found in the initial batch screening experiments. The key objective of the follow-up batch screening experiments was to further investigate the influence of pH, bromide concentration, and NOM characteristics on DXAA formation. The DXAA mass yield decreased with an increasing hydrophobic/DOC ratio or, alternatively, a decreasing hydrophilic/DOC ratio, thereby indicating that the NOM in the hydrophilic fraction was more reactive in forming DXAA than the NOM in the hydrophobic fraction.

In general, the influence of pH mirrored that of the initial batch screening experiments: increasing DXAA formation as the pH decreased. With one water at one of the three Br^-/DOC ratios, the peak DXAA molar yield occurred at pH 8, rather than pH 7. Both the general pH trend and the occasional DBP peak concentrations at pH 8 were also observed by Symons et al. (1998). These results suggest that a pH dependence of NOM reactivity, a pH dependence of haloamine speciation, a bromide dependence of haloamine speciation, or some combination of these factors accounts for the relatively high reactivity around pH 8 in some waters.

Regardless of the pH, Br^-/DOC had a minor influence on DXAA molar yield formation for the waters tested, which somewhat contradicted the results found in the initial batch screening experiments. However, the Br^-/DOC ratios used in the initial batch screening experiments were greater than in these experiments, where levels were lower and more typical of the range of commonly found in drinking water sources. Increasing Br^-/DOC ratios caused a shift toward the bromine-substituted species without significantly influencing the DXAA molar yield.

The DXAA mass yield was approximately proportional to chloramine demand; as the amount of chloramine consumed increased, the DXAA mass yield increased. Many other water quality characteristics such as DOC, UV, SUVA, SUVA x Br^-, hydrophilic DOC, and hydrophobic DOC were considered, but none yielded valuable correlations with DXAA formation.

The influence of source water on DXAA formation was investigated through the analysis of a combined ANOVA of the water sources on a molar yield basis. The water factor played a very significant role in molar DXAA yield, followed by pH, and the Br^-/DOC ratio factor was not found to be influential.

CHAPTER 6
DXAA FORMATION KINETICS

WATER CHARACTERISTICS

Both source and treated water from four different locations were studied to discern the impact of pH, bromide concentration, temperature, and treatment on DXAA formation kinetics. The water quality characteristics for each of these waters are shown in Table 6.1. DOC is often used as a measurement of NOM: Lake Austin and Metedeconk River source waters have moderate DOC concentrations (3.74 mg/L and 4.32 mg/L, respectively), Charleston source water has a higher DOC concentration (5.11 mg/L), and Biscayne Aquifer source water has a very high DOC concentration (11.95 mg/L). DOC can be further described by measuring the molecular weight ranges, hydrophobicity, and SUVA. The NOM characterizations shown in Table 6.1 suggest that the organic carbon in Biscayne Aquifer water is predominately hydrophobic. A relatively large fraction of the organic carbon in Charleston source water was hydrophobic as well, while the Lake Austin and Metedeconk River source water contained a moderate amount of hydrophobic DOC. In addition, a majority of the organic carbon in Charleston source water is in the <3000 apparent molecular weight range, whereas most of the organic carbon in the other source waters was in the >3000 apparent molecular weight fraction. Lake Austin source water had a relatively low SUVA (which can be used as a measure of the reactivity of the water) of 2.49, while the other source waters had higher SUVAs of approximately 3.5.

The reactivity of the source waters was also measured by chlorination under uniform formation conditions (UFC tests). Chlorination of Biscayne Aquifer, Metedeconk River, and Charleston source waters under these conditions produced similar concentrations of HAA_9 while Lake Austin source water formed considerably less. The organic carbon concentrations, however, vary dramatically between water sources; therefore, reactivity is better characterized on a yield basis. The Metedeconk River source water, which yielded 50.7 μg HAA_9/mg DOC, was the most reactive of the source waters studied. The Charleston source water was also rather reactive, yielding 42.3 μg HAA_9/mg DOC, while the Lake Austin source was considerably less reactive (24.1 μg HAA_9/mg DOC). Biscayne Aquifer source water, which formed the greatest concentration of HAA_9 under the UFC test, was the least reactive, producing less than 20 μg HAA_9/mg DOC.

The source waters selected also span the typical range of bromide concentrations: Metedeconk River had a low bromide concentration of 33 μg/L, while Biscayne Aquifer had a bromide concentration of 100 μg/L, which is about average, and Lake Austin and Charleston source waters had moderate to high bromide concentrations of 168 μg/L and 183 μg/L, respectively. The Br^-/DOC ratio can be used to estimate the amount of bromine that will be incorporated into the HAAs. Lake Austin and Charleston source waters had higher Br^-/DOC ratios (44.9 μg/mg and 35.8 μg/mg, respectively) than Metedeconk River and Biscayne Aquifer source waters (7.6 μg/mg and 8.4 μg/mg, respectively); therefore, Lake Austin and Charleston waters have a greater tendency to form bromine-substituted HAAs. Another significant difference among the water sources was their alkalinities. Biscayne Aquifer and Lake Austin are high alkalinity source waters, with 216 and 148 mg $CaCO_3$/L, respectively, while the Charleston and Metedeconk River source waters had much lower alkalinities of 30 and 10 mg $CaCO_3$/L, respectively.

Lake Austin source water can be summarized as a hard, high alkalinity source water with a low SUVA, moderate DOC concentration, and a high ambient bromide concentration. Metedeconk River is a high SUVA, soft, low alkalinity source water with moderate DOC and low bromide concentrations. Charleston is also a low alkalinity, high SUVA source water, but additionally has high DOC and bromide concentrations. Biscayne aquifer is a high SUVA source water as well, but also has a very high DOC concentration, high alkalinity, and a moderate bromide concentration.

The effect of treatment on DXAA formation kinetics was also studied. Treatment had a significant impact on water quality characteristics. Alum coagulation of Charleston and Metedeconk River water removed approximately 50% and 47% of the DOC concentration, respectively, whereas softening was less effective for Biscayne Aquifer and Lake Austin water, removing only 21% and 11%, respectively. Both softening and coagulation typically remove higher molecular weight NOM more effectively than other NOM constituents (Liao and Randkte, 1985; Collins et al. 1996). Coagulating Metedeconk River source water removed a significant fraction on the higher molecular weight NOM, increasing the <3K/>3K apparent molecular weight ratio from less than 0.8 to greater than 3.6. Softening Lake Austin source water, however, only slightly increased the ratio from less than 0.5 to approximately 0.6. Conversely, softening Biscayne Aquifer source water preferentially removed <3000 apparent molecular fraction, decreasing the <3K/>3K AMW ratio from 0.74 to 0.46. In addition, softening only caused a 16% SUVA reduction in Biscayne Aquifer and an 18% SUVA reduction in Lake Austin, whereas alum coagulation on Metedeconk River and Charleston waters resulted in SUVA decreases of 56% and 54%, respectively. The reactivity of the waters was also measured using the UFC or the SUFC test where appropriate. Treatment reduced the HAA_9 formation in all of the treated waters studied. Alum coagulating Metedeconk River and Charleston dramatically decreased HAA_9 formation, forming only 30% and 34% of that in the source waters, respectively. Softening Lake Austin and Biscayne Aquifer waters had a lesser impact on HAA_9 formation. The treated waters formed 64% and 73% of the source water HAA_9 formation, respectively.

Table 6.1
Water quality characteristics

Parameter	LA Source	LA Softened	MR Source	MR Coagulated	BA Source	BA Softened	Chl Source	Chl Coagulated
pH	8.2	9.7	6.78	7.25	7.53	9.61	7.2	6.5
TOC (mg/L)	3.87	3.3	n/a	n/a	12.24	9.36	5.1	n/a
DOC (mg/L)	3.74	3.32	4.32	2.31	11.95	9.45	5.11	2.55
Hydrophobic DOC (%)	50	49	51*	51*	78	74	66	54
Transphilic DOC (%)	17	16	15*	24*	23	27	ND	ND
Hydrophilic DOC (%)	20	20	15*	20*	10	11	ND	ND
AMW < 3K (mg/L DOC)	1.31	1.26	1.42*	1.7*	5.1	2.97	3.91	ND
AMW > 3K (mg/L DOC)	2.67	2.06	1.78*	0.47*	6.85	6.48	1.2	ND
AMW (<3K/>3K)	0.49	0.61	0.80	3.6	0.74	0.46	3.3	ND
UV_{254} (cm^{-1})	0.093	0.068	0.167	0.039	0.412	0.274	0.178	0.041
Alkalinity (mg $CaCO_3$/L)	148	61	10	44	216	43.6	30.1	9.5
UFC HAA_9 (µg/L)	90.2	NA	219	65.0	234	NA	216	72.5
UFC HAA_9 Yield (µg/mg)	24.1	NA	50.7	28.1	19.6	NA	42.3	28.4
SUFC HAA_9 (µg/L)	NA	58.1	NA	NA	NA	170	NA	NA
SUFC HAA_9 Yield (µg/mg)	NA	17.5	NA	NA	NA	18.1	NA	NA
Turbidity (NTU)	0.66	0.09	2.6	0.6	1	0.66	1.77	0.13
Bromide (µg/L)	168	164	33	49	100	135	183	174
SUVA (L/mg-m)	2.49	2.05	3.87	1.68	3.45	2.90	3.48	1.59
Ambient Br^- /DOC (µg/mg)	44.9	49.4	7.6	21.2	8.4	14.3	35.8	68.2
Spiked Br^- /DOC (µg/mg)	179	200	123.4	237	50.2	67.2	133	264

NA = Not Analyzed, ND = Not determined; n/a = Not Applicable, * = Data from initial batch screening experiments.
LA = Lake Austin, Austin, TX; MR = Metedeconk River, Brick Township, NJ; BA = Biscayne Aquifer, Boca Raton, FL; Chl= Bushy Park Reservoir and Edisto River, Charleston, SC

CHLORAMINATION KINETICS

Chloramination has been implemented in a wide variety of ways, which can lead to significant differences in DBP formation. The traditional approach to chloramination involves the addition of ammonia and chlorine simultaneously very early in the treatment train. However, many plants now have a significant period of free chlorination prior to ammonia addition for purposes of meeting CT requirements (Seidel et al. 2000; Connell et al. 2000). Clearly, significant exposure to free chlorine will produce greater DBP formation than with chloramines alone; therefore, the key objectives of the kinetic experiments were to discern the impacts of certain water quality and chloramination variables, as well as the impact of prechlorination, on DXAA formation.

Traditional Chloramination

Water from four different sources was used to determine the impact of pH, bromide concentration, temperature, and treatment on DXAA formation kinetics during traditional chloramination (simultaneous ammonia and chlorine addition). In these experiments, described in Chapter 3, waters were dosed to achieve a 4/1 Cl_2/N mass ratio and a target chloramine residual of 2 mg/L at 48 hours, which is typical of drinking water treatment plants in practice.

Kinetics and speciation of the DXAA formation in Metedeconk River source water are illustrated in Figure 6.1. The kinetics were rather fast, approximately 73% of the 72-hour DXAA concentration formed after only 0.5 hours and almost 82% formed after 1 hour. Similar DXAA formation kinetics were observed in the other waters studied (Appendix E). Metedeconk River source water has a low bromide concentration (33 μg/L) and thus, a low Br^-/DOC ratio (8 μg/mg). Therefore, DCAA was the predominant dihalogenated species that formed; minimal formation of bromine-substituted DXAAs (BCAA and DBAA) was observed. Source waters with higher ambient bromide concentrations, such as Lake Austin and Charleston, exhibited greater bromine-substituted DXAA formation. Lake Austin source water, for example, had a higher n_D' (0.64 vs. 0.05 for Metedeconk River) after 72 hours because of a higher ambient bromide concentration (168 μg/L) and Br^-/DOC ratio (45 μg/mg). Charleston source water had higher ambient bromide (183 μg/L) and DOC concentrations (5.11 mg/L), but a lower Br^-/DOC ratio (36 μg/mg) and n_D' (0.23) than Lake Austin source water. The lower n_D' may be attributed to the lower Br^-/DOC ratio. The effect of the Br^-/DOC ratio on DXAA speciation is presented in more detail in subsequent sections of this chapter.

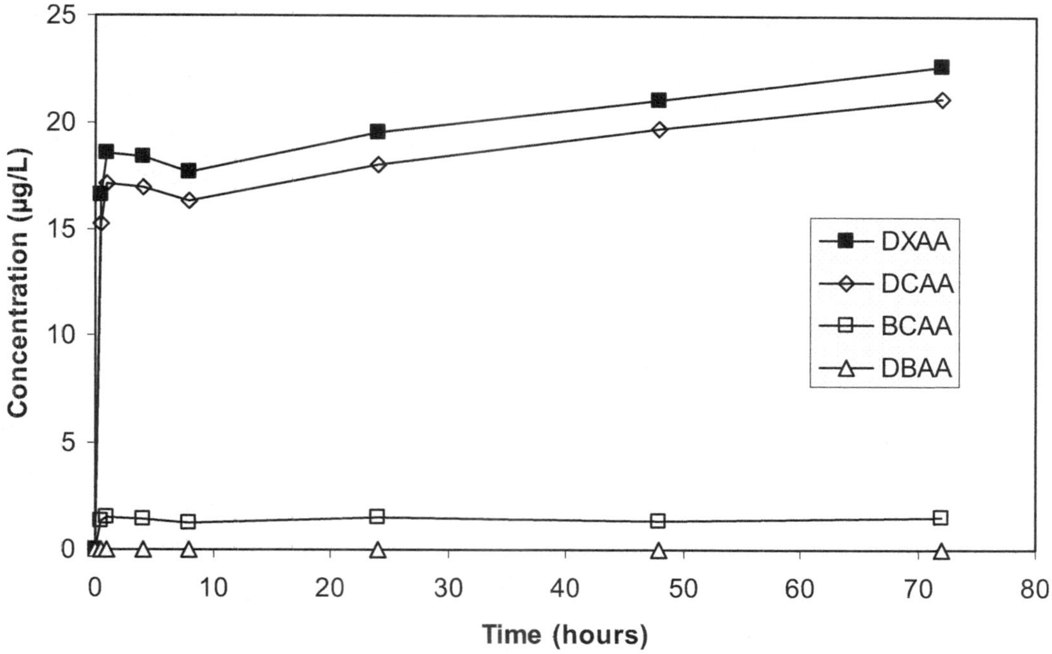

Figure 6.1 DXAA formation kinetics of Metedeconk River source water at pH 8

Effect of Bromide Concentration

The effect of bromide concentration on DXAA formation and speciation was investigated by spiking Metedeconk River source water with 0.5 mg/L of bromide (Figure 6.2). The addition of bromide caused both an increase in DXAA formation, 27 µg/L compared to 22 µg/L after a 72-hour incubation period, and a shift in speciation to the bromine-substituted species; the source water spiked with bromide formed less DCAA and more BCAA and DBAA than the source incubated at the ambient bromide concentration (Figure 6.1). However, when these data were analyzed on a molar basis (Figure 6.3), a minimal increase in HAA formation was observed in the bromide-spiked water. The discrepancy between the mass and molar analyses is explained by variations in speciation (bromine-substituted species weigh more than their chlorine-substituted counterparts). Therefore, HAA formation in Metedeconk River source water in the presence of a bromide spike (0.5 mg/L) mainly caused a shift in speciation. An increase in bromine incorporation was observed, but the presence of bromide did not significantly increase the molar yield. However, in the source waters, treatments, and NOM fractions of both Metedeconk River and Lake Austin studied in the initial batch screening experiments (Figure 4.10 and Figure 4.12) increasing bromide concentrations caused a more significant increase in DXAA formation. Symons et al. (1998) observed that the addition of bromide did not have a large effect on the HAA_6 concentration in Lake Houston water, but significantly increased HAA_6 concentrations in California Sate Project Water. Hwang et al. (2000) observed that bromide addition increased bromine incorporation, but did not significantly increase the molar yield of

HAAs in Colorado River water NOM fractions. The inconsistency of these results indicates that increased HAA formation due to the presence of bromide is water specific. Different water sources exhibit different NOM characteristics and perhaps different speciation of the various haloamines, which probably influence the reactivity with respect to DXAA formation.

The relative kinetics of each species are also shown in Figure 6.2. The bromine-substituted species (BCAA and DBAA) formed very rapidly, whereas the DCAA formation kinetics was characterized by an initial fast period of formation followed by a slower period. Only about 73% of the 72-hour DCAA concentration formed after 0.5 hours in the Metedeconk River source water in the presence of a bromide spike, whereas the bromine-substituted species essentially formed almost immediately in this water (*i.e.*, in <0.5 hr). This trend in speciation is also apparent in the investigation of the bromine incorporation ratio (Figure 6.3). As expected, n_D' decreased over time because of the relative kinetics of each species. The bromide present can react with free chlorine produced by monochloramine hydrolysis to form hypobromous acid or with the monochloramonium ion to produce monobromamine. During chloramination, bromide oxidation by monochloramonium ion usually dominates, and once hypobromous acid and monobromamine are formed, they undergo further rapid reactions to produce chlorobromamines, primarily chlorobromamine (Valentine 1986; Vikesland et al. 2001), which has a pKa of 4.9 (Bousher et al., 1989). Therefore, $NClBr^-$ is the dominant species over the typical pH range of drinking water. Valentine (1986) also concluded that the bromine atom of bromochloramine is very labile and reactive and that in reactions with organic chemicals it should produce products similar in nature to those produced by free bromine. Therefore, the observed shift in speciation, overall increase in HAA formation, and the relative kinetics of each species likely occurred because of the relatively higher reactivity of the bromine-substituted reactive species as compared to their chlorine-substituted counterparts.

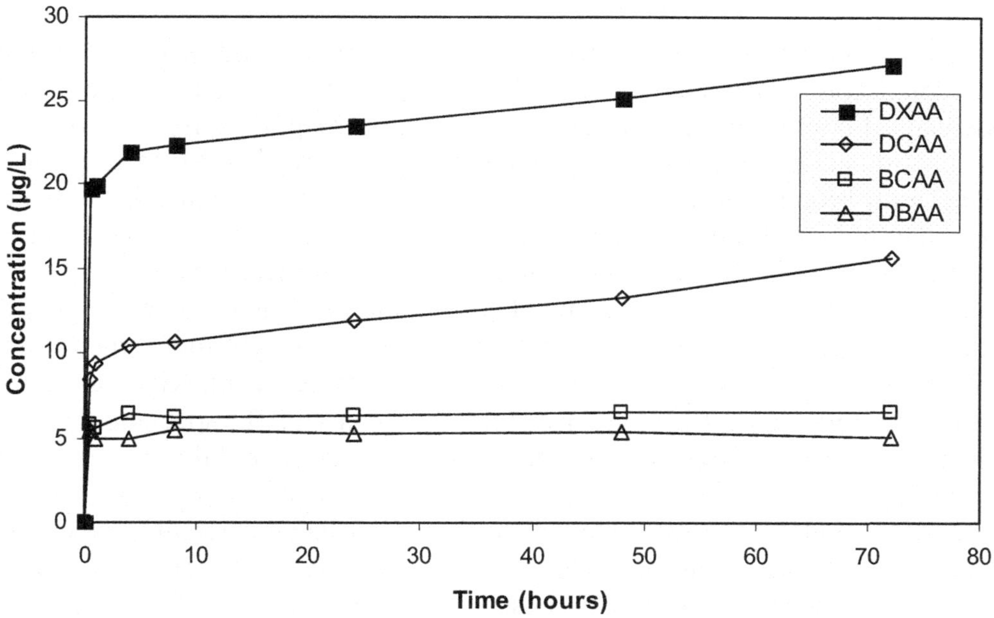

Figure 6.2 DXAA formation kinetics of Metedeconk River source water with 0.5 mg/L of bromide addition at pH 8

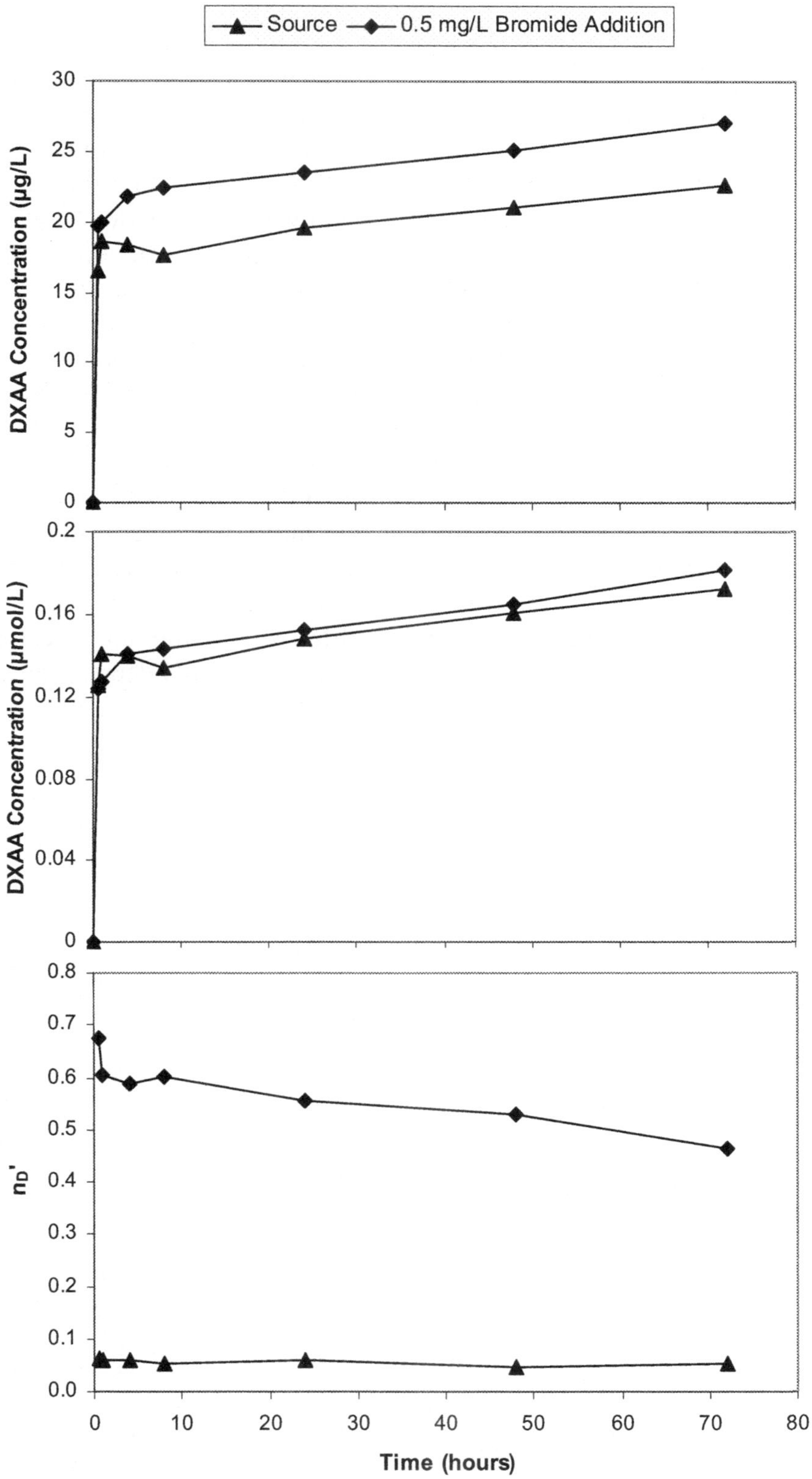

Figure 6.3 DXAA formation kinetics and bromine incorporation ratio (n_D') of Metedeconk River source and source with 0.5 mg/L Br^- addition at pH 8

Effect of pH

The impact of pH on DXAA formation kinetics was also investigated because pH was determined to significantly influence DXAA formation (Chapter 4). Lake Austin source water incubated at pH 7 exhibited greater DXAA formation than that at pH 9 (Figure 6.4), and in both cases, a majority of the DXAA formation occurred before 1 hour. Both the waters incubated at pH 7 and pH 9 had a fast period of formation followed by a slower period; however, more rapid kinetics were observed in the water incubated at pH 7 during both periods of formation. This trend indicates that acid catalyzed monochloramine reactions may be the cause of increased HAA formation as the pH decreased. As mentioned in Chapter 4, when the solution pH is decreased, greater concentrations of dichloramine, free chlorine, and monochlorammonium ion are available to react with NOM, indicating that one or more of these reactive species may be the cause of the observed increase in DXAA formation at lower pH.

The increase in pH did not significantly impact the amount of bromine incorporated into the DXAAs that formed; after 72 hours the n_D' values of the source water incubated at pH 7 and pH 9 were 0.58 and 0.64, respectively. However, the initial batch screening experiments with Lake Austin source water (Chapter 4, Figure 4.13) indicated that, on average, the n_D' increased approximately 15% as the pH decreased from 9 to 7. The follow-up batch screening experiments (Chapter 5), which were conducted on several different source and treated waters, also demonstrated that DXAA bromine incorporation increased as the pH decreased; for example, n_D' in St. Paul source water increased approximately 48%, from 0.07 to 0.11, as the pH decreased from 9 to 7.

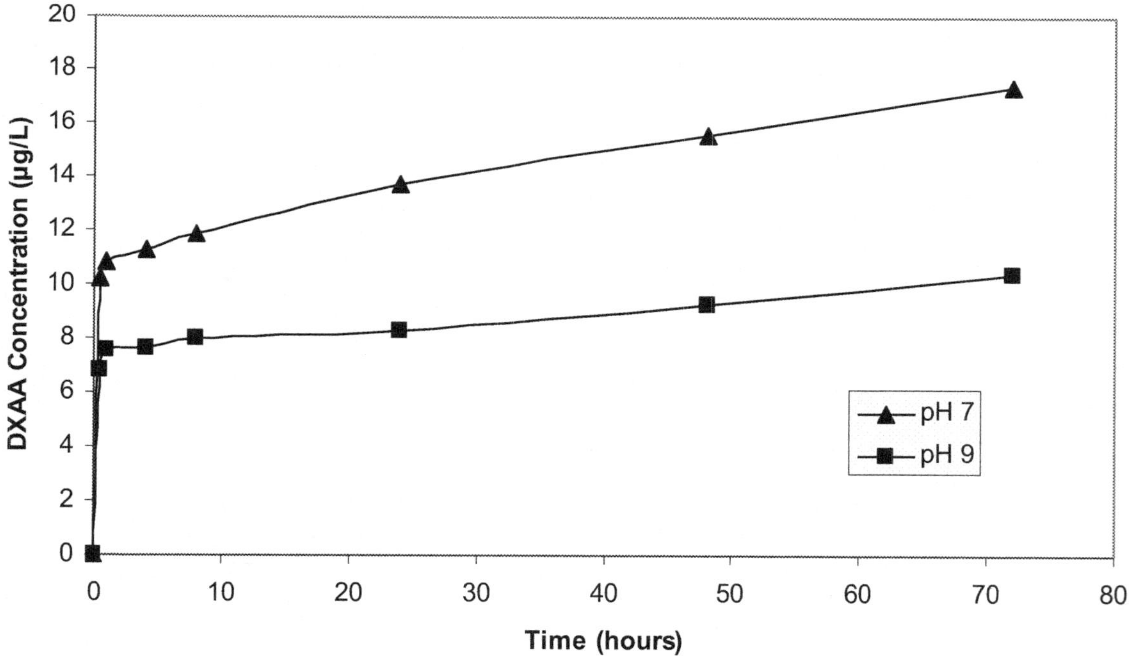

Figure 6.4 Effect of pH on DXAA formation in Lake Austin source water

Effect of Temperature

Temperature has also been shown to influence HAA formation and is especially important in southern climates where surface water temperatures can increase substantially during the summer. In addition to temperature variation, the source water NOM characteristics can vary seasonally as well. A review of ICR data (EPA, 2000) for surface water treatment plants known to be practicing chloramination shows elevated HAA concentrations in the summer months compared to the winter. The observed increase in HAA formation at higher temperatures may be due to faster reaction kinetics, a seasonal increase in the concentration or reactivity of NOM, or a combination of these factors. Temperatures of 70°F and 85°F were studied in Charleston and Lake Austin treated waters (Figure 6.5) to characterize the DXAA formation kinetics in response to temperature changes alone. Increasing temperature caused only a minimal increase in DXAA formation over the range tested in these treated waters. However, the batch screening experiments conducted on Lake Austin and Metedeconk River source waters, treatments, and fractions (Chapter 4, Figure 4.17) indicated that temperature may cause a more significant increase in HAA formation than found in these experiments.

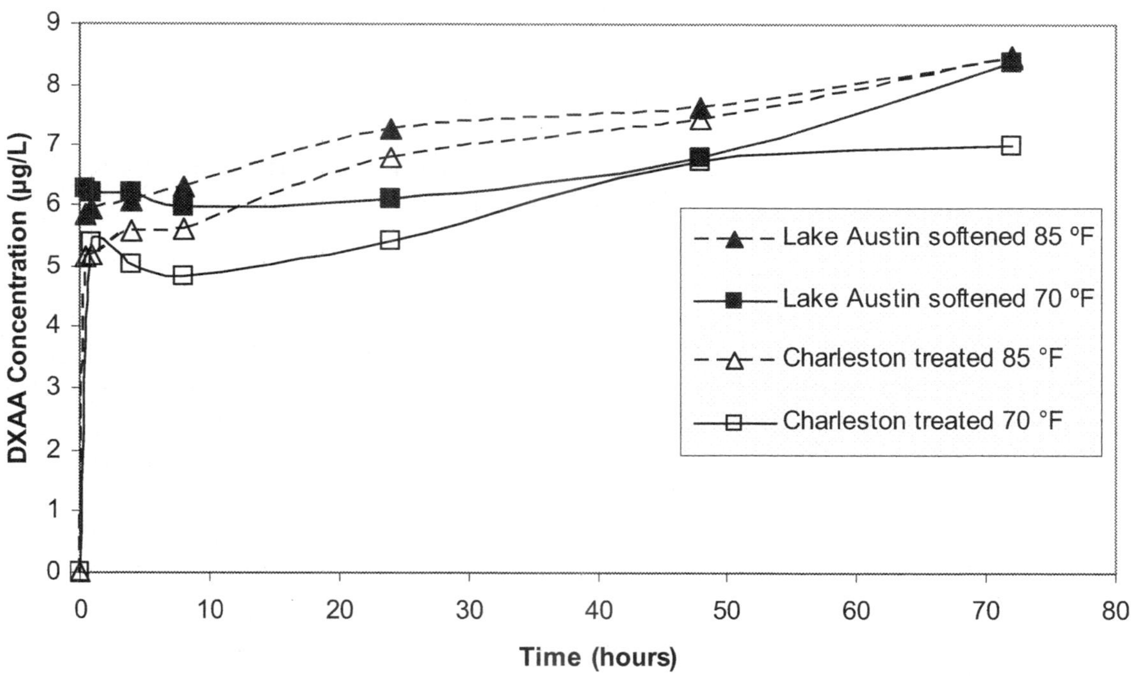

Figure 6.5 Effect of temperature on Lake Austin (pH 9) and Charleston (pH 8) treated waters (4/1 Cl_2/N mass ratio)

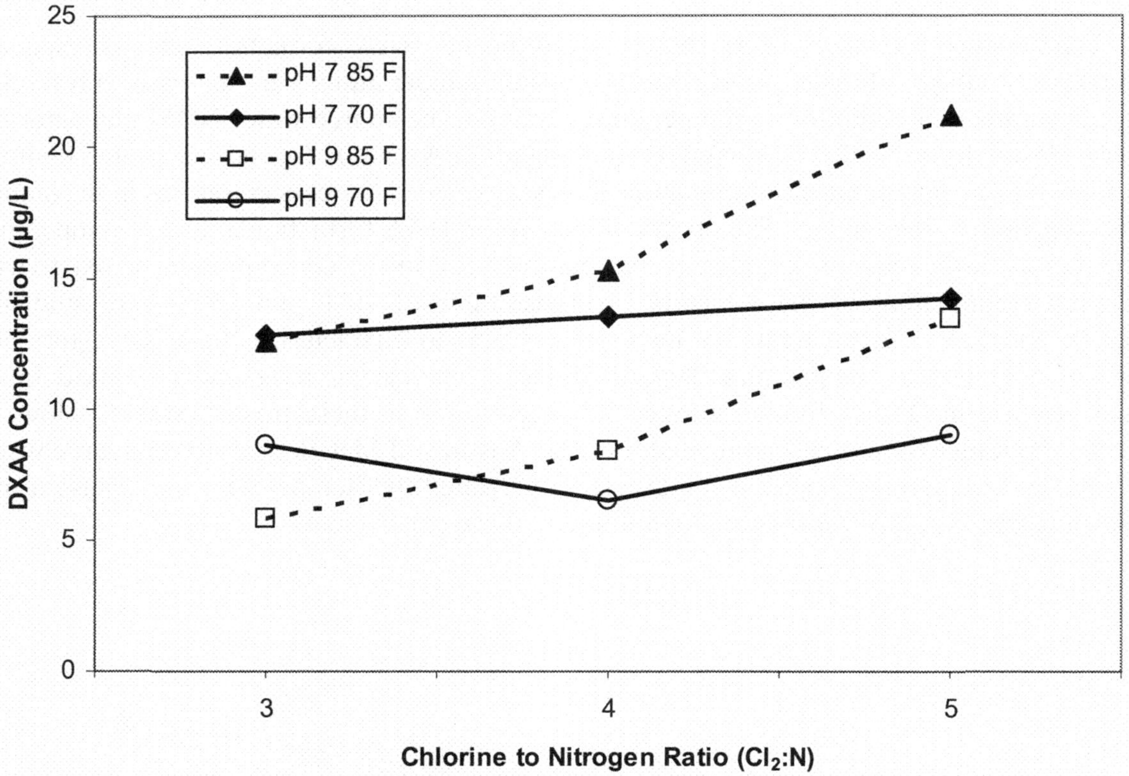

Figure 6.6 Effect of temperature and Cl_2/N ratio on DXAA formation in softened Lake Austin water (2 mg/L Cl_2 residual; 48 hour incubation)

The effect of temperature on DXAA formation at different pH and Cl_2/N ratios was investigated with Lake Austin softened water (Figure 6.6). The results indicated that, at the 3/1 Cl_2/N ratio, temperature had little effect on DXAA formation. In addition, at 70°F the Cl_2/N ratio had little effect on DXAA formation. At 85°F, however, DXAA formation increased somewhat as the Cl_2/N ratio increased; the increased DXAA formation was especially apparent at pH 7 and the 5/1 Cl_2/N ratio. Therefore, the influence of temperature on DXAA formation is dependent on other chloramination conditions such as pH and Cl_2/N ratio. As mentioned above, decreased pH promotes increased concentrations of several reactive species such as dichloramine, free chlorine, and monochlorammonium ion. In addition, larger Cl_2/N ratios cause more rapid monochloramine degradation, resulting in the formation of dichloramine and its decomposition products (e.g., free chlorine). Formation of reactive species, NOM reactions with these species, or both may be influenced by temperature. Therefore, increased HAA formation due to increased temperature is expected to be more pronounced as the pH decreases and Cl_2/N ratio increases. Nevertheless, the effect of temperature alone on DXAA formation appears to be relatively small; seasonal changes in NOM concentration or reactivity are also likely to be important in situations in practice where large changes in DXAA formation are observed with seasonal temperature changes.

Effect of Treatment

HAA formation is strongly influenced by organic carbon concentration. The effect of treatment on DXAA formation kinetics in Metedeconk River water is illustrated in Figure 6.7. As expected, the treated water exhibited lower DXAA concentrations that the source water because of DOC removal. The coagulated water formed only 34% of the DXAA concentration in the Metedeconk River source water after 72 hours of incubation. This considerable decrease is attributed to the significant DOC removal (almost 47%) attained by coagulation, which also decreased the SUVA 57% and preferentially removed the heavier (>3000 apparent molecular weight fraction) organic carbon. However, treatment did not have a significant effect on the rates of DXAA formation after the initial fast period of formation. As seen in Figure 6.7, the slopes, and thus formation rates after the period of rapid DXAA formation were similar for both the coagulated and source water. The source water formed approximately an additional 1.2 μg/L of DXAA per day, while the coagulated water formed approximately 0.7 μg/L of DXAA per day after the first hour of rapid DXAA formation.

The impact of treatment on the bromine incorporation ratios of the Metedeconk River waters is also depicted in Figure 6.7. The n_D' of both the source and coagulated waters decreased over time because of the relative kinetics of each species. However, the coagulated water had consistently higher DXAA bromine incorporation than the source water. Coagulation of Metedeconk River source water removed approximately 47% of the DOC, thereby increasing the Br^-/DOC ratio and decreasing the number of possible NOM reactive sites. The bromine-substituted reactive species are typically more reactive than their chlorine-containing counterparts. As a result, they will react more quickly and out compete the chlorine-substituted species for the available NOM reactive sites, increasing the n_D'. In contrast, the initial batch screening experiments (Chapter 4) showed that the bromine incorporation decreased as a result of coagulating this source water in a majority of the trials. These experiments used a fractional factorial experimental design to study practical ranges of several variables, which resulted in experiments at pH 7 and 9. Assuming that the response at pH 8 falls between pH 7 and pH 9, the bromine incorporation of the source water was significantly greater in the initial batch screening experiments than in the kinetics experiments. However, the bromine incorporation of the alum coagulated waters was similar in both experiments. The source waters used for both experiments had similar bromide concentrations and Br^-/DOC ratios, and coagulation removed similar percentages of SUVA. In addition, the follow-up batch screening experiments with Metedeconk River source water (Chapter 5) resulted in minimal bromine incorporation. Therefore, the larger n_D' of the source water used in the initial batch screening experiments may have resulted from temporal variations. The water used in the initial batch studies was collected in the early Fall of 2001, while the water used in the kinetics experiments and follow-up batch screening experiments was collected in the Spring of 2002. The type and concentration of NOM in source waters varies temporally (Biber et al. 1996;Volk et al. 2002), and these variations may have influenced the reactivity, the effect of treatment, and therefore the n_D', of the source waters.

Figure 6.8 shows the impact of treatment on both the DXAA formation kinetics and the bromine incorporation ratio of Charleston source water. As with Metedeconk River source water, Charleston source water was alum coagulated, which removed a significant amount of organic carbon, reducing the DOC concentration approximately 50% and the SUVA 54%. This significant DOC removal was reflected in the DXAA formation. The treated water formed only 41% of the DXAA formation of the Charleston source water after 72 hours of incubation. The

DXAA bromine incorporation of the Charleston source and coagulated waters also followed trends similar to the Metedeconk River waters studied (Figure 6.7). The n_D' of both the source and treated waters decreased over time, and coagulation resulted in more DXAA bromine incorporation than the source water, again presumably because of the greater Br^-/DOC ratio in the treated water.

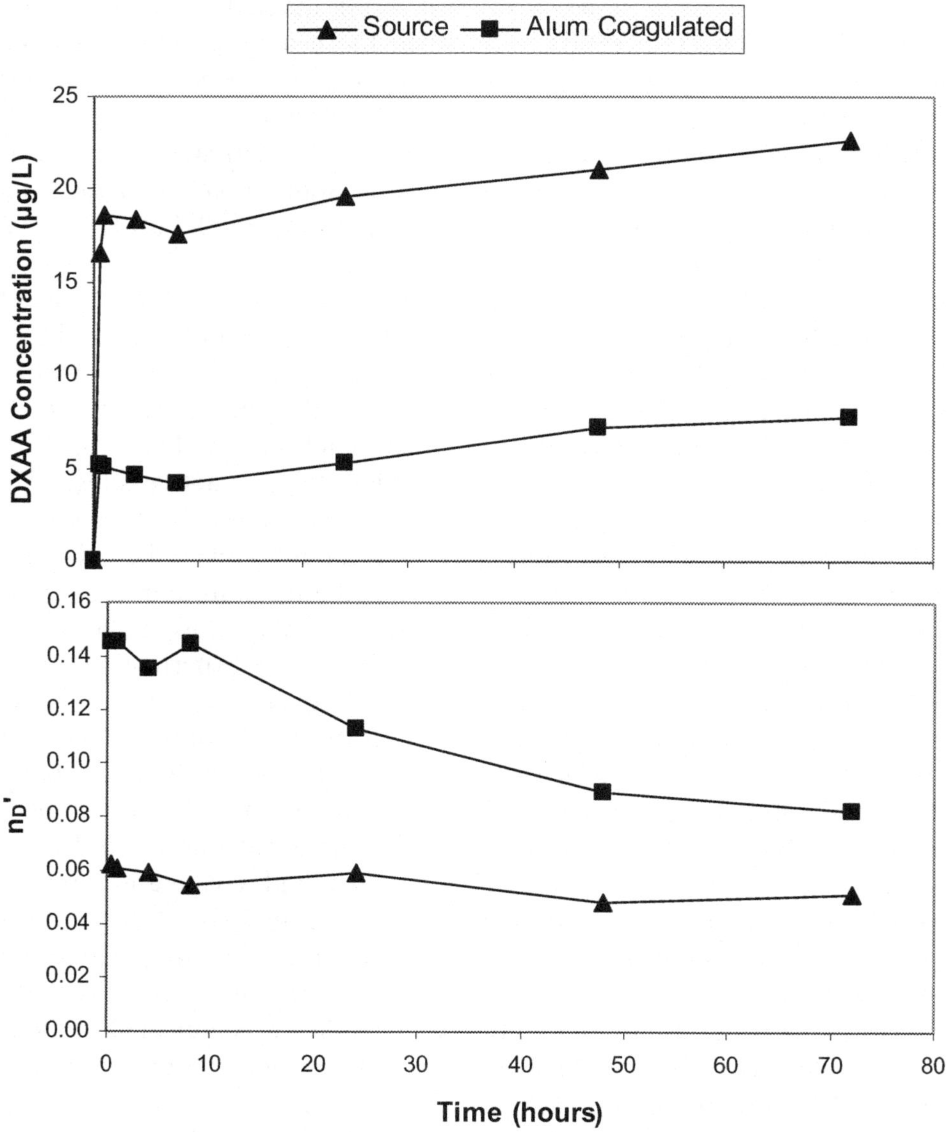

Figure 6.7 DXAA formation kinetics and bromine incorporation ratio (n_D') of Metedeconk River source and coagulated water (pH 8)

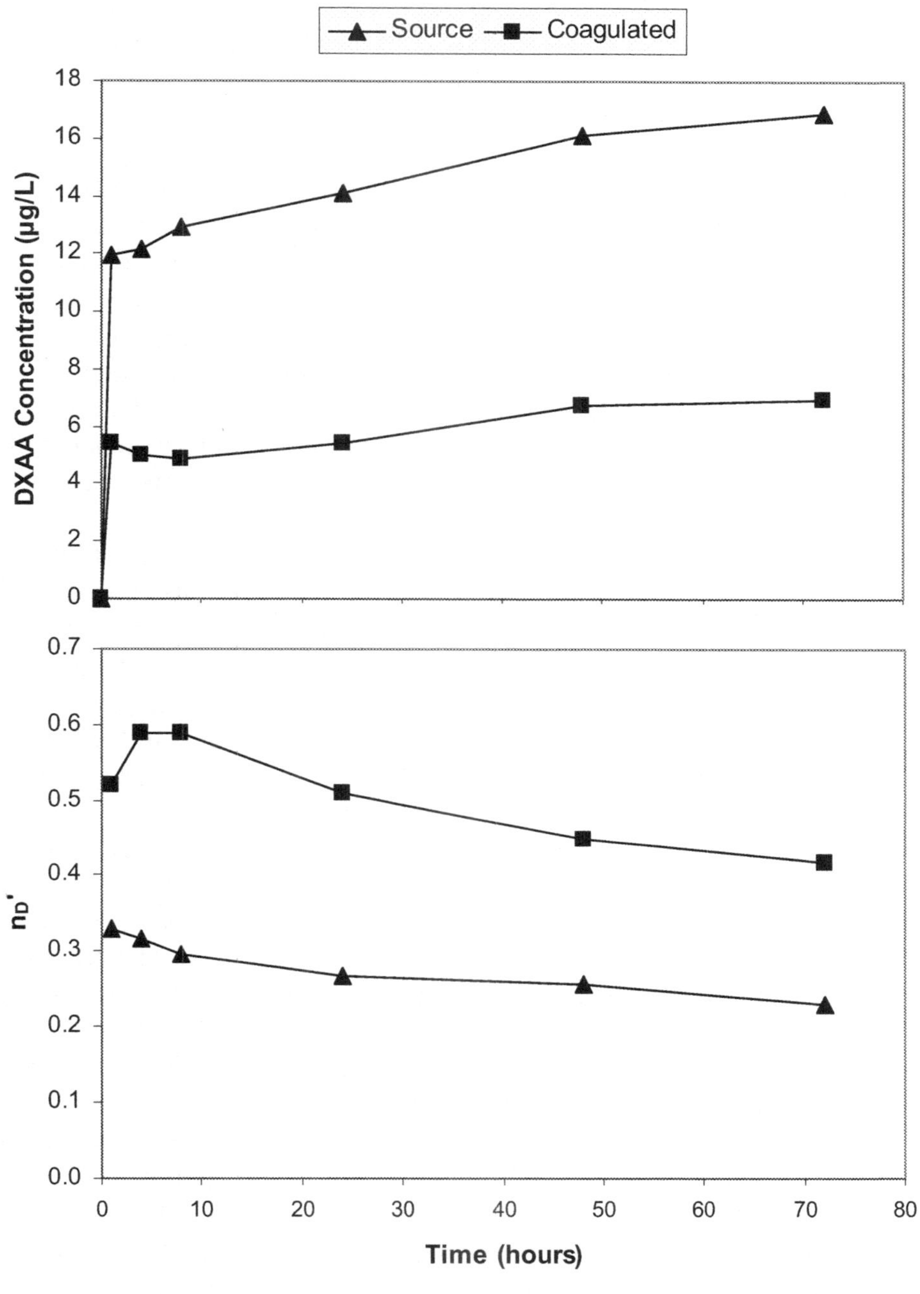

Figure 6.8 DXAA formation kinetics and bromine incoporation ratio (n_D') of Charleston source and coagulated water (pH 8)

The impact of softening Lake Austin source water on DXAA formation kinetics is illustrated in Figure 6.9. The softened water formed almost 81% of the DXAA that formed in the source water after 72 hours of incubation. Softening Lake Austin water had a minimal effect on reducing DXAA formation when compared to the coagulated Metedeconk River and Charleston source waters (Figure 6.7 and Figure 6.8). This relative difference can be related to the degree of organic carbon removal; softening Lake Austin source water only removed 11% of the DOC whereas coagulating Metedeconk River and Charleston source waters removed 47% and 50% of the DOC, respectively. Similarly, softening only reduced the SUVA 18% in Lake Austin while coagulating Metedeconk River source water, for example, decreased the SUVA 57%.

The bromine incorporation ratios of the source and softened Lake Austin waters are also illustrated in Figure 6.9. Again, as with Metedeconk River and Charleston, the n_D' of the treated water was typically greater than the source water. However, the relative increase in bromine substitution due to softening was not as significant as that of the coagulated waters. The batch screening experiments (Chapter 4, Figure 4.13) also indicated, on average, only a slight increase in bromine substitution as a result of softening Lake Austin source water. Softening only removed a small amount of organic carbon, which did not significantly decrease the number of NOM reactive sites that both the bromine-substituted and chlorine-substituted haloamines compete for. Therefore, softening had minimal influence on the degree of bromine incorporation.

The DXAA formation kinetics and bromine incorporation ratio of Biscayne Aquifer source and softened waters are shown in Figure 6.10. As with Lake Austin, softening Biscayne Aquifer source water did not remove a substantial amount of NOM, only reducing the DOC 21% from almost 12 mg/L to about 9.5 mg/L and the SUVA 18%. The Biscayne Aquifer source and softened waters, which both have relatively high SUVA values and DOC concentrations, displayed low reactivities that tracked well with the chlorination UFC and SUFC results (Table 6.1). Under chlorination UFC tests, the source water was the least reactive of all the source waters tested, and the chlorination SUFC yield in the treated water was virtually the same as in Lake Austin water, both of which were considerably less than the UFC yields in Charleston and Metedeconk River. During the chloramination experiments, these waters yielded less than 1.2 μg DXAA/mg DOC while the Lake Austin source and softened waters yielded approximately 2.5 and 2 μg DXAA/mg DOC, respectively. Similarly, Symons et al. (1998) studied Biscayne Aquifer source water from another location in the aquifer and observed a DXAA yield of only 1.85 μg/mg DOC after 48 hours of chloramination. Treatment, as well as spatial and temporal variations, can influence the composition, and therefore the reactivity the organic carbon in source waters. Lake Austin, Metedeconk River, and Charleston are surface waters, whereas Biscayne Aquifer is a groundwater. In addition, the NOM of the Biscayne Aquifer groundwater source is comprised of predominately hydrophobic DOC; only about 10% of the DOC is hydrophilic. Approximately 20% of the Lake Austin and Metedeconk River source water DOC is hydrophilic (Table 6.1), which can be substantially more reactive than the non-polar hydrophobic fractions in forming HAAs during chloramination (Hwang et al. 2000). Also, analogous to Lake Austin water, the bromine incorporation ratio was only slightly greater in the softened water due to the minimal decrease in organic carbon concentration achieved by softening this source water.

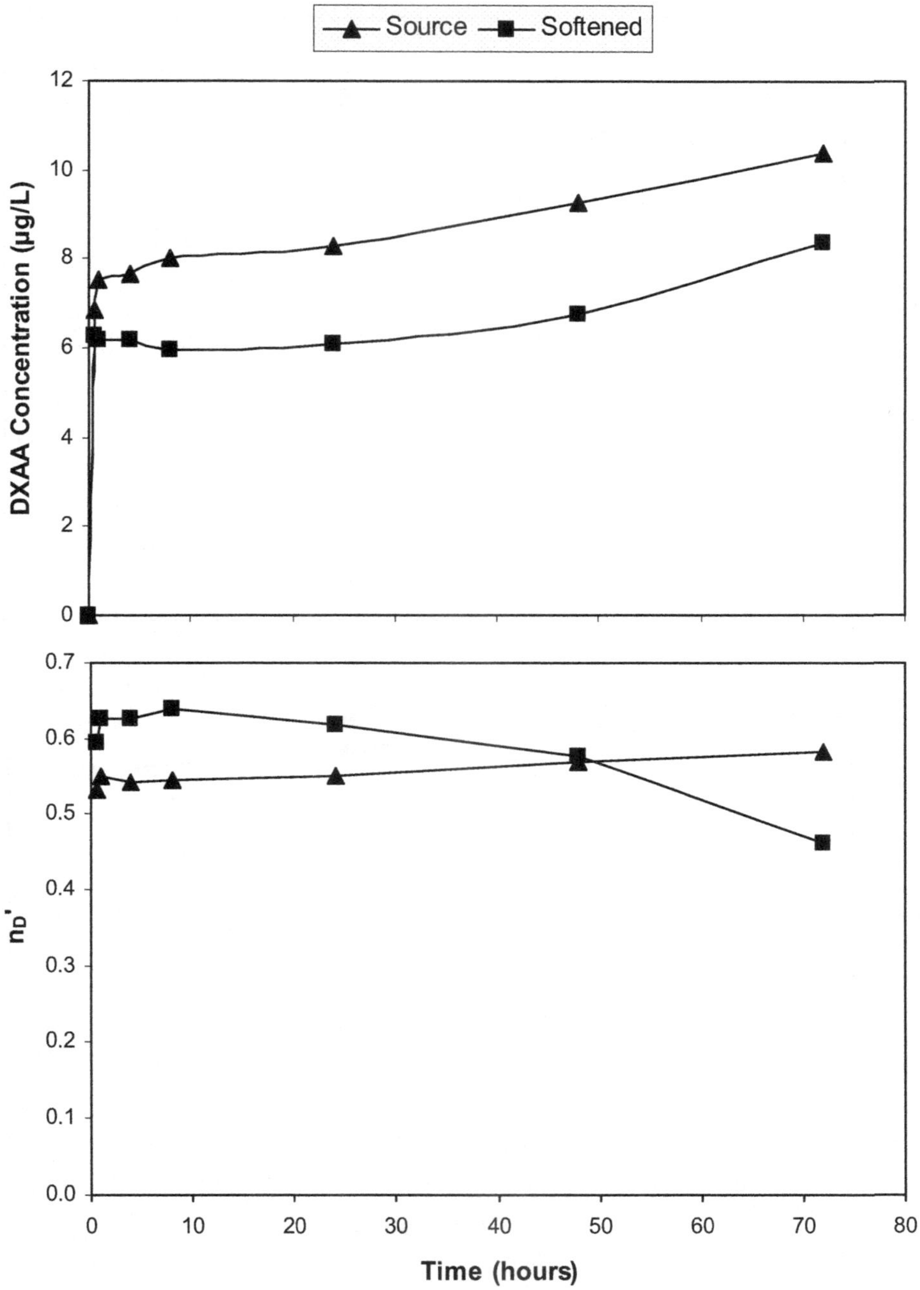

Figure 6.9 DXAA formation kinetics and bromine incorporation ratio (n_D') of Lake Austin source and softened water (pH 9)

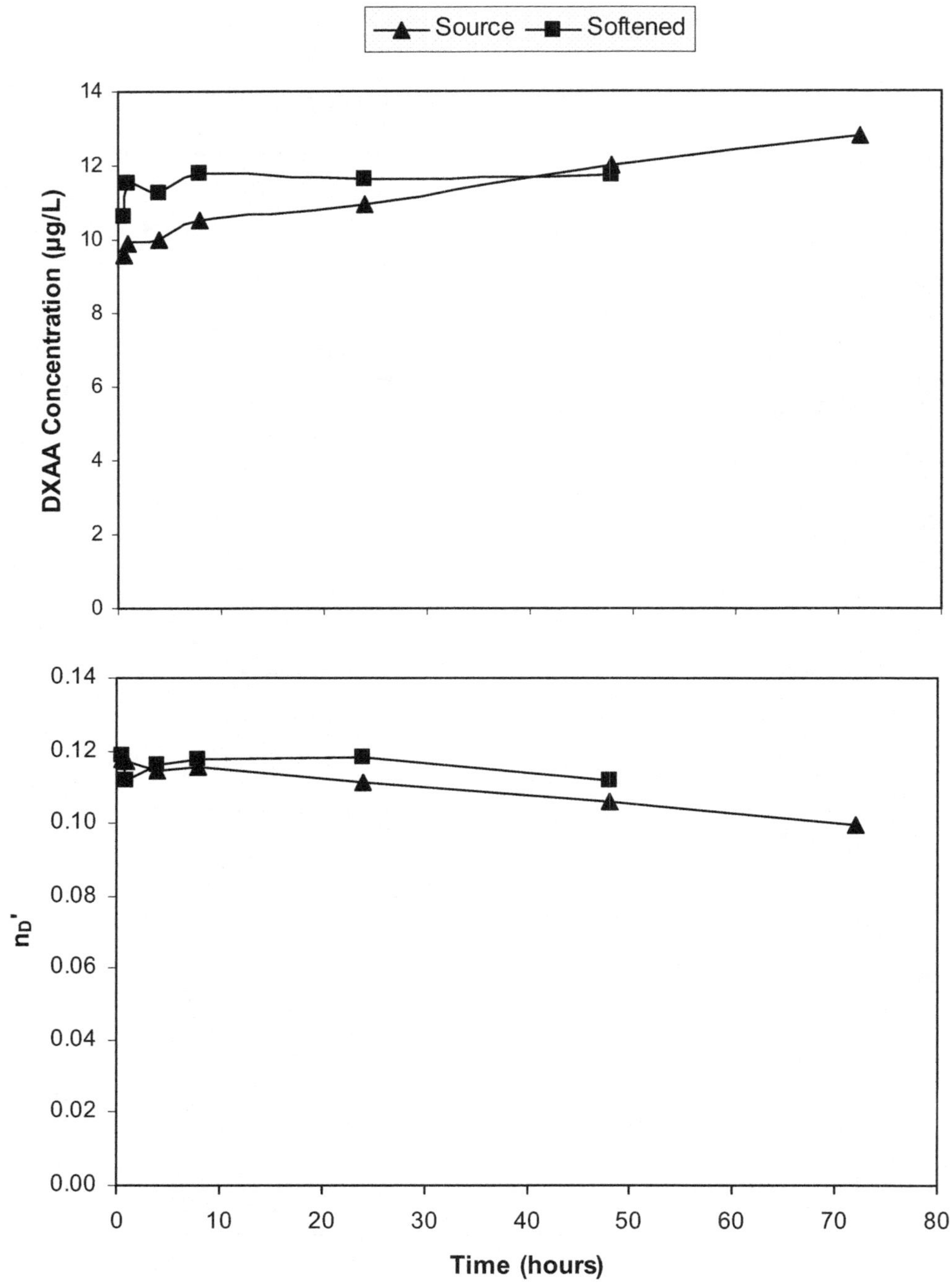

Figure 6.10 DXAA formation kinetics and bromine incorporation ratio (n_D') of Biscayne Aquifer source and softened water (pH 9)

Both the reactivity and the concentration of the organic carbon in waters can influence HAA formation. To determine if observed differences in DXAA formation were truly related to differing reactivity, or just a reflection of DOC concentration, DXAA yields were examined. Figure 6.11 and Figure 6.12 illustrate the effect of coagulation on DXAA yields in Metedeconk

River and Charleston source waters, respectively. The coagulated Metedeconk River water produced 73% of the DXAA mass yield of that of the source water after 72 hours. Similarly, the coagulated Charleston water produced about 78% of the source water DXAA mass yield. The decreased DXAA mass yields indicate this treatment strategy preferentially removed the more reactive fraction of NOM.

The impact of softening on DXAA mass yield was also investigated in Lake Austin and Biscayne Aquifer source waters (Figure 6.13 and Figure 6.14, respectively). Softening these source waters had less of an effect on DXAA yield than coagulation of the Metedeconk River and Charleston source waters. Softening Lake Austin water decreased the reactivity slightly; the yield in the softened water was approximately 90% of that in the source water. Conversely, the softened Biscayne Aquifer water was more reactive; the DXAA yield increased 24% but was quite small both before and after treatment. The follow-up batch screening experiments (Chapter 5) with Biscayne Aquifer water showed this trend in DXAA yield as well. Singer et al. (1999) studied Turnpike Aquifer water from Palm Beach County, Florida. This groundwater had 14.4 mg/L of TOC, 156 µg/L of bromide, an alkalinity of 256 mg $CaCO_3$, and a SUVA of 2.76 L/mg-m and, therefore, was similar to Biscayne Aquifer source water. Coagulation of Turnpike Aquifer source water increased the DCAA yield 65%, from 0.97 µg DCAA/mg DOC to almost 1.6 µg DCAA/mg DOC.

Based on the observed DXAA yields, softening preferentially removed the more reactive DXAA precursors in Lake Austin water and the less-reactive DXAA precursors in Biscayne Aquifer source water. Softening Lake Austin and coagulating Metedeconk River source waters principally removed the higher molecular weight fraction (>3000 AMW) of the NOM (Table 6.1) and decreased the reactivity. In contrast, softening Biscayne Aquifer source water predominantly removed the lower molecular fraction (<3000 AMW) of the NOM (Table 6.1), but increased the reactivity (Figure 6.14). These results imply that the higher molecular weight fractions of NOM may be more reactive with chloramines in forming DXAA.

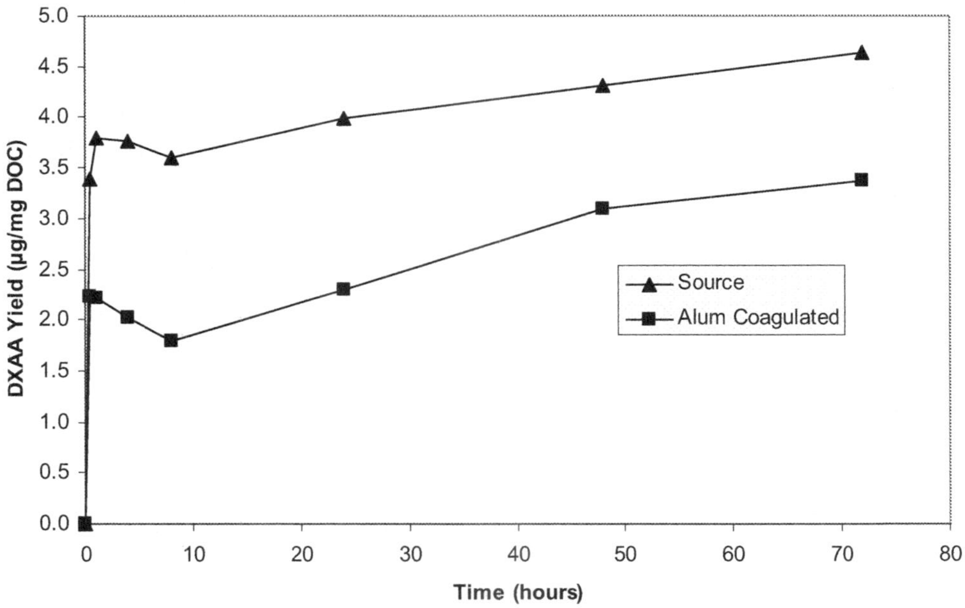

Figure 6.11 DXAA yield in Metedeconk River source and coagulated water (pH 8)

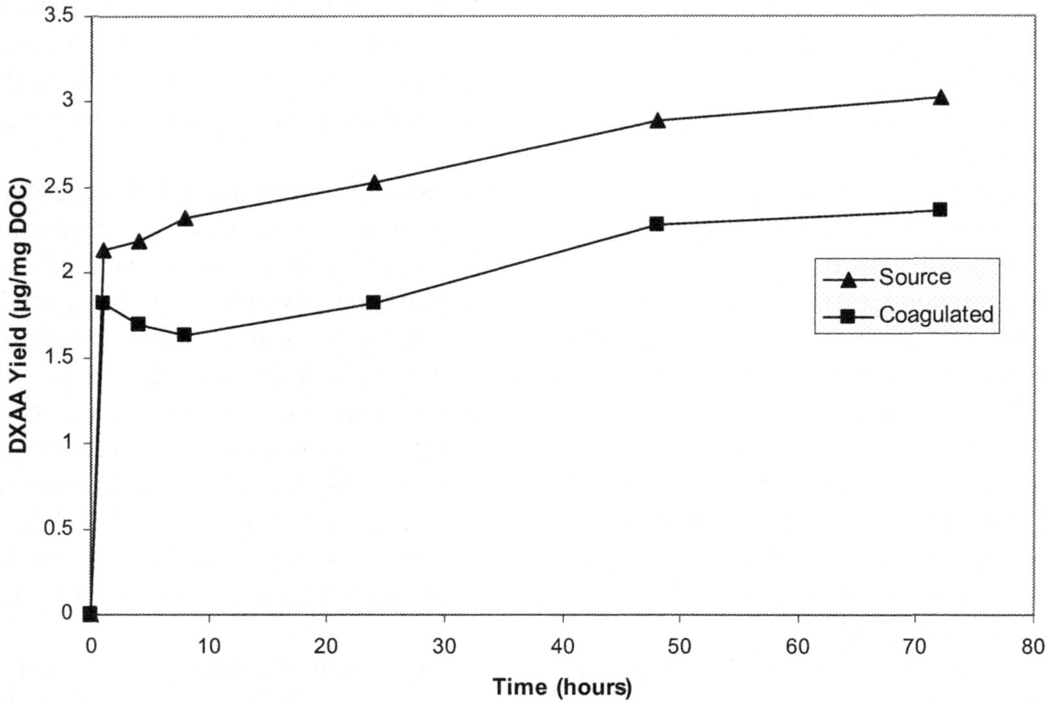

Figure 6.12 DXAA yield in Charleston source and coagulated water (pH 8)

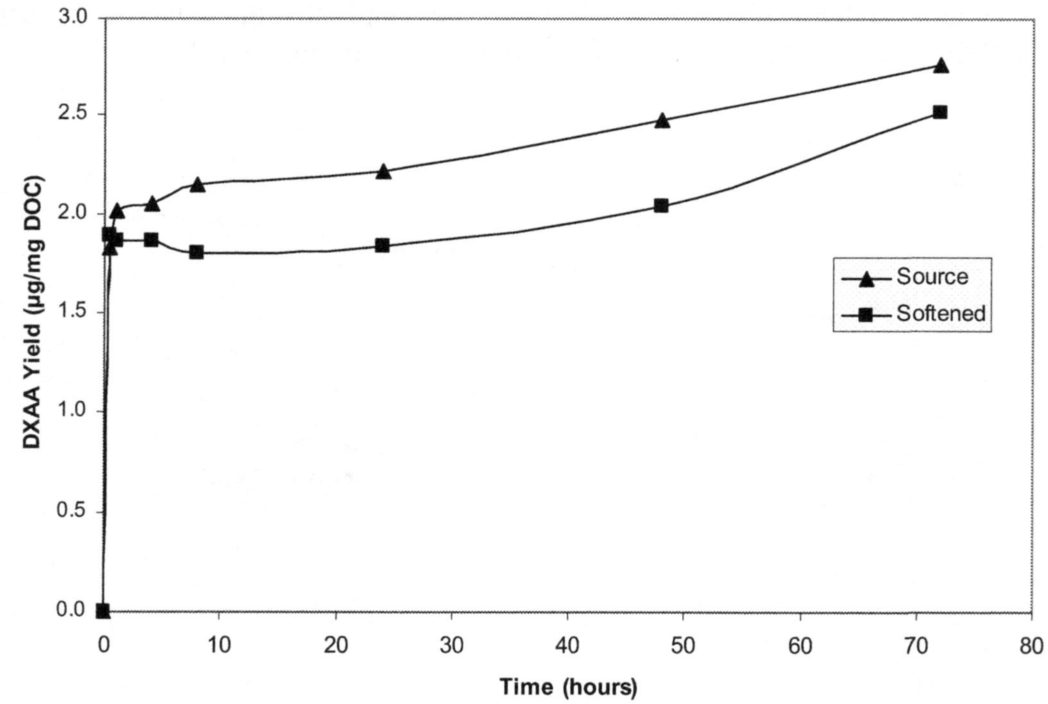

Figure 6.13 DXAA yield in Lake Austin source and softened water (pH 9)

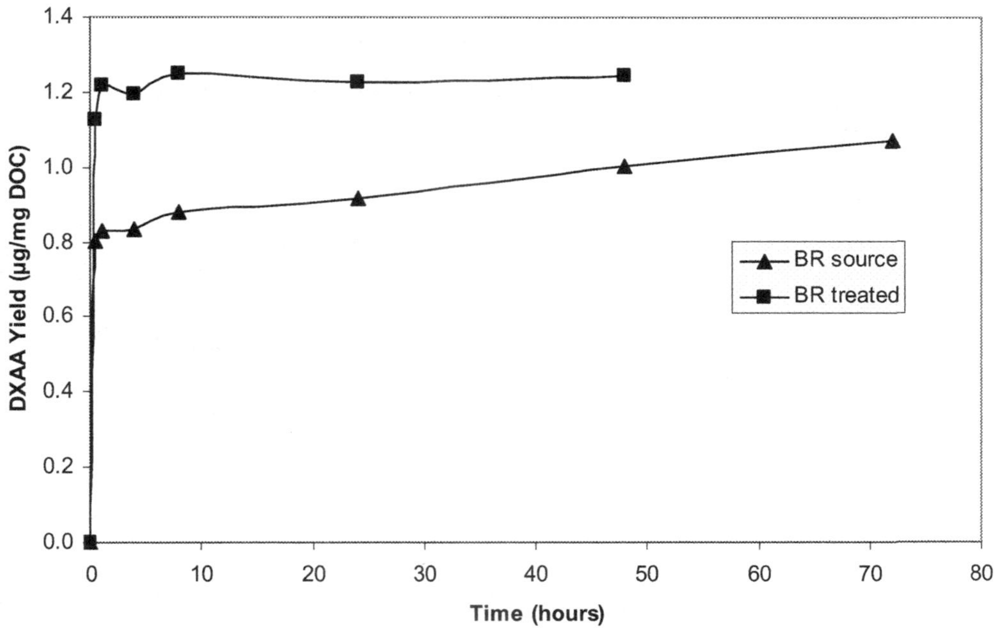

Figure 6.14 DXAA yield in Biscayne Aquifer source and softened water (pH 9)

Prechlorination

The impact of prechlorination on DXAA formation is important because many utilities have a period of free chlorination prior to ammonia addition to achieve CT requirements. Three different waters (Metedeconk River coagulated, Lake Austin source, and Biscayne Aquifer softened) were used to study the impact of prechlorination on DXAA formation. The type of water studied (*i.e.*, treated or source) corresponded to the type of water that is prechlorinated at the actual drinking water treatment plants. In these experiments, described in more detail in Chapter 3, samples were prechlorinated for either 5 or 20 minutes prior to ammonia addition. The relative doses of chlorine and ammonia were determined for each water in order to achieve both a 4/1 Cl_2/N ratio and a 2 mg/L combined chlorine residual after 48 hours of incubation.

Prechlorination increased DXAA formation as well as formation of mono- and tri-halogenated acetic acids. The relative effects of prechlorination on DXAA formation were similar in all of the waters studied (Figure 6.15). In Lake Austin source water, for example, chloramination resulted in only 9 µg/L DXAA after 48 hours, while prechlorinating for 5 and 20 minutes formed 25 µg/L and 35 µg/L of DXAA, respectively. Also, as expected, the longer the period of free chlorination the greater the DXAA formation; however, the increase was much less than proportional to the prechlorination time. In addition, a majority of the DXAA formation occurred during the prechlorination period. At least 75% of the 72-hour DXAA concentration formed during prechlorination in all the waters studied. Typically, however, greater than 90% of the DXAA formed during prechlorination. This indicates that even though prechlorination caused increased DXAA formation, very little additional DXAA formed during the chloramination period.

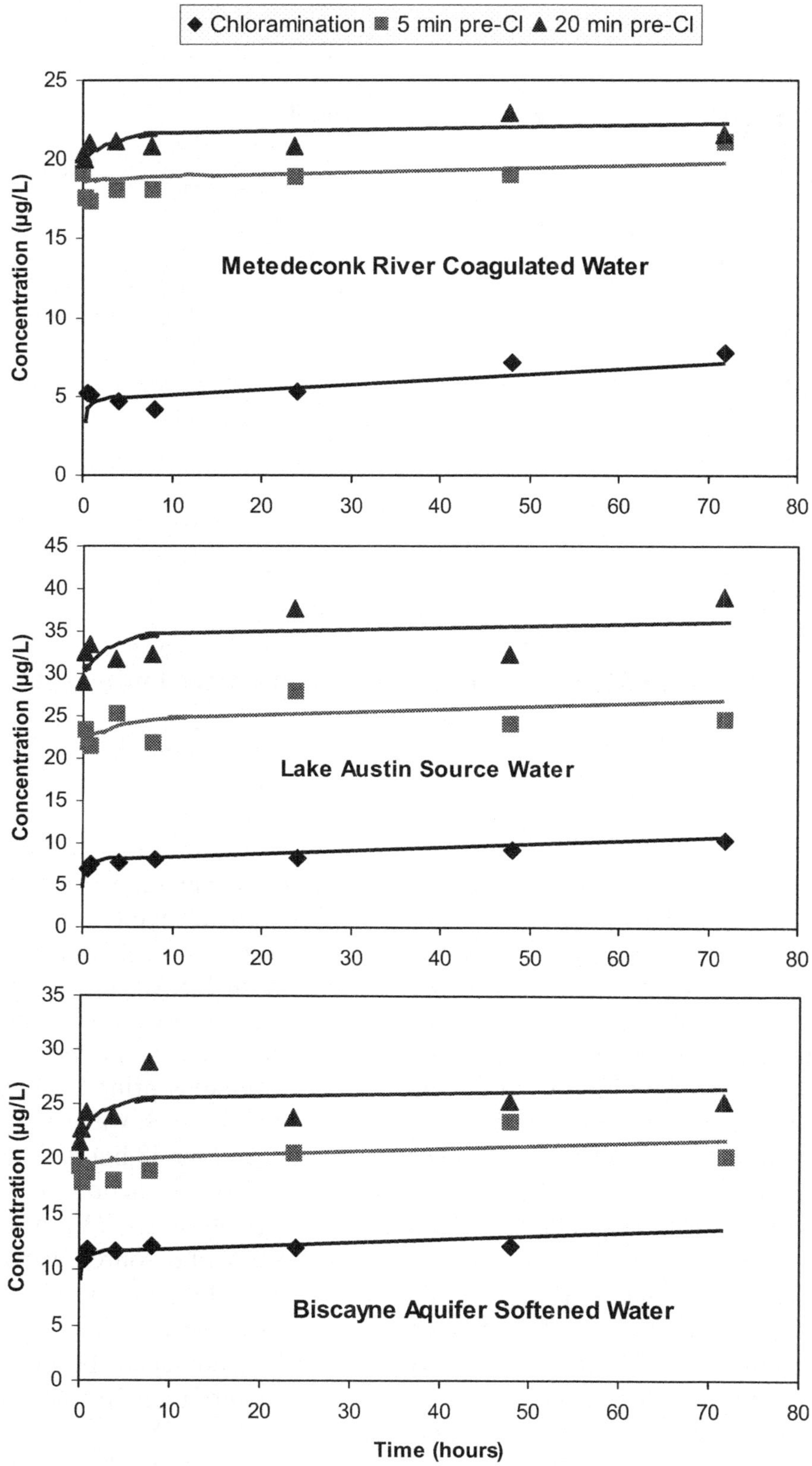

Figure 6.15 Effect of prechlorination on DXAA formation kinetics (time zero corresponds with start of chloramination)

The DXAA formation during the chloramination period was examined in greater detail by subtracting the formation as a result of prechlorination from the total DXAA concentration at each time interval (Figure 6.16). Again, the trends were similar in all three water sources. Prechlorination of Metedeconk River coagulated and Biscayne Aquifer softened waters resulted in essentially no additional DXAA formation after ammonia addition. The Lake Austin source water data, however, were more scattered and had some outliers, yet still minimal additional DXAA formation occurred during the chloramination period. These results indicate that, during the prechlorination period, free chlorine reacts with most of the fast reactive NOM sites; therefore, after subsequent ammonia addition and chloramine formation very little additional DXAA is formed. This is of practical significance to plants that prechlorinate as it indicates that most all of the HAAs form during the chlorination step. These results also suggest that both chlorine and monochloramine (or monochloramonium ion) react with the same fraction of the NOM to produce DXAAs, or that the DXAAs formed during chloramination do so as a result of the HOCl that is formed by monochloramine hydrolysis.

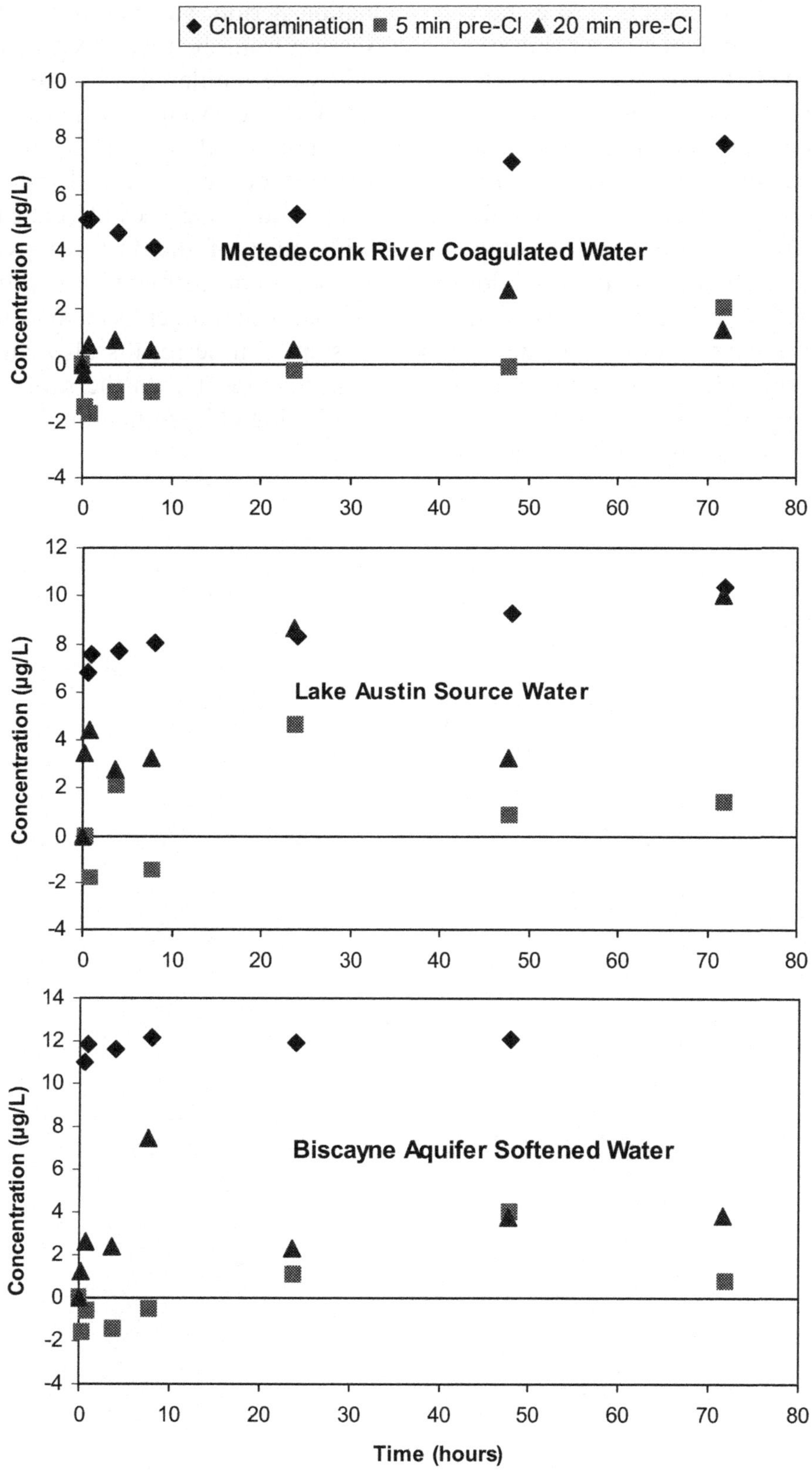

Figure 6.16 DXAA formation during the period of chloramination only (DXAA formed during chloramination = Total DXAA – DXAA formed during prechlorination)

Prechlorination also impacted THM formation (Table 6.2). Chloramination typically produces negligible THM formation, whereas periods of free chlorination cause THMs to form, as is well known. The longer the period of free chlorination the greater the THM formation, and in both Lake Austin and Biscayne Aquifer waters, once the prechlorination period was complete (ammonia was added and chloramines formed) minimal THM formation occurred. In contrast, quenching the prechlorination period with ammonia did not limit the THM formation as effectively in coagulated Metedeconk River water; THM concentrations doubled between the prechlorination periods of either 5 or 20 minutes and 72 hours. However, the total THM concentrations after 72 hours were only 15 and 23 µg/L for the 5 and 20 minute prechlorination periods, respectively. The additional THM formation in coagulated Metedeconk River water may be linked either to its NOM characteristics or to water quality characteristics that affect the concentration of the reactive halogenating species. Teng and Veenstra (1996) studied the THM formation potential of different molecular weight fractions isolated from Kaw Reservoir, Oklahoma water. They found that chloramination produced only minimal levels of THM formation potential except in the <1000 molecular weight fraction. Approximately 78% of DOC in Metedeconk River coagulated water is in the more reactive <3000 molecular weight range, whereas less than 33% of the DOC in both the Lake Austin source and Biscayne Aquifer softened waters is this range. In addition, the different treatment conditions, specifically the pH that each water was subjected to (Metedeconk River coagulated water was incubated at pH 8, while Lake Austin source and Biscayne Aquifer softened waters were incubated at pH 9), may have influenced THM formation patterns along with the differing reactivity of NOM in the water sources (Metedeconk River source water was significantly more reactive than the other source waters studied with respect to HAA_9 yield under chlorination UFC – Table 6.1).

Table 6.2
THM concentrations of kinetic experiments with prechlorination

Water	Prechlorination Time (min)	Incubation Time	AVERAGE THM Concentrations TCM Conc. (µg/L)	BDCM Conc. (µg/L)	DBCM Conc. (µg/L)	TBM Conc. (µg/L)	TTHM Conc. (µg/L)
Lake Austin	5	5 min	10.0	6.4	1.7	0.2	18.4
Lake Austin	5	72 hours	13.4	6.5	1.7	0.4	22.0
Lake Austin	20	20 min	17.0	13.8	6.5	1.1	38.3
Lake Austin	20	72 hours	22.0	12.7	6.8	0.9	42.4
Metedeconk River	5	5 min	6.9	0.4	ND	ND	7.3
Metedeconk River	5	72 hours	13.9	0.9	ND	ND	14.8
Metedeconk River	20	20 min	11.8	2.5	ND	ND	14.3
Metedeconk River	20	72 hours	19.5	3.6	ND	ND	23.1
Biscayne Aquifer	5	5 min	8.3	ND	ND	ND	8.3
Biscayne Aquifer	5	72 hours	11.5	ND	ND	ND	11.5
Biscayne Aquifer	20	20 min	12.2	0.7	ND	ND	13.0
Biscayne Aquifer	20	72 hours	14.2	ND	ND	ND	14.2

ND = Not Detected

TRENDS IN DXAA SPECIATION

As mentioned in previous chapters, bromide present in source waters may be incorporated into haloamines such as bromamines and chlorobromamines, and may also be present as HOBr during chloramination. Increasing bromide concentrations increase the bromine-substituted reactive species that are available to react with NOM. These bromine-substituted reactive species may compete with non-bromine substituted reactive species (*e.g.*, chloramines and HOCl) for the available NOM reactive sites, resulting in the formation of bromine-substituted HAAs. In addition, during the prechlorination period, HOCl and HOBr compete to react with NOM to form HAAs. To investigate the effect of prechlorination on the formation of bromine-substituted DXAAs, Br^-/DOC ratios and Br^-/average Cl^+ molar ratios were investigated after 48 hours of incubation. Symons et al. (1993) studied the influence of these two ratios on bromine substitution during chlorination; both higher Br^-/DOC and Br^-/average Cl^+ ratios led to greater bromine incorporation into DBPs.

Figure 6.17 and Figure 6.18 illustrate the influence of prechlorination and Br^-/DOC ratio on DXAA speciation. In general, higher Br^-/DOC ratios resulted in increased DXAA bromine incorporation. Experiments with Metedeconk River and Charleston source and coagulated waters (Figure 6.17) were performed at pH 8, while those with source and softened Lake Austin and Biscayne Aquifer waters (Figure 6.18) were performed at pH 9 to match the typical conditions of their respective treatment plants. In these figures, the line shows the trend for chloramination alone. The data are limited, but the results indicated that prechlorination at low Br^-/DOC ratios may increase the formation of HAAs that contain bromine. At higher Br^-/DOC ratios, prechlorination did not appear to have a significant effect on bromine incorporation.

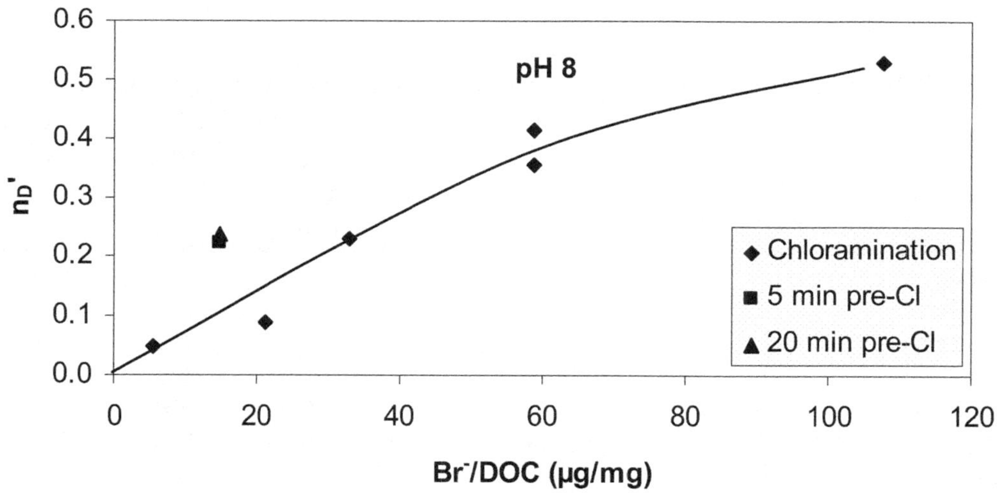

Figure 6.17 Effect of Br^-/DOC ratio on DXAA speciation in Metedeconk River and Charleston source and coagulated waters

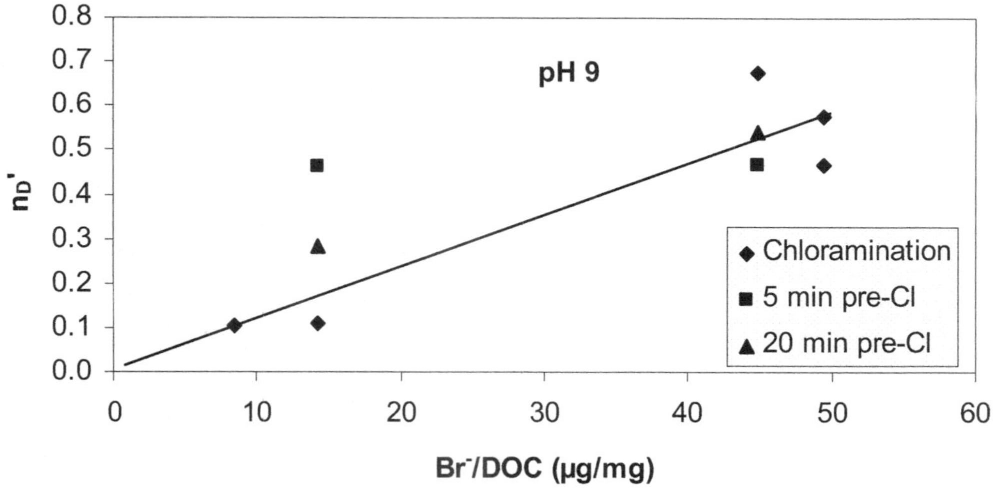

Figure 6.18 Effect of Br^-/DOC ratio on DXAA speciation in Lake Austin and Biscayne Aquifer source and softened waters

Both chlorine- and bromine-substituted reactive species are in competition for available NOM reactive sites. Symons et al. (1996) observed that the degree of bromine substitution increases as the initial bromide to average free available chlorine ratio increases. In Figure 6.19 and Figure 6.20 this ratio was extended to bromide to average total chlorine (determined as the average between the free chlorine dose and the combined chlorine residual) molar ratios to study the effect of prechlorination on DXAA speciation. Again, Metedeconk River and Charleston source and coagulated were incubated at pH 8 (Figure 6.19), and Lake Austin and Biscayne Aquifer source and softened waters were incubated at pH 9 (Figure 6.20). The line in these figures represents the trend for chloramination alone. Again, the data are limited, but increased Br^-/Cl^+ molar ratios led to increased n_D' at low Br^-/DOC ratios. The pH may also have an impact on bromine incorporation. Cowman and Singer (1996) observed that the formation of bromine-substituted HAAs was enhanced by lowering the pH from 8 to 6 during chlorination. The increase in n_D' at lower pH may be due to a greater concentration of HOBr at lower pH. The concentration of HOBr and HOCl available to react with NOM is dependent on the speciation between these more reactive forms and hypobromite and hypochlorite, respectively (Reaction 6.1 and Reaction 6.2). The acidity constants of these reactions show that at average pH values of most finished drinking waters more bromine will be in its more reactive form than will chlorine, and as the pH is increased lesser concentrations of both species will be present in their more reactive forms. Therefore, n_D' would be expected to decrease as pH increases.

$$HOCl + H_2O \leftrightarrow OCl^- + H_3O^+ \qquad K_{Cl} = 10^{-7.55} \qquad \text{(Reaction 6.1)}$$
$$HOBr + H_2O \leftrightarrow OBr^- + H_3O^+ \qquad K_{Br} = 10^{-8.68} \qquad \text{(Reaction 6.2)}$$

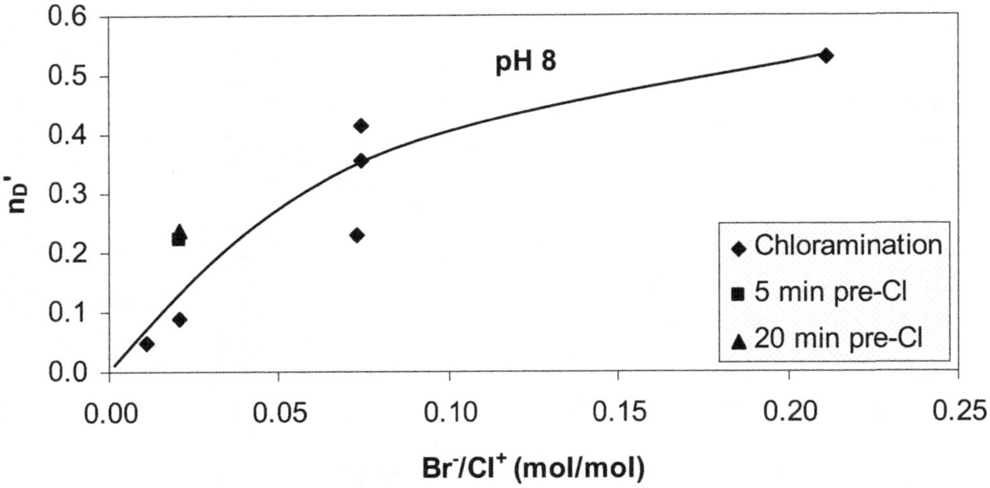

Figure 6.19 Effect of Br^-/Cl^+ on DXAA speciation in Metedeconk River and Charleston source and coagulated waters

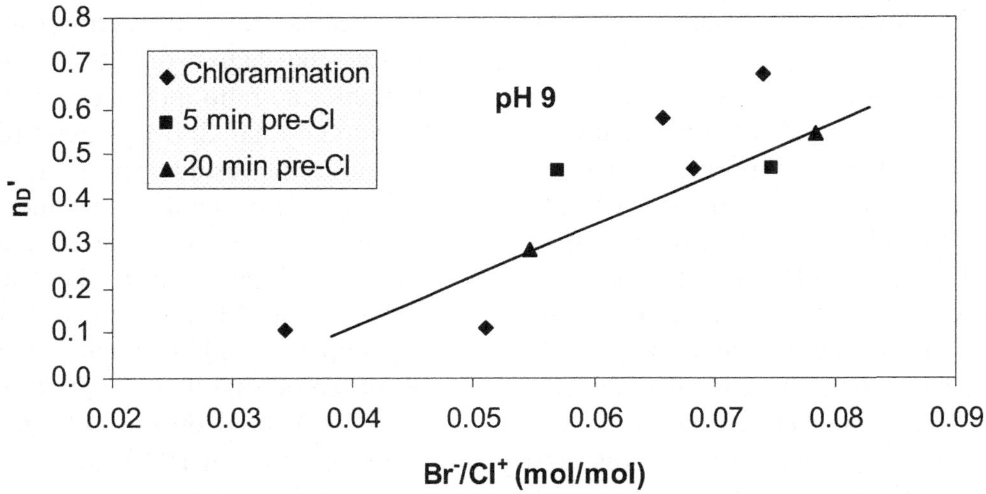

Figure 6.20 Effect of Br^-/Cl^+ on DXAA speciation in Lake Austin and Biscayne Aquifer source and softened waters

DAVIS DWTP (LAKE AUSTIN) AND DISTRIBUTION SYSTEM

The Davis Drinking Water Treatment Plant and associated distribution system were sampled to confirm the HAA formation kinetics measured in the laboratory. This plant chlorinates Lake Austin source water for approximately 6 minutes prior to ammonia addition and softening. Approximately 1.5 mg/L of free chlorine is added to the pH 7.7 source water to produce a free chlorine residual of 0.8 mg/L. Then, 1.35 mg/L of chlorine is added with ammonia and lime during rapid mix to produce a combined chlorine residual of approximately 2.18 mg/L at approximately a 4/1 Cl_2/N ratio. This converts the free chlorine to chloramines and softens the water, which increases the pH to greater than 9. Samples were taken from the plant immediately before ammonia addition, after rapid mixing (ammonia and lime addition), after settling, and after filtration. The distribution system was also sampled at three different points that provided approximate residence times of 1.1, 2.4, and 3.1 days. The DXAA formation in the Davis DWTP and distribution system is illustrated in Figure 6.21. A majority of the DXAA formation occurred during the period of free chlorination with minimal additional formation occurring after ammonia was added to form chloramines. In fact, all DXAA formation occurred within the treatment plant itself, indicating that HAA control strategies should be focused on the chlorination step in plants that prechlorinate prior to chloramination.

The Lake Austin source batch kinetic laboratory data are also shown in Figure 6.21. The trends in DXAA formation of the laboratory data parallel the formation at the drinking water treatment plant. However, the DXAA formation at the Davis DWTP was greater than the formation observed in the laboratory. This discrepancy was due to differences in pH during prechlorination. During chlorination, HAA concentrations increase as the pH is decreased (Stevens et al. 1989). Carlson and Hardy (1998) studied the impact of pH on HAA_5 formation at short contact times and observed that the pH of maximum HAA formation ranged from 6.5 to 7.5 depending on the water source. Therefore, the laboratory experiments formed less DXAA because they were conducted at pH 9, while prechlorination at the treatment plant occurred at pH 7.7.

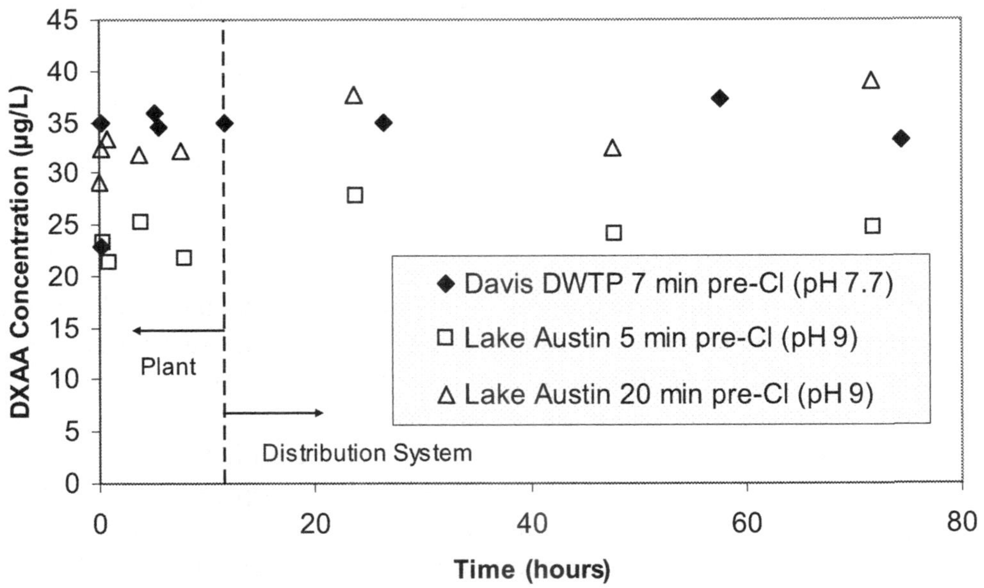

Figure 6.21 DXAA Formation in the Davis DWTP and Distribution System

The DXAA speciation in the Davis DWTP is depicted in Figure 6.22. DCAA accounted for a majority of the total DXAA, but significant concentrations of the bromine-substituted species (DBAA and BCAA) also formed because Lake Austin water had a significant bromide concentration (150 μg/L). The rate of formation of each species is also of interest. Almost all of each species formed during the prechlorination period and remained stable throughout the rest of the treatment plant and distribution system.

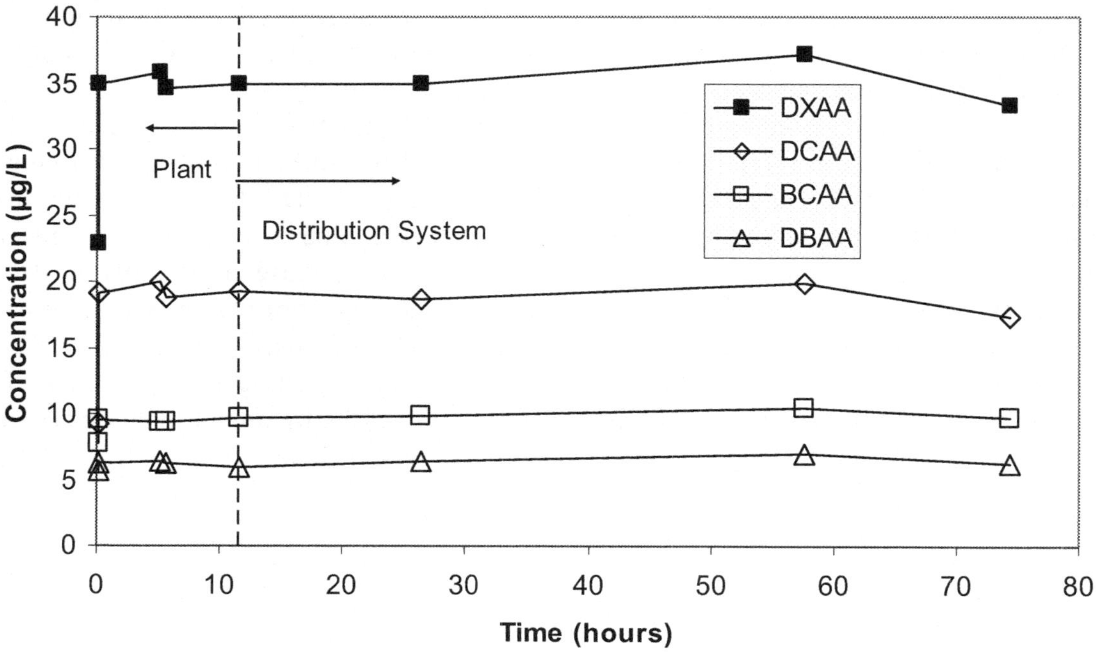

Figure 6.22 DXAA Speciation - Davis DWTP

Chloramination limits the formation of mono-and tri-halogentated HAAs (MXAAs and TXAAs, respectively), predominately forming DXAAs. During traditional chloramination, DXAAs typically accounted for approximately 90% of the HAA_9 formation. Even brief periods of prechlorination, however, may cause significant concentrations of MXAAs and TXAAs to form as well. The HAA_9 formation at the Davis water treatment plant is shown in Figure 6.23. DXAAs were the predominant species formed, accounting for approximately 60% of the total HAA_9 formation, however, significant concentrations of TXAAs and some MXAAs also formed. Prechlorination, therefore, increased the formation of the other regulated HAA species, forming approximately 60 μg/L of HAA_9 and almost 47 μg/L of HAA_5 (regulated at 60 μg/L). Again, as with the DXAAs, almost all of the formation occured in the prechlorination step, and the species remained stable throughout the treatment plant and distribution system.

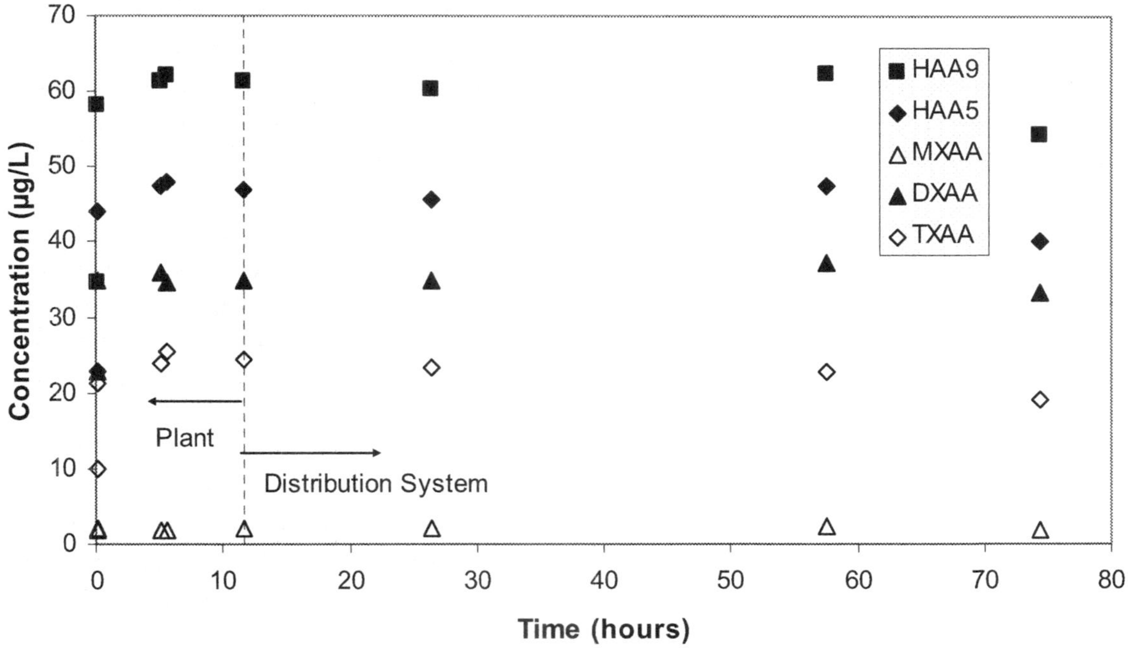

Figure 6.23 Total HAA Formation in the Davis DWTP and Distribution System

SUMMARY

Experiments were conducted on both source and treated waters from four different locations to study the impact of pH, bromide concentration, temperature, and treatment on DXAA formation kinetics during chloramination. The waters selected, Lake Austin, Metedeconk River, Biscayne Aquifer, and Charleston, spanned the typical range of bromide and DOC concentrations, as well as alkalinity and SUVA, common in source waters. The traditional approach to chloramination involves the addition of ammonia and chlorine simultaneously very early in the treatment train. However, many plants have a significant period of free chlorination prior to ammonia addition for purposes of meeting CT requirements (Seidel et al.2000; Connell et al. 2000); and therefore, the key objectives of the kinetic experiments was to study the impact of prechlorination on DXAA formation in addition to the aforementioned water quality and chloramination variables.

The kinetics of DXAA formation during chloramination are rather fast. For example, approximately 73% of the 72-hour DXAA concentration formed after only 0.5 hours and almost 82% formed after 1 hour in Metedeconk River source water. Since chloramination often occurs at the beginning of treatment plants, a vast majority of the DXAA formation will occur within the water treatment plant itself and only a small amount of formation will occur in the distribution system.

DXAA speciation is also important because the health effects of bromine-substituted HAAs may be of greater concern. In general, higher Br^-/DOC ratios promoted greater bromine incorporation into DXAAs. In order to evaluate the impact of bromide on DXAA formation, 0.5 mg/L of bromide was added to Metedeconk River source water, which initially had a low ambient bromine concentration (27 µg/L). The addition of bromide caused only a slight increase in molar DXAA formation, but significantly shifted the speciation to the bromine-substituted species. In addition, the bromine-substituted species (BCAA and DBAA) formed very rapidly (< 0.5 hr);in contrast, DCAA formation kinetics were characterized by an initial period (~ 1 hour) of rapid formation followed by slow formation over the remainder of the 72-hour incubation period..

The impact of pH on DXAA formation kinetics was also investigated, and as expected, DXAA formation increased as the pH decreased. In Lake Austin source water, for example, a majority of the DXAA formation occurred before 1 hour; both the waters incubated at pH 7 and pH 9 had a fast period of formation followed by a slower period. However, more rapid kinetics was observed in the water incubated at pH 7 during both periods of formation. This trend indicates that acid catalyzed monochloramine reactions may be the cause of increased HAA formation as the pH decreased.

Temperature has also been shown to influence HAA formation and is especially important in southern climates where surface water temperatures can increase substantially during the summer. In addition to temperature variations, source water NOM characteristics also vary seasonally. Increasing temperature caused only a minimal increase in DXAA formation over the range tested in Charleston and Lake Austin treated waters. However, batch experiments with Lake Austin softened water indicated that increased HAA formation as a result of increased temperature may be more pronounced at lower pH and higher Cl_2/N ratios. Nevertheless, the effect of temperature alone on DXAA formation appears to be relatively small; seasonal changes in NOM concentration or reactivity are also likely to be important in situations in practice where large changes in DXAA formation are observed with seasonal temperature changes.

Treatment can also have a significant effect on DXAA formation. The effectiveness of treatment on DXAA formation is related to DOC removal; softening Lake Austin and Biscayne Aquifer source water resulted in minimal DOC removal, and therefore only a slight decrease in DXAA formation. However, alum coagulation of Metedeconk River and Charleston source water removed a larger fraction of the DOC, resulting in a more significant decrease in DXAA formation. Alum coagulation of both Metedeconk River and Charleston source water was beneficial in decreasing DXAA yield, indicating preferential removal of DXAA precursors, whereas softening Lake Austin and Biscayne Aquifer source water had a minimal effect on yield. In addition, treatment did not have a significant effect on the rate of DXAA formation after 0.5 hours (the rapid period of DXAA formation); the rates during the slow period were similar for both the treated and source waters.

Prechlorination increased DXAA formation as well as formation of mono- and tri-halogenated acetic acid species. Also, as expected, the longer the period of free chlorination the greater the DXAA formation; however, the increase was much less than proportional to the prechlorination time. During the prechlorination period, free chlorine reacts with most of the fast reactive NOM sites; therefore, after subsequent ammonia addition and chloramine formation very little additional DXAA is formed. In addition, prechlorination at low Br^-/DOC ratios may increase the formation of HAAs that contain bromine relative to chloramination alone. At higher Br^-/DOC ratios, however, prechlorination did not appear to have a significant effect on bromine incorporation.

Prechlorination also impacted THM formation; the longer the period of free chlorination the greater the THM formation. Once the prechlorination period was complete (ammonia was added and chloramines formed) minimal THM formation occurred in the Lake Austin and Biscayne Aquifer waters studied. In contrast, quenching the prechlorination period with ammonia did not limit the THM formation as effectively in coagulated Metedeconk River water; THM concentrations doubled between the prechlorination periods of either 5 or 20 minutes and 72 hours. These results indicate the differing NOM characteristics (approximately 78% of DOC in Metedeconk River coagulated water is in the more reactive <3000 molecular weight range, whereas less than 33% of the DOC in both the Lake Austin source and Biscayne Aquifer softened waters is this range), treatment conditions (e.g., pH), or water quality characteristics that affect the concentration of the reactive halogenating species may have influenced THM formation patterns.

The Davis Drinking Water Treatment Plant, which chlorinates Lake Austin source water for approximately 6 minutes prior to ammonia addition and softening, and associated distribution system were sampled to confirm the HAA formation kinetics measured in the laboratory. All of the DXAA formation occurred within the treatment plant itself, indicating that HAA control strategies should be focused on the chlorination step in plants that prechlorinate prior to chloramination.

CHAPTER 7
MODELING

MODEL DEVELOPMENT

The objective of this task was to develop a simple and practical model to predict the DXAA concentration from a number of selected water quality characteristics. Modeling can be useful for engineers, researchers or plant operators to anticipate DXAA formation with seasonal changes, for optimizing treatment conditions, or when modifying treatment practice. Numerous analytical measurements of water characteristics, chloramination conditions, and DXAA concentrations were gathered throughout the research to create an extensive database for model development through linear regression analysis (SAS Institute Inc., 2001). The water quality variables and chloramination conditions were identified as the predictors, and the DXAA data were identified as the response.

The selection of the water characteristics and chloramination variables was based on previous research and conclusions from the initial and follow-up batch screening experiments reported in Chapters 4 and 5. The pH, bromide concentration, and water source were determined to be the most significant variables from these previous experiments. The Cl_2/N mass ratio was the next most significant contributor to DXAA formation in the initial batch screening experiments. Temperature and chloramine residual also impacted DXAA formation, but were much less influential than the aforementioned factors. Due to the minimal influence of the temperature parameter, it was not included in the model development, which helped to decrease the total number of terms. Other parameters considered for the model development were chloramine demand and the water quality characteristics, SUVA, UV_{254}, DOC, hydrophobic DOC, hydrophobic/DOC, and hydrophilic/DOC.

Similar parameters were considered by Westerhoff et al. (2001), such as pH, bromide concentration, temperature, chlorine dose, and reaction time, for the prediction of DXAA during chlorination disinfection. Instead of DOC concentration, their model used fixed concentrations of two types of DOC sites, fast and slow reactive fractions. Another difference between models was the kinetic component, which for simplicity was not included in the models generated in this study because the DXAA formation reactions were complete after a short period of time (as described in Chapter 6). The selected models were intended to simulate general trends rather than case specific results.

The response was expressed in various ways; however, DXAA mass concentration is appropriate as the regulations for HAA_5 are also expressed on a mass unit basis. DXAA mass and molar yields were also considered as a way of normalizing the DXAA formation to the DOC content of the water, thus providing a measure of the reactivity of a water source. In addition, the natural log, Ln, of the DXAA response was also selected because it provided better correlations between the model and actual data.

The models were developed from data generated during the initial batch screening experiments (112 points) with Lake Austin and Metedeconk River, the follow-up batch screening experiments (35 points) with Metedeconk River, Biscayne Aquifer source and treated, and St. Paul source and treated waters, and finally Charleston source water (32 points), which was not included in either of the screening experiments.

Regression analysis was used to create a linear model. At first, DXAA could not be accurately predicted without including a large number of interaction terms, which caused serious multicolinearity problems. Multicollinearity (or collinearity) refers to a condition in multiple linear regression analysis where two or more regressors carry redundant information about the response. If two or more regressors are too similar in their information about the response, a least squares solution for the parameter estimates may not be possible. Although this scenario is rare, it is common to have regressors similar enough that the least squares solution may be unstable and even misleading. The optimal scenario is when all the regressors are orthogonal to each other. Orthogonality is the highest level of independent information between two or more regressors. Two or more regressors are defined to be orthogonal if they carry no redundant information about the response.

A transformation of the DXAA response using the natural log significantly improved the general model including all the selected factors. Then, the regression model was optimized to best fit the data through the "Forward-Stepwise" function, by which the model terms are added one by one in order of highest to lowest significance. Afterward, a "Backward-Stepwise" approach was considered where all the possible model terms were included from the start and the less significant model terms were removed one by one.

The final regression model was selected based on the R^2 value, the Mallow's Cp, referred to as Cp, the distribution of the residuals, collinearity issues, and the Variance Inflation Factor (VIF). The Cp is a statistic that measures the variance and the bias in a model. Ideally, the Cp should be small and close in value to the number of regressors (terms) in the model. The residuals are defined as the actual measured DXAA minus the predicted DXAA. The plot of the residuals versus the predicted value should be randomly distributed and show no remarkable pattern. The VIF provides information on how multicollinearities in the data inflate the coefficients and which coefficients are being impacted the most. In theory, the VIF should be equal to one, implying that all the regressors are orthogonal to each other.

Three different regression models were developed. The first model, Ln DXAA Molar Yield, was developed using the natural log of the DXAA molar yield. The second model, Ln DXAA Mass Yield was based on the Ln DXAA Molar Yield, and the last model, Ln DXAA Mass Concentration, was developed from the Ln DXAA Mass Yield. The three final regression models were quite similar to each other with respect to the parameters and their associated coefficients and signs as shown in Table 7.1. The Ln DXAA Mass Concentration model differed from Ln DXAA Mass and Molar Yield models solely by the inclusion of the Ln DOC term. Because the DOC component was not included in the response for the Ln DXAA Mass Concentration model as it was in the yield models, a DOC term was found to be significant and had to be included as a predictor. In addition, the signs and magnitudes of nearly all of the coefficients were equivalent, especially for the yield models. Plots of measured versus predicted DXAA formation are shown in Figure 7.1 for each model; the solid lines represent perfect agreement between measurements and predictions, while the dashed lines indicate the 95% confidence interval of the model.

Table 7.1
Parameters and coefficients for each multi-linear regression model

Parameter	Coefficient		
	Ln DXAA Molar Yield*	Ln DXAA Mass Yield[†]	Ln DXAA Mass Concentration[‡].
Intercept	-5.281	-0.250	0.021
Cl_2/N Ratio, mg/mg	0.097	0.099	0.095
NH_2Cl Demand, mg/L	0.082	0.064	0.075
NH_2Cl Residual, mg/L	0.082	0.087	0.082
pH	-0.228	-0.257	-0.254
Bromide, mg/L	0.484	0.734	0.735
SUVA, L/m-mg	0.361	0.388	0.310
UV_{254}, cm^{-1}	-1.782	-2.117	-1.805
Hydrophobic DOC, mg/L	-0.220	-0.234	-0.283
Hydrophobic/DOC	1.928	1.967	1.947
Hydrophilic/DOC	3.451	3.763	3.328
Ln DOC	----	----	1.155

* Ln DXAA Molar Yield = -5.281 + 0.097 x Cl_2/N Ratio + 0.082 x NH_2Cl Demand + 0.082 x NH_2Cl Residual – 0.228 x pH + 0.484 x Bromide + 0.361 x SUVA – 1.782 x UV_{254} – 0.220 x Hydrophobic + 1.928 x Hydrophobic/DOC + 3.451 x Hydrophilic/DOC

† Ln DXAA Mass Yield = -0.250 + 0.099 x Cl_2/N Ratio + 0.064 x NH_2Cl Demand + 0.087 x NH_2Cl Residual – 0.257 x pH + 0.734 x Bromide + 0.388 x SUVA – 2.117 x UV_{254} – 0.234 x Hydrophobic + 1.967 x Hydrophobic/DOC + 3.763 x Hydrophilic/DOC

‡ Ln DXAA Mass Concentration = 0.021 + 0.095 x Cl_2/N Ratio + 0.075 x NH_2Cl Demand + 0.082 x NH_2Cl Residual – 0.254 x pH + 0.735 x Bromide + 0.310 x SUVA – 1.805 x UV_{254} – 0.283 x Hydrophobic + 1.947 x Hydrophobic/DOC + 3.328 x Hydrophilic/DOC + 1.155 Ln DOC

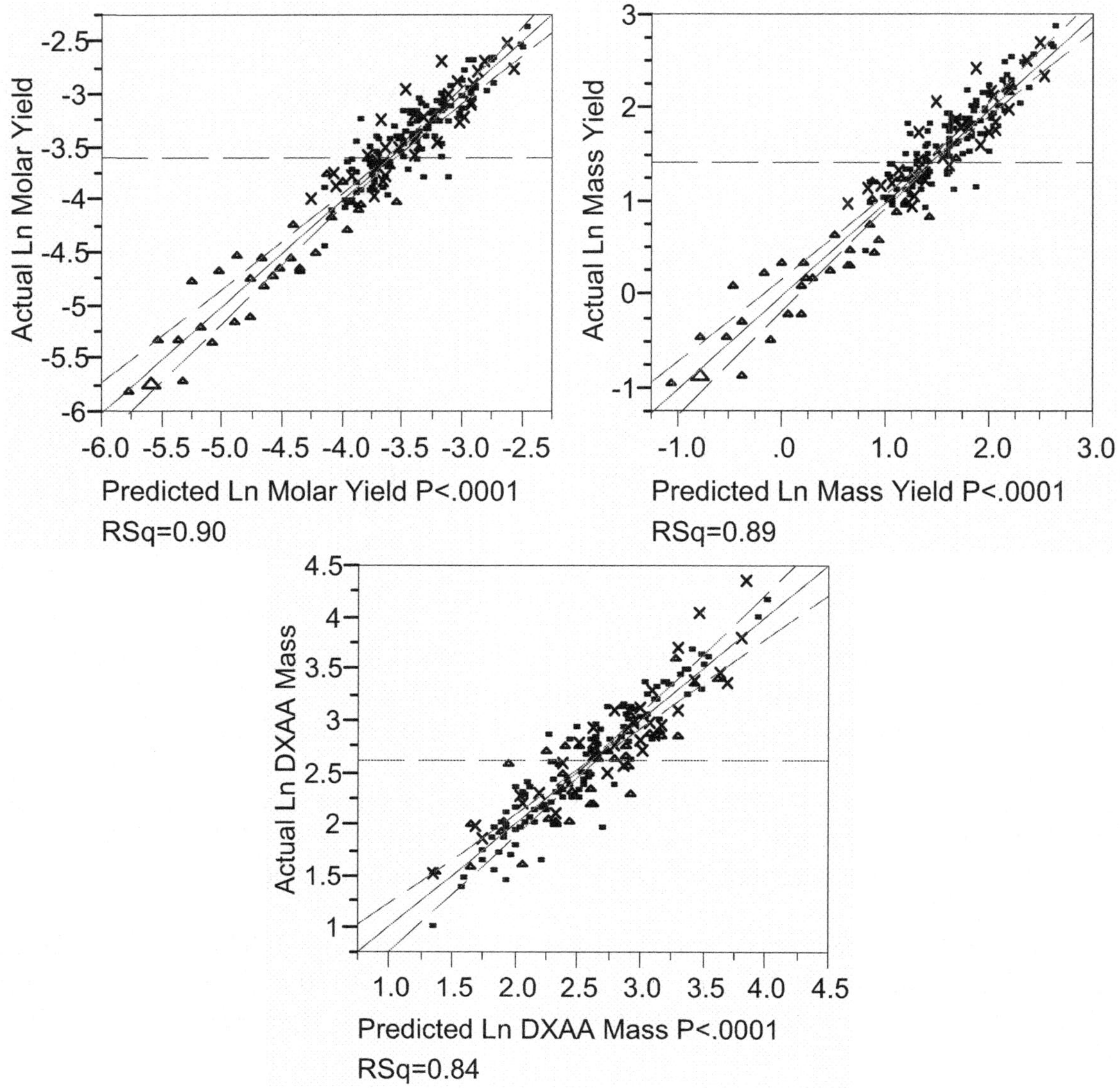

▪ Initial batch screening experiments (112 pts)
Δ Follow-up batch screening experiments (35 pts)
x Charleston (32 pts)

Figure 7.1 Measured data vs. chloramination model predictions for Ln DXAA Molar Yield, Ln DXAA Mass Yield, and Ln DXAA Mass Concentration using 179 points

The coefficient, Ln DOC, in the DXAA Mass Concentration model was slightly larger than one due to the small multicollinearities that existed in the model. As expected, a few model components were correlated (*e.g.*, hydrophobic content was a function of hydrophilic content since the latter was found by difference between total and hydrophilic DOC content). Overall, the three models had acceptable multicolinearity values based on a statistical analysis called

Principal Components, which indicated that most of the principal components of the model were not significantly correlated with other components.

The R^2 values for Ln DXAA Molar and Mass Yield models were 0.90 and 0.89, respectively, suggesting that a relatively good model fit was observed. The incorporation of DOC into both the response, as molar and mass yield, as well as in the predictor variables, as SUVA (UV_{254}/DOC), hydrophobic/DOC, and hydrophilic/DOC, resulted in inflation of the R^2 values. Nevertheless, the yield model can be used for evaluating relationships between the different predictors and the response as well as for qualitative DXAA prediction. The Ln DXAA Mass Concentration model had a lower R^2 value, 0.84, than the two other models, possibly because the Ln DXAA Mass Concentration model did not have the DOC parameter incorporated simultaneously in the response and in the predictor terms, and thus did not create any artificial inflation in the R^2 value. Overall, the three models were considered statistically consistent and reliable for qualitatively predicting DXAA formation based on their high R^2 values, low multicollinearity, low Cp value, acceptable VIF, and the use of a significant number of data points generated from various source waters and chloramination conditions.

In addition, normal distribution of the residuals (residuals = actual - predicted values) was assessed. When the residuals are normally distributed a good balance is implied between the points that are over-predicting versus those that are under-predicting. Therefore, if the residuals are normally distributed they should be represented as a bell-shaped curve on a histogram and follow a linear pattern on a plot based on percentiles or probability (normal quantile plot). If the residuals are not normally distributed, no balance would be reached between points of overestimation and underestimation, which is often demonstrated graphically by curvature in the normal quantile plot, or skewness in the histogram of residuals (SAS Institute Inc., 2001). As illustrated in Figure 7.2, the residuals were considered normally distributed and were within the 95% confidence interval for all models.

A satisfactory model depends upon how well the regressors in the model predict the response, which is not driven by the distribution of the response or residuals. One can have a poor model with a normal response and an excellent model with a non-normal response. Although normality is not a requirement for a satisfactory model, it can be useful to know as it is an assumption made when performing t-tests, F-tests, ANOVA, forming confidence intervals, etc. The least squares fitting method used in regression analysis does not require any specific assumptions about the distribution of the response beyond its need to be continuous. However, any statistical test comparison necessitating a normal distribution, such as a t-test or F-test, can be performed because the residuals were normally distributed for all three models in this study. For example, the ANOVA results from the batch screening experiments can easily be compared to the regression modeling results since all data sets were considered normally distributed.

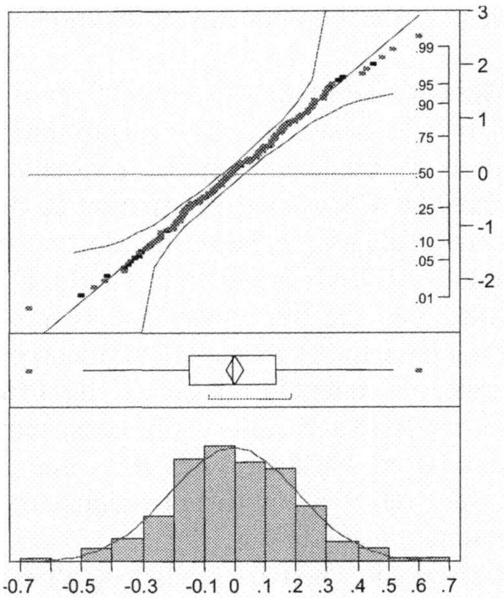

Ln DXAA Molar Yield Model

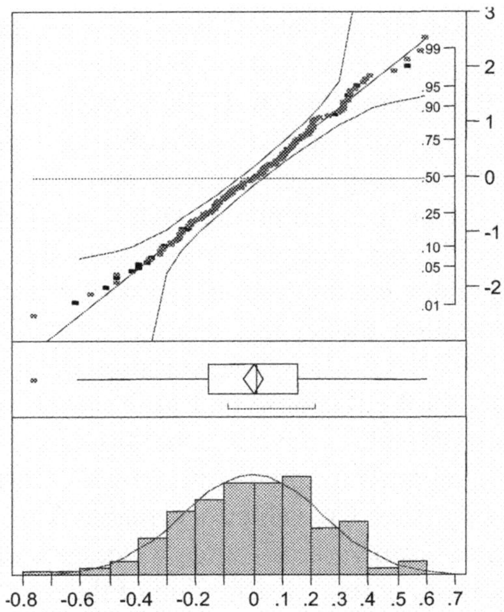

Ln DXAA Mass Yield Model

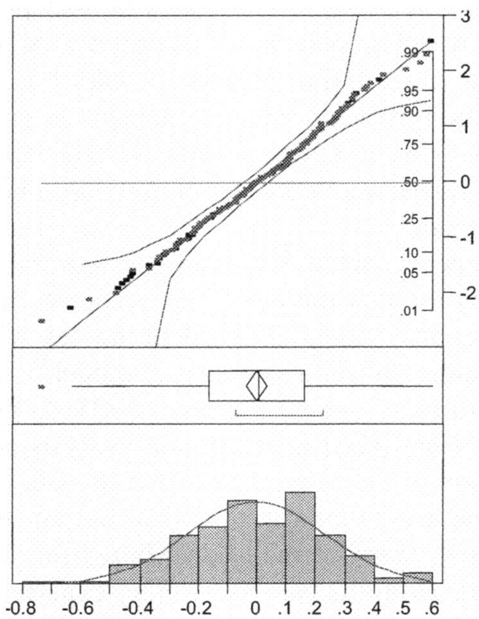

Ln DXAA Mass Concentration Model

Figure 7.2 Normal quantile plots

Based on the scaled estimates analysis, provided by the program JPM IN (Table 7.2), the most influential parameters in the DXAA formation models developed in this study were evaluated. The scaled estimates analysis allows comparison of the relative influence of each term in the model by normalizing them on a dimensionless scale. The most influential model terms for the two DXAA Yield models were hydrophobic DOC, hydrophobic/DOC, hydrophilic/DOC, SUVA, UV_{254}, pH, bromide concentration, NH_2Cl demand, NH_2Cl residual, and chlorine to ammonia mass ratio (Cl_2/N). It is interesting to note that the first six most influential terms are all related to characteristics of the NOM. The most influential model terms for the Ln DXAA Mass Concentration model were almost in the same order except that Ln DOC was first, the influence of SUVA and UV_{254} was slightly diminished, and the influence of pH and bromide slightly increased. In general, the most influential model terms are the NOM characteristics of the water followed by pH, bromide concentration, and chloramination conditions.

Table 7.2
Model parameters ranking based on the Scaled Estimates Analysis for the models Ln DXAA Molar Yield, Ln DXAA Mass Yield, and Ln DXAA Mass Concentration

Rank	Ln DXAA Molar Yield	Ln DXAA Mass Yield	Ln DXAA Mass Conc.
1	N/A	N/A	N/A
2	Hydrophobic, mg/L	Hydrophobic, mg/L	Hydrophobic, mg/L
3	Hydrophobic/DOC	Hydrophobic/DOC	Hydrophobic/DOC
4	Hydrophilic/DOC	Hydrophilic/DOC	Hydrophilic/DOC
5	SUVA, L/mg-m	SUVA, L/mg-m	pH
6	UV_{254}, cm^{-1}	UV_{254}, cm^{-1}	SUVA, L/mg-m
7	pH	pH	Bromide, mg/L
8	Bromide, mg/L	Bromide, mg/L	UV_{254}, cm^{-1}
9	NH_2Cl demand, mg/L	NH_2Cl demand, mg/L	NH_2Cl demand, mg/L
10	NH_2Cl residual, mg/L	NH_2Cl residual, mg/L	NH_2Cl residual, mg/L
11	Cl_2/N ratio, mg/mg	Cl_2/N ratio, mg/mg	Cl_2/N ratio, mg/mg

N/A = Not applicable

The ranking based on the scaled estimates in the modeling effort coincided with the findings from the initial and follow-up batch experiments as summarized in Table 7.3. The most influential factors in DXAA molar yield formation were NOM characteristics (DOC, hydrophobic DOC, hydrophobic/DOC, hydrophilic/DOC, SUVA, UV_{254}), pH, bromide and chloramination conditions (NH_2Cl Demand, NH_2Cl Residual, and Cl_2/N ratio). However, pH was more influential than NOM characteristics in the initial batch screening experiments, which could be attributed to the fact that only two source waters were considered: Lake Austin and Metedeconk River.

The molar yield model was chosen as a basis for comparison of the factors influencing DXAA formation between the two batch screening experiments and the modeling effort for two reasons. First, the yield is more appropriate than concentration when comparing different source waters to determine if observed differences are truly related to differing reactivity rather than just a reflection of the DOC concentration of the waters. Second, a molar basis was preferred to eliminate the influence of the varying molecular weights of the chlorine- and bromine-substituted DXAA species (DCAA, BCAA, DBAA).

The ANOVA was used to rank the various factors studied, and because it assumes a normal distribution of the data, normality was verified for each batch screening experiment as well as on all the data pooled together. The DXAA formation was determined to be normally distributed when the analysis was done using all the data and when the initial and follow-up batch experiments were considered separately.

Table 7.3
Ranking of factors influencing DXAA molar yield formation

Batch Screening Experiments (144 experiments)	Follow-up Batch Screening Experiments (35 experiments)	Modeling (179 experiments)*
pH	Water/NOM characteristics	Water/NOM characteristics
Water/NOM characteristics	pH	pH
Bromide	Bromide	Bromide
Chloramination conditions†	Chloramination conditions†	Chloramination conditions†

* 32 experiments could not be included in the modeling effort because of incomplete water quality measurements.
† Includes chloramine demand, chloramine residual, and chloramine to ammonia ratio

Overall, the NOM and water quality characteristics were more important in predicting DXAA formation than chloramination conditions. Similarly, the important parameters considered for the dual sites model of chloramination by Duirk et al. (2001) were DOC concentration, chloramine demand, and pH. However, Aldrich humic acid was used in place of natural aquatic organic matter.

FULL-SCALE SYSTEM EVALUATION

Duplicate sampling of three different treatment plants and corresponding distribution systems was undertaken to compare the DXAA predictions from the models generated in this study to actual measured values in the field. As the models were generated based on chloramination disinfection, chloramination-only plants were targeted for the full-scale evaluation, which limited the geographic diversity for plant selection. In order to meet CT without a prechlorination period, the options were restricted to utilities located in the southern United States where the climate is warmer. The field sites selected for this phase included Manatee (FL), Port Orange (FL), and Chalmette (LA). However, one of the utilities participating in the study practiced prechlorination. At the time of sampling, Manatee was adding a low dose (1.5 mg/L) of free chlorine to one of its water sources. This utility was included in the study because it was challenging to find utilities that were strictly chloraminating and because independent DBP studies conducted by the utility indicated prechlorination contributed less than 10% to the total DBP formation (MacLeod 2003).

WATER CHARACTERISTICS

The water characteristics for Manatee, FL, Port Orange, FL, and Chalmette, LA are summarized in Table 7.4. Manatee County treatment plant draws its source water from a mix of ground water (Floridian aquifer) and surface water (Lake Manatee Reservoir). Manatee treated water has a moderate DOC concentration with a low aromatic content, and low Br^-/DOC ratio. Port Orange is a softened groundwater characterized by a high DOC concentration, low SUVA and moderate Br^-/DOC ratio. Chalmette primarily draws its water from the Mississippi River and performs coagulation. Chalmette coagulated water may be characterized as having a low DOC concentration, higher aromaticity (SUVA = 3.58) and moderate Br^-/DOC ratio.

Table 7.4
Summary of water characteristics for the sources used in model validation

	Manatee, Florida		Port Orange, Florida		Chalmette, Louisiana	
Parameter	Source	Treated	Source	Softened	Source	Coagulated
pH	7.23	6.49	7.33	9.82	7.95	7.56
TOC (mg/L)	11.81	3.75	8.4	6.43	3.9	2.54
DOC (mg/L)	11.78	3.93	8.17	6.44	3.8	2.57
Hydrophobic DOC (%)	NR	65	NR	51	NR	73
Hydrophilic (XAD-8) DOC (%)	NR	51	NR	29	NR	62
UV_{254} (cm^{-1})	0.463	0.060	0.227	0.140	0.116	0.092
SUVA (L/mg-m)	3.93	1.53	2.78	2.17	3.05	3.58
Bromide (µg/L)	117	87	203	195	78	86.6
Ambient Br^-/DOC (µg/mg)	9.89	29.67	24.85	30.28	20.53	33.70
Alkalinity (mg $CaCO_3$/L)	29	3	251	47	103	95

NR = Not Run

MODEL VALIDATION

The comparison of the DXAA predictions from the models generated in this study to actual measured conditions in the field are shown for each model in Figure 7.3. The data used to generate the models and the actual field data exhibited similar trends with regression lines nearly parallel to each other. However, the models slightly over-predicted the response measured in the field by 7-26% (Ln DXAA Mass Concentration), 7-34% (Ln DXAA Molar Yield), and 10-50% (Ln DXAA Mass Yield). The straight line going through the origin, in Figure 7.3, represents a perfect fit where the prediction of the models would be equal to the actual measured field data.

As the prediction of the model, on the abscissa, was greater than the actual measured field data, on the ordinate, the six points from the full-scale evaluation appeared below the line of perfect fit.

Over-prediction of the DXAA formation by the models could be due to a number of reasons. The small number of DXAA data points (6) collected from the full-scale systems may not be sufficient to establish a significant trend. Only utilities from the southern United States could be sampled because a limited number of chloramination plants were available to participate in this study. In contrast, waters from both the northern and southern United States were used in developing the models. Moreover, most of the waters used in model development were from surface waters, while Port Orange and a portion of Manatee are groundwater sources, which may have lower DXAA yields. In addition, DXAA biodegradation in the distribution system (Motosugi et al. 1982; Williams et al. 1998; Xie and Zhou 2002; McRae et al. 2004) of Manatee, Port Orange and Chalmette could also have been the cause of the lower DXAA concentration found in the field compared to the model predictions. The models predict DXAA formation when entering the distribution system and as biodegradation would occur after that point throughout the distribution system, it could not have been incorporated in the models. The addition of a biodegradation term would be a way of adjusting for the DXAA overestimation from the models; however, a calibration of each model would have to be performed with a larger number of locations first. In addition, an adjustment factor for DXAA overestimation due to biodegradation would only be appropriate for locations showing evidence of biodegradation. Overall, the comparison of the model predictions to actual measured field samples was considered satisfactory and no further attempt was made to adjust the models.

Future modeling efforts based on water quality and chloramination conditions should be improved by reducing the number of parameters and predicting the individual species of DXAA. In addition, calibration should be performed using a larger number of water sources with varying water qualities. More work is needed for identification of specific NOM precursors in DXAA formation during chloramination. It would be useful to separate the transphilics, hydrophilic acids plus neutrals, and hydrophilic bases, and characterize the NOM using elemental analysis, molecular weight measurements, and ^{13}C Nuclear Magnetic Resonance (NMR) spectroscopy to provide more functional and structural information to correlate with reactivity.

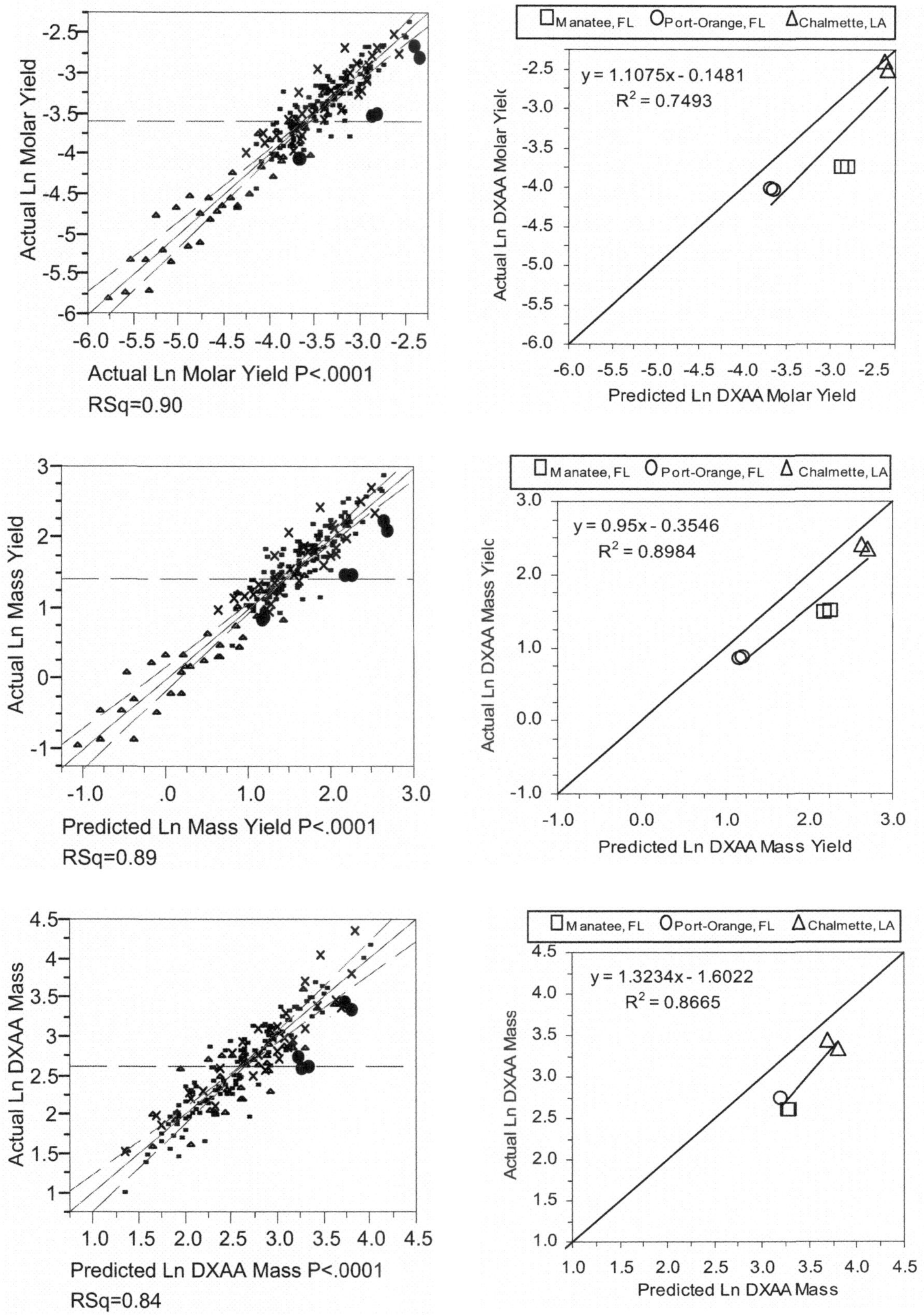

Figure 7.3 Full-scale system evaluation of Ln Molar Yield, Mass Yield, and Mass Concentration models

SUMMARY

Based on percent contributions from ANOVA analyses performed with batch screening experiments I and II as well as from the scaled estimates from the model development, the most important parameters were the NOM characteristics (DOC, hydrophobic DOC, hydrophobic/DOC, and SUVA), pH, bromide concentration, and lastly chloramination process conditions (i.e., Cl_2/N ratio and residual concentration). The models developed in this study, Ln DXAA Molar Yield, Ln DXAA Mass Yield, and Ln DXAA Mass Concentration adequately predicted DXAA formation (R^2 of 0.90, 0.89, and 0.84, respectively). In general, the models somewhat over-predicted the trends observed in practice; however, the full-scale system evaluation of the models was considered satisfactory.

CHAPTER 8
CONCLUSIONS

FACTORS AFFECTING DXAA FORMATION

An initial round of batch screening experiments was conducted on two diverse water sources: Lake Austin water, a low SUVA, hard, high alkalinity source treated by lime softening, and Metedeconk River water, a high-SUVA, soft, low alkalinity source treated by alum coagulation. Source waters, treated waters, and NOM fractions were studied to determine the influence of pH, bromide concentration, Cl_2/N ratio, temperature, and chloramine residual on DXAA formation during chloramination. A fractional factorial experimental design was used to assess the significance and relative contribution of each factor, as well as water type, through ANOVA.

To complement the initial batch screening experiments, five source and treated waters were studied from a variety of locations to broaden the range of water quality and NOM characteristics studied. The two most influential factors attributed to DXAA formation in the initial batch screening experiments, the bromide concentration and pH, were studied in a fractional factorial experimental design. Three levels rather than two were investigated for each factor to assess possible non-linear effects. The findings from both sets of batch screening experiments are summarized below.

pH

The ANOVA analyses over a broad variety of experimental conditions usually pointed to pH as being the most significant factor in DXAA formation. DXAA formation decreased as the pH increased. Therefore, utilities interested in decreasing DXAA formation should first examine the possibility of raising the pH of their water.

Bromide

The evaluation of bromide's impact on DXAA formation was studied in two ways. In the initial batch screening experiments two levels of bromide concentration were considered: ambient and ambient plus 0.5 mg/L. In the follow-up batch experiments, the Br^-/DOC ratio was selected as the factor for studying the impact of bromide after accounting for varying DOC content; ratios ranged from less than 1 to 87.5 µg/mg and were grouped into three levels: <15, 40, and 87.5 µg/mg, respectively. The Br^-/DOC ratio was not a controlled factor in the initial batch screening experiments; therefore, a wide variety of Br^-/DOC ratios prevailed in these experiments, ranging from 9 to 415. The impact of bromide can be viewed from several perspectives: the DXAA mass concentration or yield, DXAA molar concentration or yield, and DXAA speciation, which can be described by the molar bromine incorporation ratio (n_D'). Generally, bromide impacts the DXAA speciation and DXAA mass concentration or yield more so than DXAA molar concentration or yield. Therefore, the results are discussed below from each of these perspectives to provide a more complete picture.

The ANOVA analyses on the initial batch screening experiments showed that pH and bromide concentration were the most significant factors in DXAA formation. DXAA formation increased and a shift in speciation to the bromine-substituted species occurred as the bromide concentration increased. It should be noted, however, that some experimental conditions in these batch experiments resulted in relatively high bromide concentrations and Br^-/DOC ratios in comparison to most natural waters. Therefore, the wide range of bromide concentrations studied may have influenced the relative significance of bromide as factor in DXAA formation.

The follow-up batch experiments used Br^-/DOC ratios that were more representative of the range seen in typical practice. Under these conditions, the Br^-/DOC ratio had a minor influence on DXAA molar yield for the waters tested, which somewhat contradicted the results found in the initial batch screening experiments. Increasing Br^-/DOC ratios caused a shift toward the bromine-substituted species without significantly influencing the DXAA molar yield. Because of the higher molecular weight of bromine in comparison to chlorine, the shift in speciation implies an increase in the DXAA mass yield or concentration. Therefore, the overall results suggest that the DXAA molar yield generally will not increase over the range of Br^-/DOC ratios typical of practice. A source water experiencing a large increase in bromide concentration, however, may exhibit an increase in the DXAA molar yield. An increase in the Br^-/DOC ratio will cause a shift toward the bromine-substituted species and increase the DXAA mass yield and concentration. An increase in the Br^-/DOC ratio is undesirable from two perspectives. First, the regulations are based on mass concentration, so the likelihood of violating the HAA MCL increases. Second, the bromine-substituted HAAs may be of greater health concern. In general, higher Br^-/DOC ratios and lower pH led to greater bromine incorporation into DXAAs. At Br^-/DOC ratios of less than 200 μg/mg (the typical range of most natural waters), increases in the Br^-/DOC ratio caused an increase in DXAA bromine incorporation. Values of n_D', which characterizes the bromine substitution of DXAAs on a molar basis, varied considerably among the water sources studied, but ranged from a low of 0 at Br^-/DOC ratios around 10 μg/mg to a high of 1.4 at a Br^-/DOC ratio of 200 μg/mg. At Br^-/DOC ratios above 200 μg/mg, further increases caused no discernable impact on bromine incorporation into DXAAs.

Other Factors

The Cl_2/N mass ratio was the next most significant contributor to DXAA formation after pH and bromide. DXAA formation decreased as the Cl_2/N ratio decreased. Temperature and chloramine residual also impacted DXAA formation, but were much less influential than the aforementioned factors. DXAA formation increased as the temperature increased, and the impact of temperature was more pronounced as the Cl_2/N ratio increased. Increasing the chloramine residual concentration caused an increase in DXAA formation in Lake Austin water, but had a minimal impact on DXAA formation in Metedeconk River water.

Impact of Treatment

Treatment can also have a significant effect on DXAA formation. The effectiveness of treatment on DXAA formation is largely related to overall DOC removal, although preferential removal of more reactive NOM fractions also may contribute to reduced DXAA formation. Comparison of DXAA yield before and after treatment helps distinguish whether decreased DXAA formation is related to differing reactivity of the waters or is just a reflection of the lower DOC concentration after treatment. Softening Lake Austin and Biscayne Aquifer source water resulted in minimal DOC removal and therefore only slightly decreased DXAA formation.

Alum coagulation of Metedeconk River and Charleston source water removed a larger fraction of the DOC, resulting in a more significant decrease in DXAA formation. Treatment reduced the DXAA yield in all of the source waters studied except Biscayne Aquifer. This decrease in yield indicates that treatment generally removes the more reactive DXAA precursors. Treatment processes that remove DOC can also affect DXAA formation and speciation through the resulting increase in the Br^-/DOC ratio.

NOM Characteristics

The follow-up batch experiments showed that DXAA mass yield decreased with an increasing hydrophobic/DOC ratio or, alternatively, a decreasing hydrophilic/DOC ratio, thereby indicating that the NOM in the hydrophilic fraction was more reactive in forming DXAA than the NOM in the hydrophobic fraction. In addition, the DXAA mass yield was approximately proportional to chloramine demand; as the chloramine demand increased, the DXAA mass yield increased. Many other water quality characteristics such as DOC, UV, SUVA, SUVA x Br^-, hydrophilic DOC, and hydrophobic DOC were considered, but none yielded valuable correlations with DXAA formation when considered individually.

The most influential terms in the DXAA mass and molar yield models (discussed below) were, in order, hydrophobic DOC concentration, hydrophobic/DOC ratio, hydrophilic/DOC ratio, SUVA, and UV_{254}. The models highlighted the importance of multiple NOM characteristics in predicting DXAA formation, although it was not generally possible to correlate individual NOM characteristics with DXAA formation. The need to consider multiple NOM characteristics simultaneously illustrates the complexity of the relationship between NOM and DXAA formation.

DXAA FORMATION KINETICS

Experiments were conducted on both source and treated waters from four different locations to study the impact of pH, bromide concentration, temperature, and treatment on DXAA formation kinetics during chloramination. In addition, the impact of prechlorination on DXAA formation kinetics was studied because of the widespread use of short periods of chlorination prior to chloramination. The waters selected, Lake Austin, Metedeconk River, Biscayne Aquifer, and surface impoundments near Charleston, SC, spanned the typical range of bromide and DOC concentrations, as well as alkalinity and SUVA, common in source waters. The conclusions drawn from these experiments are:

1. The kinetics of DXAA formation during chloramination are rather fast. For example, approximately 73% of the 72-hour DXAA concentration formed after only 0.5 hours and almost 82% formed after 1 hour in Metedeconk River source water. Therefore, the vast majority of the DXAA formation will occur within the water treatment plant itself.
2. The bromine-substituted species (BCAA and DBAA) formed very rapidly, with the formation reactions typically being completed in less than 0.5 hours. In contrast, DCAA formation kinetics were characterized by an initial period (~ 1 hour) of rapid formation followed by slow formation over the remainder of the 72-hour incubation period.

3. More rapid DXAA formation kinetics were observed at pH 7 than at pH 9 during both the rapid and slow periods of formation. This pH trend indicates that acid catalyzed monochloramine reactions may be the cause of increased HAA formation as the pH decreases.
4. Increasing temperature caused only a minimal increase in DXAA formation over the range tested in Charleston and Lake Austin treated waters. However, additional experiments with Lake Austin softened water indicated that increased HAA formation as a result of increased temperature may be more pronounced at lower pH and higher Cl_2/N ratios. Nevertheless, the effect of temperature alone on DXAA formation appears to be relatively small; seasonal changes in NOM concentration or reactivity are also likely to be important in situations in practice where large changes in DXAA formation are observed with seasonal temperature changes.
5. During short periods of prechlorination (5 or 20 minutes), significantly more DXAA formation, as well as formation of mono- and tri-halogenated acetic acids, occurred relative to chloramination alone. Also, as expected, the longer the period of free chlorination the greater the DXAA formation; however, the increase was much less than proportional to the prechlorination time.
6. During the prechlorination period, free chlorine reacts with most of the fast-reactive NOM sites. As a result, after subsequent ammonia addition and chloramine formation very little additional DXAA is formed. At least 75% of the 72-hour DXAA concentration formed during prechlorination in all the waters studied, and greater than 90% of the 72-hour DXAA concentration typically formed during prechlorination. Therefore, the general expectation is that very little of the DXAA formation in chlorination/chloramination plants is associated with the chloramination step, despite the dramatic difference in the chloramine and chlorine contact times that is typical of such plants. Thus, HAA control strategies should be focused on the chlorination step in plants that prechlorinate prior to chloramination.
7. The absence of significant DXAA formation during chloramination after prechlorination suggests that both chlorine and monochloramine (or monochloramonium ion) react with the same fraction of the NOM to produce DXAAs, or that the DXAAs formed during chloramination do so as a result of the HOCl that is formed by monochloramine hydrolysis.

PREDICTIVE MODEL OF DXAA FORMATION

The final step of this study consisted of the development of simple and practical models for predicting DXAA formation during chloramination based on water characteristics and chloramination conditions. Linear regression modeling was performed on the data generated during the initial batch screening experiments (112 points) with Lake Austin and Metedeconk River, the follow-up batch screening experiments (35 points) with Metedeconk River, Biscayne Aquifer source and treated, and St. Paul source and treated waters, and finally Charleston source water (32 points), which was not included in either of the screening experiments. Models for DXAA mass yield, DXAA molar yield, and DXAA mass concentration were developed. The major findings are summarized as follows:

1. The modeling effort was a culmination of the work accomplished throughout this study and assumed fast kinetics based on the findings from the kinetics experiments.
2. The models developed in this study, Ln DXAA Molar Yield, Ln DXAA Mass Yield, and Ln DXAA Mass Concentration adequately predicted DXAA formation (R^2 of 0.90, 0.89, and 0.84, respectively). In general, the models over-predicted the levels observed in practice; however, the full-scale system evaluation of the models was considered satisfactory, especially since the trends between model predictions and measurements were consistent.
3. The modeling effort allowed the ranking of factors through scaled-estimate analysis, which was compared to the factor ranking of the initial and follow-up batch screening experiments based ANOVA analyses. The most influential model terms for the two DXAA Yield models were, in rank order, hydrophobic DOC concentration, the hydrophobic/DOC ratio, the hydrophilic/DOC ratio, SUVA, UV_{254}, pH, bromide concentration, NH_2Cl Demand, NH_2Cl Residual, and chlorine to ammonia ratio (Cl_2/N). It is interesting to note that the first six most influential terms are all related to water/NOM characteristics. The most influential model terms for the DXAA Mass model were almost in the same order except that Ln DOC concentration was first, the influence of SUVA and UV_{254} was slightly diminished, and the influence of pH and bromide concentration was slightly increased. In general, the most influential model terms were the NOM characteristics of the water followed by pH, bromide concentration and chloramination conditions. This ranking of model factors based on scaled estimates coincided with the factor rankings from the initial and follow-up batch experiments, which were based on percent contributions from ANOVA analyses.

CHAPTER 9
RECOMMENDATIONS TO UTILITIES

The use of models can be helpful for utilities to evaluate general tendencies in their distribution system and anticipate possible difficulties with respect to DBP regulation compliance. The development of simple and practical models predicting DXAA formation during chloramination based on select water characteristics and chloramination conditions was the final stage of this research. A number of kinetic models predicting monochloramine loss and/or DBP formation have been presented in the literature. The models developed in this present study were not based on kinetics considerations as most of the DXAAs were formed in less than 1 hour during chloramination. Numerous analytical measurements of water characteristics, chloramination conditions, and DXAA concentrations were gathered throughout the research to create an extensive database for model development through linear regression analysis. The water quality variables and chloramination conditions were identified as the predictors and the DXAA data (in mass, mass yield, and molar yield) were identified as the response. The multi-linear regression models are listed in Table 7.1.

The use of these models will allow utilities to qualitatively assess their distribution system DXAAs based on treated water characteristics and chloramination conditions. The DXAAs calculated from the model will tend to over-predict the actual DXAAs found in the distribution system, but the trends should still be applicable to actual treated water quality and chloramination conditions. Thus, the influence from changes in the NOM and inorganic characteristics of the treated water (due to changing conditions in the source water quality from watershed influences or treatment process variations) may be qualitatively assessed to determine any problematic issues in satisfying the HAA regulations. In addition, changes or modifications to chloramination conditions may also be qualitatively evaluated in a similar fashion.

In addition to the multi-linear regression models for assessing problematic HAA by-product issues, this research also provided guidance to utilities on an approach for systematically implementing the most influential changes to the treatment and/or chloramination process for reducing DXAAs in the distribution system. Based on the statistical experimental approach used in this study (plus the personal experiences of the authors) the corrective treatment changes to minimize distribution system DXAAs from chloramination should be implemented in the following order:

- **Increase pH**: The pH during chloramination should be raised above pH 7. Results from this study suggested that approximately a 50% reduction in DXAAs can be achieved by increasing pH from 7 to 9. This reduction was independent of all other factors and conditions.
- **Decrease free chlorination interval**: The time interval between chlorine and ammonia additions should be reduced as much as possible with preference given to simultaneous additions. As much as 95% of the 72-hour DXAA concentration can be formed with as little as 5 minutes of contact time between free chlorination and ammonia addition.
- **Decrease NOM/DOC precursors**: As expected, as much NOM/DOC precursor material as possible should be removed before chloramination. Sole dependence on softening to reduce organic precursors sufficiently should be carefully evaluated;

enhanced softening, involving significant magnesium precipitation, may be needed to achieve substantial (> 20%) DOC removal (Smith et al. 2004). Both hydrophobic and hydrophilic NOM fractions should be removed, but preference should be given to the hydrophilics (XAD-8 eluate). Consequently, treatment processes that are known to favor hydrophilic DOC removal (e.g., biological treatment) may need to be considered. In addition, utilities need to be mindful that DOC removal will increase the Br^-/DOC ratio, which will shift the DXAA speciation more toward the bromine-substituted species and in some cases may increase the DXAA yield.

- **Decrease Cl_2/N ratio**: Although not as significant as the aforementioned changes, decreasing the Cl_2/N ratio will decrease DXAA formation somewhat in most cases.
- **Decrease chloramine residuals**: Although reductions in DXAAs are expected for lower chloramine residuals, the amount of reduction was small in this research. Managing DXAA reduction by limiting chloramine residuals should be used only as a last resort.

APPENDIX A
INITIAL BATCH SCREENING EXPERIMENTS

Table A.1
Lake Austin source water (mass HAA concentrations)

Trial No.	Cl/N ratio	pH	Br^- (mg/L)	Total Cl Residual ($mg Cl_2/L$) actual	Total Cl Residual ($mg Cl_2/L$) target	Chloramine Demand (mg/L)	Temperature (°F)	AVERAGE HAA CONCENTRATIONS (µg/L) MCAA	MBAA	DCAA	BCAA	DBAA	TCAA	BDCAA	DBCAA	TBAA	Total HAA	HAA5	DXAA	MXAA	DXAA/HAA9	DXAA Yield (DXAA/DOC)
1	3:1	7	Ambient	1.89	2	0.51	70	ND	0.60	7.08	4.72	4.60	0.97	ND	ND	ND	17.96	13.24	16.39	0.60	0.91	4.38
2	3:1	7	Ambient + 0.5	3.79	4	1.31	70	ND	1.86	5.75	7.38	14.09	1.06	ND	2.14	1.02	33.31	22.77	27.23	1.86	0.82	7.28
3	3:1	7	Ambient	3.41	4	1.59	85	ND	1.14	10.51	7.36	6.68	1.13	ND	1.10	ND	27.92	19.46	24.55	1.14	0.88	6.56
4	3:1	7	Ambient + 0.5	1.96	2	1.44	85	ND	2.31	5.93	8.39	18.27	0.94	ND	2.38	2.39	40.60	27.44	32.59	2.31	0.80	8.71
5	5:1	7	Ambient	3.95	4	1.95	70	ND	BDL	10.36	9.13	6.33	0.79	1.25	1.46	0.95	30.25	17.47	25.81	0.00	0.85	6.90
6	5:1	7	Ambient + 0.5	1.93	2	0.97	70	ND	1.33	5.27	9.74	19.32	BDL	BDL	0.96	1.16	37.79	25.92	34.34	1.33	0.91	9.18
7	5:1	7	Ambient	1.79	2	2.31	85	ND	0.53	10.21	10.38	7.76	BDL	1.07	1.26	0.92	32.13	18.50	28.36	0.53	0.88	7.58
8	5:1	7	Ambient + 0.5	4.37	4	5.13	85	ND	3.44	9.28	19.66	36.59	2.51	BDL	1.84	1.49	74.81	51.82	65.52	3.44	0.88	17.52
9	5:1	9	Ambient	2.19	4	0.31	70	ND	ND	5.84	2.94	2.91	BDL	ND	ND	0.91	12.61	8.75	11.70	0.00	0.93	3.13
10	5:1	9	Ambient + 0.5	1.92	2	0.48	70	ND	ND	4.60	3.01	11.70	BDL	1.08	1.17	1.45	23.01	16.30	19.32	0.00	0.84	5.17
11	5:1	9	Ambient	1.94	2	0.56	85	ND	ND	6.93	2.72	2.45	BDL	ND	1.06	0.90	14.08	9.39	12.11	0.00	0.86	3.24
12	5:1	9	Ambient + 0.5	3.89	4	1.01	85	ND	BDL	7.96	4.97	7.51	BDL	1.04	1.33	1.35	24.15	15.46	20.44	0.00	0.85	5.46
13	3:1	9	Ambient	1.97	2	0.03	70	ND	BDL	4.21	2.36	3.88	0.88	ND	2.16	ND	13.50	8.98	10.46	0.00	0.77	2.80
14	3:1	9	Ambient + 0.5	3.96	4	0.34	70	ND	0.97	4.81	2.73	15.34	1.07	ND	2.17	2.20	29.29	22.18	22.89	0.97	0.78	6.12
15	3:1	9	Ambient	3.90	4	0.70	85	ND	0.52	7.20	3.13	4.14	1.06	ND	ND	ND	16.06	12.93	14.47	0.52	0.90	3.87
16	3:1	9	Ambient + 0.5	2.03	2	0.27	85	ND	0.86	5.30	2.56	9.07	0.91	ND	2.15	1.17	22.01	16.13	16.93	0.86	0.77	4.53

DXAA = DCAA + DBAA + BCAA
MXAA = MCAA + MBAA
ND = Not Detected
BDL = Below Detection Limit of 0.5 µg/L
DOC (mg/L) = 3.74

Table A.2
Lake Austin source water (molar HAA concentrations)

Trial No.	Cl/N ratio	pH	Br^- (mg/L)	Total Cl Residual ($mg Cl_2/L$) actual	Total Cl Residual ($mg Cl_2/L$) target	Chloramine Demand (mg/L)	Temperature (°F)	AVERAGE HAA CONCENTRATIONS (µmol/L) MCAA	MBAA	DCAA	BCAA	DBAA	TCAA	BDCAA	DBCAA	TBAA	Total HAA	HAA5	DXAA	MXAA	DXAA/HAA9	DXAA Yield (DXAA/DOC)	Bromine incorporation factor (n_D')
1	3:1	7	Ambient	1.89	2	0.51	70	ND	0.0043	0.0549	0.0272	0.0211	0.0059	ND	ND	ND	0.1134	0.0862	0.1032	0.0043	0.9096	0.0276	0.673
2	3:1	7	Ambient + 0.5	3.79	4	1.31	70	ND	0.0134	0.0446	0.0426	0.0647	0.0065	ND	0.0085	0.0034	0.1837	0.1292	0.1519	0.0134	0.8267	0.0406	1.132
3	3:1	7	Ambient	3.41	4	1.59	85	ND	0.0082	0.0815	0.0424	0.0306	0.0069	ND	0.0044	ND	0.1741	0.1273	0.1546	0.0082	0.8881	0.0413	0.671
4	3:1	7	Ambient + 0.5	1.96	2	1.44	85	ND	0.0166	0.0460	0.0484	0.0839	0.0057	ND	0.0094	0.0081	0.2181	0.1522	0.1782	0.0166	0.8174	0.0477	1.213
5	5:1	7	Ambient	3.95	4	1.95	70	ND	BDL	0.0803	0.0526	0.0291	0.0048	0.0060	0.0058	0.0032	0.1818	0.1142	0.1620	0.0000	0.8911	0.0433	0.684
6	5:1	7	Ambient + 0.5	1.93	2	0.97	70	ND	0.0071	0.0409	0.0562	0.0879	BDL	BDL	0.0023	0.0037	0.1980	0.1358	0.1849	0.0071	0.9340	0.0495	1.254
7	5:1	7	Ambient	1.79	2	2.31	85	ND	0.0038	0.0792	0.0599	0.0356	BDL	0.0051	0.0050	0.0031	0.1917	0.1186	0.1747	0.0038	0.9113	0.0467	0.751
8	5:1	7	Ambient + 0.5	4.37	4	5.13	85	ND	0.0248	0.0719	0.1134	0.1691	0.0168	BDL	0.0055	0.0048	0.4063	0.2826	0.3544	0.0248	0.8723	0.0948	1.274
9	5:1	9	Ambient	2.19	4	0.31	70	ND	ND	0.0453	0.0170	0.0134	BDL	ND	ND	0.0031	0.0787	0.0587	0.0756	0.0000	0.9610	0.0202	0.578
10	5:1	9	Ambient + 0.5	1.92	2	0.48	70	ND	ND	0.0357	0.0174	0.0537	BDL	0.0052	0.0046	0.0049	0.1215	0.0894	0.1068	0.0000	0.8790	0.0286	1.169
11	5:1	9	Ambient	1.94	2	0.56	85	ND	ND	0.0538	0.0157	0.0113	BDL	ND	0.0042	0.0030	0.0880	0.0650	0.0807	0.0000	0.9175	0.0216	0.473
12	5:1	9	Ambient + 0.5	3.89	4	1.01	85	ND	BDL	0.0617	0.0287	0.0345	BDL	0.0050	0.0053	0.0046	0.1397	0.0962	0.1248	0.0000	0.8939	0.0334	0.782
13	3:1	9	Ambient	1.97	2	0.03	70	ND	BDL	0.0327	0.0136	0.0178	0.0054	ND	0.0086	ND	0.0781	0.0559	0.0641	0.0000	0.8211	0.0171	0.768
14	3:1	9	Ambient + 0.5	3.96	4	0.34	70	ND	0.0069	0.0373	0.0158	0.0704	0.0065	ND	0.0086	0.0074	0.1530	0.1212	0.1235	0.0069	0.8073	0.0330	1.268
15	3:1	9	Ambient	3.90	4	0.70	85	ND	0.0038	0.0558	0.0181	0.0190	0.0065	ND	ND	ND	0.1032	0.0851	0.0929	0.0038	0.9003	0.0248	0.603
16	3:1	9	Ambient + 0.5	2.03	2	0.27	85	ND	0.0062	0.0411	0.0148	0.0416	0.0055	ND	0.0085	0.0039	0.1217	0.0944	0.0975	0.0062	0.8014	0.0261	1.005

DXAA = DCAA + DBAA + BCAA
MXAA = MCAA + MBAA
ND = Not Detected
BDL = Below Detection Limit of 0.5 µg/L
DOC (mg/L) = 3.74

Table A.3

Lake Austin algal source water (mass concentration)

Trial				Total Cl Residual ($mg\ Cl_2/L$)		Chloramine Demand	Temperature	AVERAGE HAA CONCENTRATIONS (µg/L)															DXAA Yield
No.	Cl/N ratio	pH	Br^- (mg/L)	actual	target	(mg/L)	(°F)	MCAA	MBAA	DCAA	BCAA	DBAA	TCAA	BDCAA	DBCAA	TBAA	Total HAA	HAA5	DXAA	MXAA	DXAA/HAA9		(DXAA/DOC)
1	3:1	7	Ambient	2.26	2	0.14	70	ND	BDL	7.02	4.50	2.21	BDL	BDL	ND	BDL	13.73	9.23	13.73	0.00	1.00		3.71
2	3:1	7	Ambient + 0.5	4.44	4	0.36	70	ND	3.85	6.70	11.27	20.55	BDL	0.58	BDL	0.68	43.63	31.11	38.52	3.85	0.88		10.41
3	3:1	7	Ambient	4.16	4	0.84	85	ND	2.93	10.91	7.93	5.67	BDL	BDL	BDL	BDL	27.44	19.51	24.52	2.93	0.89		6.63
4	3:1	7	Ambient + 0.5	2.17	2	0.63	85	ND	1.31	5.30	8.70	17.39	BDL	BDL	BDL	0.76	33.45	24.00	31.39	1.31	0.94		8.48
5	5:1	7	Ambient	4.06	4	2.34	70	ND	1.65	9.09	11.18	8.07	0.57	1.04	0.92	ND	32.52	19.39	28.34	1.65	0.87		7.66
6	5:1	7	Ambient + 0.5	1.93	2	1.27	70	ND	1.41	4.37	10.68	22.23	BDL	0.90	1.04	1.09	41.71	28.00	37.28	1.41	0.89		10.08
7	5:1	7	Ambient	2.18	2	1.22	85	ND	0.70	11.67	10.28	7.58	BDL	0.72	0.71	ND	31.65	19.94	29.53	0.70	0.93		7.98
8	5:1	7	Ambient + 0.5	3.52	4	4.88	85	ND	2.40	9.90	13.89	30.84	1.12	1.19	1.35	1.00	61.69	44.26	54.63	2.40	0.89		14.76
9	5:1	9	Ambient	4.41	4	0.29	70	ND	1.39	10.22	4.51	2.01	BDL	0.63	ND	ND	18.77	13.62	16.74	1.39	0.89		4.53
10	5:1	9	Ambient + 0.5	2.35	2	0.05	70	ND	ND	6.78	5.64	9.72	0.75	ND	1.08	0.99	24.96	17.25	22.14	0.00	0.89		5.98
11	5:1	9	Ambient	2.23	2	0.27	85	ND	BDL	9.21	4.31	1.83	BDL	ND	ND	ND	15.36	11.05	15.36	0.00	1.00		4.15
12	5:1	9	Ambient + 0.5	4.18	4	0.32	85	ND	1.03	9.37	6.93	11.88	BDL	0.77	0.86	0.95	31.80	22.29	28.18	1.03	0.89		7.62
13	3:1	9	Ambient	2.09	2	0.05	70	ND	BDL	5.00	3.23	2.14	BDL	ND	ND	ND	10.38	7.14	10.38	0.00	1.00		2.80
14	3:1	9	Ambient + 0.5	4.16	4	0.05	70	ND	1.28	4.98	3.47	13.92	10.59	BDL	BDL	1.00	35.24	30.77	22.37	1.28	0.63		6.05
15	3:1	9	Ambient	4.01	4	0.05	85	ND	1.13	6.75	2.24	1.89	0.66	BDL	ND	ND	12.67	10.44	10.88	1.13	0.86		2.94
16	3:1	9	Ambient + 0.5	1.87	2	0.13	85	ND	0.62	3.98	2.33	16.57	BDL	BDL	BDL	0.96	24.45	21.17	22.87	0.62	0.94		6.18

DXAA = DCAA + DBAA + BCAA
MXAA = MCAA + MBAA
ND = Not Detected
BDL = Below Detection Limit of 0.5 µg/L
DOC (mg/L) = 3.70

Table A.4

Lake Austin algal source water (molar concentration)

Trial				Total Cl Residual ($mg\ Cl_2/L$)		Chloramine Demand	Temperature	AVERAGE HAA CONCENTRATIONS (µmol/L)														DXAA Yield	Bromine incorporation factor
No.	Cl/N ratio	pH	Br^- (mg/L)	actual	target	(mg/L)	(°F)	MCAA	MBAA	DCAA	BCAA	DBAA	TCAA	BDCAA	DBCAA	TBAA	Total HAA	HAA5	DXAA	MXAA	DXAA/HAA9	(DXAA/DOC)	(n_D')
1	3:1	7	Ambient	2.26	2	0.14	70	ND	BDL	0.0544	0.0259	0.0102	BDL	BDL	ND	BDL	0.0905	0.0646	0.0905	0.0000	1.0000	0.0245	0.511
2	3:1	7	Ambient + 0.5	4.44	4	0.36	70	ND	0.0277	0.0520	0.0650	0.0943	BDL	0.0028	BDL	0.0023	0.2441	0.1740	0.2113	0.0277	0.8657	0.0571	1.200
3	3:1	7	Ambient	4.16	4	0.84	85	ND	0.0211	0.0846	0.0457	0.0260	BDL	BDL	BDL	BDL	0.1775	0.1317	0.1564	0.0211	0.8813	0.0423	0.625
4	3:1	7	Ambient + 0.5	2.17	2	0.63	85	ND	0.0094	0.0411	0.0502	0.0798	BDL	BDL	BDL	0.0025	0.1831	0.1304	0.1711	0.0094	0.9345	0.0462	1.226
5	5:1	7	Ambient	4.06	4	2.34	70	ND	0.0119	0.0705	0.0645	0.0371	0.0035	0.0050	0.0037	ND	0.1960	0.1229	0.1720	0.0119	0.8773	0.0465	0.806
6	5:1	7	Ambient + 0.5	1.93	2	1.27	70	ND	0.0101	0.0339	0.0616	0.1020	BDL	0.0043	0.0041	0.0037	0.2197	0.1460	0.1975	0.0101	0.8989	0.0534	1.345
7	5:1	7	Ambient	2.18	2	1.22	85	ND	0.0050	0.0905	0.0593	0.0348	BDL	0.0035	0.0028	ND	0.1959	0.1303	0.1846	0.0050	0.9424	0.0499	0.698
8	5:1	7	Ambient + 0.5	3.52	4	4.88	85	ND	0.0173	0.0768	0.0801	0.1416	0.0069	0.0057	0.0054	0.0034	0.3370	0.2425	0.2984	0.0173	0.8855	0.0807	1.217
9	5:1	9	Ambient	4.41	4	0.29	70	ND	0.0100	0.0793	0.0260	0.0092	BDL	0.0030	ND	ND	0.1276	0.0985	0.1145	0.0100	0.8976	0.0310	0.389
10	5:1	9	Ambient + 0.5	2.35	2	0.05	70	ND	ND	0.0525	0.0325	0.0446	0.0046	ND	0.0043	0.0033	0.1419	0.1018	0.1297	0.0000	0.9138	0.0351	0.939
11	5:1	9	Ambient	2.23	2	0.27	85	ND	BDL	0.0715	0.0249	0.0084	BDL	ND	ND	ND	0.1048	0.0799	0.1048	0.0000	1.0000	0.0283	0.398
12	5:1	9	Ambient + 0.5	4.18	4	0.32	85	ND	0.0074	0.0727	0.0399	0.0545	BDL	0.0037	0.0034	0.0032	0.1850	0.1347	0.1672	0.0074	0.9039	0.0452	0.891
13	3:1	9	Ambient	2.09	2	0.05	70	ND	BDL	0.0388	0.0187	0.0098	BDL	ND	ND	ND	0.0673	0.0486	0.0673	0.0000	1.0000	0.0182	0.569
14	3:1	9	Ambient + 0.5	4.16	4	0.05	70	ND	0.0092	0.0386	0.0200	0.0639	0.0648	BDL	BDL	0.0034	0.1999	0.1766	0.1225	0.0092	0.6129	0.0331	1.206
15	3:1	9	Ambient	4.01	4	0.05	85	ND	0.0081	0.0524	0.0129	0.0087	0.0041	BDL	ND	ND	0.0861	0.0732	0.0739	0.0081	0.8586	0.0200	0.409
16	3:1	9	Ambient + 0.5	1.87	2	0.13	85	ND	0.0045	0.0309	0.0134	0.0760	BDL	BDL	BDL	0.0032	0.1280	0.1114	0.1203	0.0045	0.9399	0.0325	1.375

DXAA = DCAA + DBAA + BCAA
MXAA = MCAA + MBAA
ND = Not Detected
BDL = Below Detection Limit of 0.5 µg/L
DOC (mg/L) = 3.70

Table A.5
Lake Austin softened water (mass concentration)

Trial No.	Cl/N ratio	pH	Br^- (mg/L)	Total Cl Residual ($mg\ Cl_2/L$) actual	Total Cl Residual ($mg\ Cl_2/L$) target	Chloramine Demand (mg/L)	Temperature (°F)	MCAA	MBAA	DCAA	BCAA	DBAA	TCAA	BDCAA	DBCAA	TBAA	Total HAA	HAA5	DXAA	MXAA	DXAA/HAA9	DXAA Yield (DXAA/DOC)
								AVERAGE HAA CONCENTRATIONS (μg/L)														
1	3:1	7	Ambient	1.91	2	0.79	70	ND	BDL	5.51	3.98	4.80	1.18	ND	ND	ND	15.48	11.49	14.30	0.00	0.92	4.31
2	3:1	7	Ambient + 0.5	3.65	4	1.40	70	ND	1.74	4.83	6.37	17.84	2.94	BDL	0.55	0.86	35.12	27.35	29.03	1.74	0.83	8.75
3	3:1	7	Ambient	3.67	4	1.63	85	ND	0.98	8.47	6.61	7.94	2.19	ND	BDL	BDL	26.18	19.57	23.02	0.98	0.88	6.93
4	3:1	7	Ambient + 0.5	1.87	2	1.28	85	ND	1.86	4.72	6.12	18.66	1.39	BDL	BDL	0.85	33.62	26.64	29.50	1.86	0.88	8.89
5	5:1	7	Ambient	3.86	4	1.44	70	ND	2.63	7.95	6.09	7.01	2.63	BDL	BDL	BDL	26.31	20.22	21.04	2.63	0.80	6.34
6	5:1	7	Ambient + 0.5	1.82	2	1.48	70	ND	1.88	4.19	7.16	17.29	0.74	0.65	0.97	0.91	33.81	24.10	28.64	1.88	0.85	8.63
7	5:1	7	Ambient	2.04	2	1.46	85	ND	0.69	7.59	6.63	6.95	0.73	ND	ND	ND	22.58	15.96	21.17	0.69	0.94	6.38
8	5:1	7	Ambient + 0.5	2.84	4	2.56	85	ND	3.07	5.93	8.08	25.60	1.97	ND	0.68	1.96	47.29	36.57	39.62	3.07	0.84	11.93
9	5:1	9	Ambient	1.93	2	0.17	70	ND	BDL	4.90	1.83	2.32	1.72	ND	ND	ND	10.77	8.94	9.05	0.00	0.84	2.73
10	5:1	9	Ambient + 0.5	1.92	2	0.18	70	ND	0.67	4.99	2.00	9.94	3.14	ND	BDL	2.00	22.73	18.74	16.92	0.67	0.74	5.10
11	5:1	9	Ambient	2.17	2	0.33	85	ND	BDL	7.56	3.07	2.79	2.19	ND	BDL	1.71	17.32	12.54	13.42	0.00	0.77	4.04
12	5:1	9	Ambient + 0.5	4.08	4	0.72	85	ND	2.12	8.61	4.93	9.92	2.26	ND	BDL	2.09	29.93	22.91	23.47	2.12	0.78	7.07
13	3:1	9	Ambient	2.03	2	0.22	70	ND	BDL	4.77	2.21	4.01	1.72	ND	ND	ND	12.71	10.50	10.99	0.00	0.86	3.31
14	3:1	9	Ambient + 0.5	4.00	4	0.45	70	ND	0.76	5.84	2.94	10.41	2.63	BDL	BDL	0.86	23.45	19.65	19.20	0.76	0.82	5.78
15	3:1	9	Ambient	3.88	4	0.67	85	ND	BDL	6.81	2.56	4.17	2.99	ND	ND	ND	16.53	13.97	13.54	0.00	0.82	4.08
16	3:1	9	Ambient + 0.5	2.12	2	0.33	85	ND	0.62	5.09	2.83	10.79	1.57	ND	0.50	0.99	22.39	18.07	18.71	0.62	0.84	5.64

DXAA = DCAA + DBAA + BCAA
MXAA = MCAA + MBAA
ND = Not Detected
BDL = Below Detection Limit of 0.5 μg/L

Table A.6
Lake Austin softened water (molar concentration)

Trial No.	Cl/N ratio	pH	Br^- (mg/L)	Total Cl Residual ($mg\ Cl_2/L$) actual	Total Cl Residual ($mg\ Cl_2/L$) target	Chloramine Demand (mg/L)	Temperature (°F)	MCAA	MBAA	DCAA	BCAA	DBAA	TCAA	BDCAA	DBCAA	TBAA	Total HAA	HAA5	DXAA	MXAA	DXAA/HAA9	DXAA Yield (DXAA/DOC)	Bromine incorporation factor (n_D')
								AVERAGE HAA CONCENTRATIONS (μmol/L)															
1	3:1	7	Ambient	1.91	2	0.79	70	ND	BDL	0.0427	0.0230	0.0220	0.0072	ND	ND	ND	0.09	0.07	0.09	0.00	0.92	0.03	0.764
2	3:1	7	Ambient + 0.5	3.65	4	1.40	70	ND	0.0125	0.0375	0.0367	0.0819	0.0180	BDL	0.0022	0.0029	0.19	0.15	0.16	0.01	0.81	0.05	1.285
3	3:1	7	Ambient	3.67	4	1.63	85	ND	0.0070	0.0657	0.0381	0.0364	0.0134	ND	BDL	BDL	0.16	0.12	0.14	0.01	0.87	0.04	0.791
4	3:1	7	Ambient + 0.5	1.87	2	1.28	85	ND	0.0134	0.0366	0.0353	0.0857	0.0085	BDL	BDL	0.0029	0.18	0.14	0.16	0.01	0.86	0.05	1.311
5	5:1	7	Ambient	3.86	4	1.44	70	ND	0.0190	0.0616	0.0351	0.0322	0.0161	BDL	BDL	BDL	0.16	0.13	0.13	0.02	0.79	0.04	0.771
6	5:1	7	Ambient + 0.5	1.82	2	1.48	70	ND	0.0135	0.0325	0.0413	0.0794	0.0045	0.0031	0.0039	0.0031	0.18	0.13	0.15	0.01	0.84	0.05	1.306
7	5:1	7	Ambient	2.04	2	1.46	85	ND	0.0049	0.0589	0.0382	0.0319	0.0044	ND	ND	ND	0.14	0.10	0.13	0.00	0.93	0.04	0.791
8	5:1	7	Ambient + 0.5	2.84	4	2.56	85	ND	0.0221	0.0460	0.0466	0.1175	0.0120	ND	0.0027	0.0066	0.25	0.20	0.21	0.02	0.83	0.06	1.340
9	5:1	9	Ambient	1.93	2	0.17	70	ND	BDL	0.0380	0.0106	0.0107	0.0105	ND	ND	ND	0.07	0.06	0.06	0.00	0.85	0.02	0.539
10	5:1	9	Ambient + 0.5	1.92	2	0.18	70	ND	0.0048	0.0387	0.0115	0.0456	0.0192	ND	BDL	0.0067	0.13	0.11	0.10	0.00	0.76	0.03	1.072
11	5:1	9	Ambient	2.17	2	0.33	85	ND	BDL	0.0587	0.0177	0.0128	0.0134	ND	BDL	0.0058	0.11	0.08	0.09	0.00	0.82	0.03	0.485
12	5:1	9	Ambient + 0.5	4.08	4	0.72	85	ND	0.0153	0.0668	0.0284	0.0455	0.0138	ND	BDL	0.0070	0.18	0.14	0.14	0.02	0.80	0.04	0.849
13	3:1	9	Ambient	2.03	2	0.22	70	ND	BDL	0.0370	0.0127	0.0184	0.0105	ND	ND	ND	0.08	0.07	0.07	0.00	0.87	0.02	0.727
14	3:1	9	Ambient + 0.5	4.00	4	0.45	70	ND	0.0055	0.0453	0.0170	0.0478	0.0161	BDL	BDL	0.0029	0.13	0.11	0.11	0.01	0.82	0.03	1.023
15	3:1	9	Ambient	3.88	4	0.67	85	ND	BDL	0.0528	0.0148	0.0191	0.0183	ND	ND	ND	0.11	0.09	0.09	0.00	0.83	0.03	0.611
16	3:1	9	Ambient + 0.5	2.12	2	0.33	85	ND	0.0045	0.0395	0.0163	0.0495	0.0096	ND	0.0020	0.0033	0.12	0.10	0.11	0.00	0.84	0.03	1.096

DXAA = DCAA + DBAA + BCAA
MXAA = MCAA + MBAA
ND = Not Detected
BDL = Below Detection Limit of 0.5 μg/L
DOC (mg/L) = 3.32

Table A.7
Lake Austin source hydrophilic fraction (mass concentration)

Trial No.	Cl/N ratio	pH	Br^- (mg/L)	Total Cl Residual ($mg\ Cl_2/L$) actual	target	Chloramine Demand (mg/L)	Temperature (°F)	AVERAGE HAA CONCENTRATIONS (µg/L) MCAA	MBAA	DCAA	BCAA	DBAA	TCAA	BDCAA	DBCAA	TBAA	Total HAA	HAA5	DXAA	MXAA	DXAA/HAA9	DXAA Yield (DXAA/DOC)
1	3:1	7	Ambient	1.99	2	0.26	70	0.54	0.57	3.25	2.62	2.98	BDL	BDL	ND	0.57	10.52	7.34	8.85	1.11	0.84	4.28
2	3:1	7	Ambient + 0.5	3.98	4	0.62	70	BDL	2.01	3.39	4.48	9.50	1.17	BDL	0.78	1.54	22.88	16.07	17.38	2.01	0.76	8.39
3	3:1	7	Ambient	3.99	4	0.49	85	BDL	2.14	4.69	3.33	3.07	BDL	BDL	ND	BDL	13.23	9.89	11.09	2.14	0.84	5.36
4	3:1	7	Ambient + 0.5	2.07	2	0.48	85	0.79	0.57	2.99	3.92	8.84	0.77	BDL	ND	0.99	18.86	13.96	15.74	1.37	0.83	7.60
5	5:1	7	Ambient	4.03	4	1.58	70	BDL	2.00	5.32	5.84	5.93	BDL	BDL	0.62	0.93	20.63	13.25	17.08	2.00	0.83	8.25
6	5:1	7	Ambient + 0.5	1.94	2	1.19	70	BDL	0.72	2.87	5.05	12.78	0.95	BDL	0.86	1.34	24.56	17.31	20.69	0.72	0.84	10.00
7	5:1	7	Ambient	2.22	2	0.88	85	0.71	0.83	5.10	5.14	5.31	BDL	BDL	ND	0.92	18.01	11.95	15.55	1.54	0.86	7.51
8	5:1	7	Ambient + 0.5	3.52	4	2.32	85	BDL	2.70	5.14	7.22	13.83	0.96	BDL	1.01	1.32	32.19	22.63	26.20	2.70	0.81	12.66
9	5:1	9	Ambient	4.23	4	0.33	70	BDL	2.00	6.12	2.08	2.32	BDL	BDL	ND	0.95	13.47	10.44	10.52	2.00	0.78	5.08
10	5:1	9	Ambient + 0.5	2.35	2	0.17	70	0.95	ND	4.10	2.29	5.06	0.55	BDL	0.54	1.21	14.68	10.65	11.44	0.95	0.78	5.53
11	5:1	9	Ambient	2.19	2	0.21	85	0.77	BDL	5.18	1.84	2.10	BDL	0.76	0.86	1.39	12.90	8.05	9.12	0.77	0.71	4.41
12	5:1	9	Ambient + 0.5	4.28	4	0.42	85	0.54	1.49	6.25	4.11	5.52	0.55	0.94	1.10	1.61	22.11	14.35	15.88	2.03	0.72	7.67
13	3:1	9	Ambient	2.03	2	0.02	70	0.56	BDL	2.47	1.55	2.19	BDL	0.71	ND	1.37	8.84	5.22	6.21	0.56	0.70	3.00
14	3:1	9	Ambient + 0.5	3.93	4	0.21	70	BDL	1.21	4.11	2.23	5.53	0.97	0.94	1.50	1.78	18.27	11.83	11.87	1.21	0.65	5.73
15	3:1	9	Ambient	4.06	4	0.18	85	BDL	0.93	4.89	1.78	2.21	BDL	0.77	ND	1.02	11.61	8.04	8.88	0.93	0.77	4.29
16	3:1	9	Ambient + 0.5	1.96	2	0.18	85	BDL	0.53	3.32	1.86	5.20	0.56	0.75	1.19	1.91	15.33	9.61	10.39	0.53	0.68	5.02

DXAA = DCAA + DBAA + BCAA
MXAA = MCAA + MBAA
ND = Not Detected
BDL = Below Detection Limit of 0.5 µg/L
DOC (mg/L) = 2.07

Table A.8
Lake Austin source hydrophilic fraction (molar concentration)

Trial No.	Cl/N ratio	pH	Br^- (mg/L)	Total Cl Residual ($mg\ Cl_2/L$) actual	target	Chloramine Demand (mg/L)	Temperature (°F)	AVERAGE HAA CONCENTRATIONS (µmol/L) MCAA	MBAA	DCAA	BCAA	DBAA	TCAA	BDCAA	DBCAA	TBAA	Total HAA	HAA5	DXAA	MXAA	DXAA/HAA9	DXAA Yield (DXAA/DOC)	Bromine incorporation factor (n_D')
1	3:1	7	Ambient	1.99	2	0.26	70	0.0057	0.0041	0.0252	0.0151	0.0137	BDL	BDL	ND	0.0019	0.0657	0.0487	0.0540	0.0098	0.82	0.0261	0.786
2	3:1	7	Ambient + 0.5	3.98	4	0.62	70	BDL	0.0145	0.0263	0.0258	0.0436	0.0072	BDL	0.0031	0.0052	0.1257	0.0916	0.0958	0.0145	0.76	0.0463	1.180
3	3:1	7	Ambient	3.99	4	0.49	85	BDL	0.0154	0.0363	0.0192	0.0141	BDL	BDL	ND	BDL	0.0850	0.0658	0.0697	0.0154	0.82	0.0337	0.681
4	3:1	7	Ambient + 0.5	2.07	2	0.48	85	0.0084	0.0041	0.0232	0.0226	0.0406	0.0047	BDL	ND	0.0033	0.1069	0.0809	0.0863	0.0125	0.81	0.0417	1.202
5	5:1	7	Ambient	4.03	4	1.58	70	BDL	0.0144	0.0413	0.0337	0.0272	BDL	BDL	0.0025	0.0031	0.1221	0.0829	0.1021	0.0144	0.84	0.0493	0.862
6	5:1	7	Ambient + 0.5	1.94	2	1.19	70	BDL	0.0051	0.0223	0.0291	0.0586	0.0058	BDL	0.0034	0.0045	0.1289	0.0919	0.1100	0.0051	0.85	0.0531	1.331
7	5:1	7	Ambient	2.22	2	0.88	85	0.0075	0.0059	0.0395	0.0297	0.0244	BDL	BDL	ND	0.0031	0.1101	0.0774	0.0936	0.0135	0.85	0.0452	0.838
8	5:1	7	Ambient + 0.5	3.52	4	4.88	85	BDL	0.0194	0.0398	0.0417	0.0635	0.0059	BDL	0.0040	0.0045	0.1788	0.1287	0.1450	0.0194	0.81	0.0701	1.163
9	5:1	9	Ambient	4.23	4	0.33	70	BDL	0.0144	0.0475	0.0120	0.0106	BDL	BDL	ND	0.0032	0.0877	0.0725	0.0701	0.0144	0.80	0.0339	0.475
10	5:1	9	Ambient + 0.5	2.35	2	0.05	70	0.0100	ND	0.0318	0.0132	0.0232	0.0034	BDL	0.0021	0.0041	0.0877	0.0684	0.0682	0.0100	0.78	0.0329	0.874
11	5:1	9	Ambient	2.19	2	0.21	85	0.0081	BDL	0.0402	0.0106	0.0096	BDL	0.0036	0.0034	0.0047	0.0803	0.0580	0.0605	0.0081	0.75	0.0292	0.494
12	5:1	9	Ambient + 0.5	4.28	4	0.42	85	0.0057	0.0107	0.0485	0.0237	0.0253	0.0034	0.0045	0.0043	0.0054	0.1317	0.0936	0.0975	0.0165	0.74	0.0471	0.762
13	3:1	9	Ambient	2.03	2	0.02	70	0.0059	BDL	0.0192	0.0089	0.0100	BDL	0.0034	ND	0.0046	0.0521	0.0351	0.0381	0.0059	0.73	0.0184	0.761
14	3:1	9	Ambient + 0.5	3.93	4	0.21	70	BDL	0.0087	0.0319	0.0128	0.0254	0.0059	0.0045	0.0059	0.0060	0.1012	0.0719	0.0701	0.0087	0.69	0.0339	0.907
15	3:1	9	Ambient	4.06	4	0.18	85	BDL	0.0067	0.0380	0.0102	0.0102	BDL	0.0037	ND	0.0035	0.0722	0.0548	0.0584	0.0067	0.81	0.0282	0.524
16	3:1	9	Ambient + 0.5	1.96	2	0.18	85	BDL	0.0038	0.0258	0.0108	0.0239	0.0034	0.0036	0.0047	0.0064	0.0824	0.0569	0.0604	0.0038	0.73	0.0292	0.969

DXAA = DCAA + DBAA + BCAA
MXAA = MCAA + MBAA
ND = Not Detected
BDL = Below Detection Limit of 0.5 µg/L
DOC (mg/L) = 2.07

Table A.9
Metedeconk River source water (mass concentration)

Trial No.	Cl/N ratio	pH	Br^- (mg/L)	Total Cl Residual ($mg\ Cl_2/L$) actual	target	Chloramine Demand (mg/L)	Temperature (ºF)	HAA CONCENTRATIONS (µg/L) TCAA	DCAA	BCAA	DBAA	DXAA	DXAA Yield (DXAA/DOC)
1	3:1	7	Ambient	2.21	2	0.79	70	BDL	6.40	2.20	2.20	10.80	3.87
2	3:1	7	Ambient + 0.5	3.79	4	1.61	70	BDL	5.30	8.50	6.80	20.60	7.38
3	3:1	7	Ambient	3.90	4	1.10	85	BDL	11.00	3.30	2.50	16.80	6.02
4	3:1	7	Ambient + 0.5	2.14	2	2.26	85	ND	5.00	8.20	6.90	20.10	7.20
5	5:1	7	Ambient	4.18	4	1.52	70	BDL	11.00	3.60	2.60	17.20	6.16
6	5:1	7	Ambient + 0.5	1.90	2	1.90	70	ND	6.50	9.20	10.40	26.10	9.35
7	5:1	7	Ambient	2.04	2	1.66	85	BDL	12.00	3.90	2.50	18.40	6.59
8	5:1	7	Ambient + 0.5	4.17	4	3.63	85	0.52	7.30	14.00	12.00	33.30	11.94
9	5:1	9	Ambient	2.14	4	0.46	70	ND	3.70	1.50	2.40	7.60	2.72
10	5:1	9	Ambient + 0.5	2.03	2	0.47	70	ND	2.90	2.80	4.10	9.80	3.51
11	5:1	9	Ambient	1.98	2	0.62	85	ND	6.40	1.30	2.20	9.90	3.55
12	5:1	9	Ambient + 0.5	3.71	4	1.49	85	0.60	8.10	3.80	4.20	16.10	5.77
13	3:1	9	Ambient	2.26	2	0.34	70	ND	3.10	1.80	2.60	7.50	2.69
14	3:1	9	Ambient + 0.5	3.76	4	0.37	70	ND	4.80	2.50	2.30	9.60	3.44
15	3:1	9	Ambient	3.69	4	0.91	85	BDL	6.20	1.60	2.40	10.20	3.66
16	3:1	9	Ambient + 0.5	1.90	2	0.70	85	BDL	8.10	4.30	5.00	17.40	6.24

DXAA = DCAA + DBAA + BCAA
ND = Not Detected
BDL = Below Detection Limit of 0.5 µg/L
DOC (mg/L) = 2.79

Table A.10
Metedeconk River source water (molar concentration)

Trial No.	Cl/N ratio	pH	Br^- (mg/L)	Total Cl Residual ($mg\ Cl_2/L$) actual	target	Chloramine Demand (mg/L)	Temperature (ºF)	HAA CONCENTRATIONS (µmol/L) TCAA	DCAA	BCAA	DBAA	DXAA	DXAA Yield (DXAA/DOC)	Bromine incorporation factor (n_D')
1	3:1	7	Ambient	2.21	2	0.79	70	BDL	0.0496	0.0127	0.0101	0.0724	0.02596	0.454
2	3:1	7	Ambient + 0.5	3.79	4	1.61	70	BDL	0.0411	0.0490	0.0312	0.1213	0.04349	0.919
3	3:1	7	Ambient	3.90	4	1.10	85	BDL	0.0853	0.0190	0.0115	0.1158	0.04151	0.363
4	3:1	7	Ambient + 0.5	2.14	2	2.26	85	ND	0.0388	0.0473	0.0317	0.1177	0.04220	0.940
5	5:1	7	Ambient	4.18	4	1.52	70	BDL	0.0853	0.0208	0.0119	0.1180	0.04230	0.378
6	5:1	7	Ambient + 0.5	1.90	2	1.90	70	ND	0.0504	0.0531	0.0477	0.1512	0.05420	0.982
7	5:1	7	Ambient	2.04	2	1.66	85	BDL	0.0931	0.0225	0.0115	0.1270	0.04553	0.358
8	5:1	7	Ambient + 0.5	4.17	4	3.63	85	0.52	0.0566	0.0807	0.0551	0.1924	0.06898	0.992
9	5:1	9	Ambient	2.14	4	0.46	70	ND	0.0287	0.0087	0.0110	0.0484	0.01733	0.634
10	5:1	9	Ambient + 0.5	2.03	2	0.47	70	ND	0.0225	0.0161	0.0188	0.0575	0.02060	0.936
11	5:1	9	Ambient	1.98	2	0.62	85	ND	0.0496	0.0075	0.0101	0.0672	0.02410	0.412
12	5:1	9	Ambient + 0.5	3.71	4	1.49	85	0.60	0.0628	0.0219	0.0193	0.1040	0.03728	0.581
13	3:1	9	Ambient	2.26	2	0.34	70	ND	0.0240	0.0104	0.0119	0.0464	0.01662	0.739
14	3:1	9	Ambient + 0.5	3.76	4	0.37	70	ND	0.0372	0.0144	0.0106	0.0622	0.02230	0.571
15	3:1	9	Ambient	3.69	4	0.91	85	BDL	0.0481	0.0092	0.0110	0.0683	0.02449	0.458
16	3:1	9	Ambient + 0.5	1.90	2	0.70	85	BDL	0.0628	0.0248	0.0230	0.1106	0.03963	0.639

DXAA = DCAA + DBAA + BCAA
ND = Not Detected
BDL = Below Detection Limit of 0.5 µg/L
DOC (mg/L) = 2.79

Table A.11
Metedeconk River alum coagulated source water (mass concentration)

Trial No.	Cl/N ratio	pH	Br^- (mg/L)	Total Cl Residual ($mg Cl_2/L$) actual	target	Chloramine Demand (mg/L)	Temperature (167F)	HAA CONCENTRATIONS (µg/L) TCAA	DCAA	BCAA	DBAA	DXAA	DXAA Yield (DXAA/DOC)
1	3:1	7	Ambient	1.47	2	0.96	70	BDL	3.7	0.6	ND	4.30	2.53
2	3:1	7	Ambient + 0.5	3.63	4	0.60	70	BDL	3.0	3.0	3.5	9.50	5.59
3	3:1	7	Ambient	2.61	4	1.29	85	BDL	5.1	1.2	0.4	6.70	3.94
4	3:1	7	Ambient + 0.5	2.00	2	0.73	85	BDL	2.9	3.1	4.2	10.20	6.00
5	5:1	7	Ambient	2.72	4	1.52	70	ND	5.5	3.6	5.0	14.10	8.29
6	5:1	7	Ambient + 0.5	2.12	2	1.02	70	BDL	2.8	3.6	5.0	11.40	6.71
7	5:1	7	Ambient	1.44	2	1.57	85	BDL	5.5	1.5	0.5	7.50	4.41
8	5:1	7	Ambient + 0.5	3.01	4	2.51	85	BDL	3.6	4.7	6.6	14.90	8.76
9	5:1	9	Ambient	2.23	4	1.26	70	BDL	3.3	0.4	0.2	3.90	2.29
10	5:1	9	Ambient + 0.5	2.01	2	0.23	70	BDL	2.7	1.1	1.7	5.50	3.24
11	5:1	9	Ambient	1.21	2	1.11	85	ND	4.0	0.4	ND	4.40	2.59
12	5:1	9	Ambient + 0.5	3.50	4	0.65	85	BDL	4.5	1.8	1.6	7.90	4.65
13	3:1	9	Ambient	1.12	2	0.38	70	ND	2.4	0.3	ND	2.70	1.59
14	3:1	9	Ambient + 0.5	3.34	4	0.10	70	BDL	2.6	1.2	1.8	5.60	3.29
15	3:1	9	Ambient	2.39	4	1.39	85	BDl	3.3	0.4	0.1	3.80	2.24
16	3:1	9	Ambient + 0.5	1.57	2	0.16	85	BDL	2.8	1.0	2.0	5.80	3.41

DXAA = DCAA + DBAA + BCAA
ND = Not Detected
BDL = Below Detection Limit of 0.5 µg/L
DOC (mg/L) = 1.70

Table A.12
Metedeconk River alum coagulated source water (molar concentration)

Trial No.	Cl/N ratio	pH	Br^- (mg/L)	Total Cl Residual ($mg Cl_2/L$) actual	target	Chloramine Demand (mg/L)	Temperature (°F)	HAA CONCENTRATIONS (µmol/L) TCAA	DCAA	BCAA	DBAA	DXAA	DXAA Yield (DXAA/DOC)	Bromine incorporation factor (n_D')
1	3:1	7	Ambient	1.47	2	0.96	70	BDL	0.0287	0.0035	ND	0.0322	0.01892	0.108
2	3:1	7	Ambient + 0.5	3.63	4	0.60	70	BDL	0.0233	0.0173	0.0161	0.0566	0.03332	0.873
3	3:1	7	Ambient	3.84	4	2.09	85	BDL	0.0396	0.0069	0.0018	0.0483	0.02842	0.219
4	3:1	7	Ambient + 0.5	2.00	2	0.73	85	BDL	0.0225	0.0179	0.0193	0.0597	0.03509	0.946
5	5:1	7	Ambient	4.02	4	3.29	70	ND	0.0427	0.0208	0.0230	0.0864	0.05081	0.772
6	5:1	7	Ambient + 0.5	2.12	2	1.02	70	BDL	0.0217	0.0208	0.0230	0.0654	0.03849	1.019
7	5:1	7	Ambient	1.44	2	1.57	85	BDL	0.0427	0.0087	0.0023	0.0536	0.03153	0.247
8	5:1	7	Ambient + 0.5	3.01	4	2.51	85	BDL	0.0279	0.0271	0.0303	0.0853	0.05019	1.028
9	5:1	9	Ambient	3.89	4	2.46	70	BDL	0.0256	0.0023	0.0009	0.0288	0.01695	0.144
10	5:1	9	Ambient + 0.5	2.01	2	0.23	70	BDL	0.0209	0.0063	0.0078	0.0351	0.02064	0.626
11	5:1	9	Ambient	1.21	2	1.11	85	ND	0.0310	0.0023	ND	0.0333	0.01961	0.069
12	5:1	9	Ambient + 0.5	3.50	4	0.65	85	BDL	0.0349	0.0104	0.0073	0.0526	0.03096	0.476
13	3:1	9	Ambient	1.12	2	0.38	70	ND	0.0186	0.0017	ND	0.0203	0.01197	0.085
14	3:1	9	Ambient + 0.5	3.34	4	0.10	70	BDL	0.0202	0.0069	0.0083	0.0353	0.02079	0.663
15	3:1	9	Ambient	4.28	4	2.11	85	BDl	0.0256	0.0023	0.0005	0.0284	0.01668	0.114
16	3:1	9	Ambient + 0.5	1.57	2	0.16	85	BDL	0.0217	0.0058	0.0092	0.0367	0.02157	0.658

DXAA = DCAA + DBAA + BCAA
ND = Not Detected
BDL = Below Detection Limit of 0.5 µg/L
DOC (mg/L) = 1.70

Table A.13
Metedeconk River biodegraded source water (mass concentration)

Trial No.	Cl/N ratio	pH	Br^- (mg/L)	Total Cl Residual ($mg\ Cl_2/L$) actual	Total Cl Residual ($mg\ Cl_2/L$) target	Chloramine Demand (mg/L)	Temperature (°F)	HAA CONCENTRATIONS (µg/L) TCAA	DCAA	BCAA	DBAA	DXAA	DXAA Yield (DXAA/DOC)
1	3:1	7	Ambient	2.36	2	0.59	70	BDL	5.00	1.70	1.90	8.60	3.82
2	3:1	7	Ambient + 0.5	4.80	4	0.34	70	BDL	3.30	3.50	4.00	10.80	4.80
3	3:1	7	Ambient	4.33	4	0.73	85	BDL	7.50	2.00	1.90	11.40	5.07
4	3:1	7	Ambient + 0.5	2.68	2	1.02	85	BDL	4.20	4.60	5.50	14.30	6.36
5	5:1	7	Ambient	3.70	4	2.64	70	ND	5.80	1.20	0.20	7.20	3.20
6	5:1	7	Ambient + 0.5	1.68	2	0.70	70	BDL	5.40	4.90	6.30	16.60	7.38
7	5:1	7	Ambient	1.57	2	2.78	85	BDL	7.80	2.40	1.90	12.10	5.38
8	5:1	7	Ambient + 0.5	3.88	4	2.33	85	0.5	5.70	7.60	7.90	21.20	9.42
9	5:1	9	Ambient	1.73	4	1.37	70	ND	3.90	1.40	1.80	7.10	3.16
10	5:1	9	Ambient + 0.5	2.09	2	1.04	70	BDL	6.60	2.00	2.70	11.30	5.02
11	5:1	9	Ambient	2.34	2	1.76	85	BDL	6.40	1.90	2.20	10.50	4.67
12	5:1	9	Ambient + 0.5	4.45	4	1.23	85	BDL	7.20	3.80	4.40	15.40	6.84
13	3:1	9	Ambient	2.22	2	1.76	70	ND	3.20	1.50	1.80	6.50	2.89
14	3:1	9	Ambient + 0.5	3.66	4	0.58	70	BDL	4.20	2.10	2.40	8.70	3.87
15	3:1	9	Ambient	4.09	4	2.39	85	BDL	5.60	1.40	1.80	8.80	3.91
16	3:1	9	Ambient + 0.5	2.59	2	0.58	85	BDL	3.60	2.60	4.30	10.50	4.67

DXAA = DCAA + DBAA + BCAA
ND = Not Detected
BDL = Below Detection Limit of 0.5 µg/L
DOC (mg/L) = 2.25

Table A.14
Metedeconk River biodegraded source water (molar concentration)

Trial No.	Cl/N ratio	pH	Br^- (mg/L)	Total Cl Residual ($mg\ Cl_2/L$) actual	Total Cl Residual ($mg\ Cl_2/L$) target	Chloramine Demand (mg/L)	Temperature (°F)	HAA CONCENTRATIONS (µmol/L) TCAA	DCAA	BCAA	DBAA	DXAA	DXAA Yield (DXAA/DOC)	Bromine incorporation factor (n_D')
1	3:1	7	Ambient	2.36	2	0.59	70	BDL	0.0388	0.0098	0.0087	0.0573	0.02547	0.476
2	3:1	7	Ambient + 0.5	4.80	4	0.34	70	BDL	0.0256	0.0202	0.0184	0.0641	0.02851	0.887
3	3:1	7	Ambient	4.33	4	0.73	85	BDL	0.0582	0.0115	0.0087	0.0784	0.03485	0.370
4	3:1	7	Ambient + 0.5	2.68	2	1.02	85	BDL	0.0326	0.0265	0.0252	0.0844	0.03749	0.913
5	5:1	7	Ambient	3.70	4	2.64	70	ND	0.0450	0.0069	0.0009	0.0528	0.02348	0.166
6	5:1	7	Ambient + 0.5	1.68	2	0.70	70	BDL	0.0419	0.0283	0.0289	0.0991	0.04403	0.869
7	5:1	7	Ambient	1.57	2	2.78	85	BDL	0.0605	0.0138	0.0087	0.0831	0.03691	0.377
8	5:1	7	Ambient + 0.5	3.88	4	2.33	85	0.5	0.0442	0.0438	0.0363	0.1243	0.05525	0.936
9	5:1	9	Ambient	1.73	4	1.37	70	ND	0.0302	0.0081	0.0083	0.0466	0.02070	0.528
10	5:1	9	Ambient + 0.5	2.09	2	1.04	70	BDL	0.0512	0.0115	0.0124	0.0751	0.03338	0.484
11	5:1	9	Ambient	2.34	2	1.76	85	BDL	0.0496	0.0110	0.0101	0.0707	0.03142	0.441
12	5:1	9	Ambient + 0.5	4.45	4	1.23	85	BDL	0.0558	0.0219	0.0202	0.0980	0.04354	0.636
13	3:1	9	Ambient	2.22	2	1.76	70	ND	0.0248	0.0087	0.0083	0.0417	0.01855	0.603
14	3:1	9	Ambient + 0.5	3.66	4	0.58	70	BDL	0.0326	0.0121	0.0110	0.0557	0.02476	0.613
15	3:1	9	Ambient	4.09	4	2.39	85	BDL	0.0434	0.0081	0.0083	0.0598	0.02656	0.412
16	3:1	9	Ambient + 0.5	2.59	2	0.58	85	BDL	0.0279	0.0150	0.0197	0.0627	0.02785	0.869

DXAA = DCAA + DBAA + BCAA
ND = Not Detected
BDL = Below Detection Limit of 0.5 µg/L
DOC (mg/L) = 2.25

Table A.15
Metedeconk River source hydrophilic fraction (mass concentration)

Trial No.	Cl/N ratio	pH	Br^- (mg/L)	Total Cl Residual ($mg Cl_2/L$) actual	Total Cl Residual ($mg Cl_2/L$) target	Chloramine Demand (mg/L)	Temperature (ºF)	HAA CONCENTRATIONS (µg/L) TCAA	DCAA	BCAA	DBAA	DXAA	DXAA Yield (DXAA/DOC)
1	3:1	7	Ambient	1.24	2	1.56	70	BDL	4.0	ND	ND	4.0	3.64
2	3:1	7	Ambient + 0.5	4.21	4	0.27	70	ND	3.8	1.9	1.6	7.3	6.64
3	3:1	7	Ambient	2.46	4	1.74	85	BDL	5.7	0.8	0.2	6.7	6.09
4	3:1	7	Ambient + 0.5	2.04	2	0.34	85	ND	4.2	2.4	2.9	9.5	8.64
5	5:1	7	Ambient	3.17	4	1.52	70	BDL	5.8	1.3	0.4	7.5	6.82
6	5:1	7	Ambient + 0.5	2.12	2	0.54	70	ND	4.1	3.2	4.2	11.5	10.45
7	5:1	7	Ambient	1.83	2	1.46	85	BDL	6.4	1.2	0.4	8.0	7.27
8	5:1	7	Ambient + 0.5	3.96	4	1.50	85	ND	4.2	4.7	5.0	13.9	12.64
9	5:1	9	Ambient	1.06	4	1.46	70	ND	3.5	0.4	0.2	4.1	3.73
10	5:1	9	Ambient + 0.5	2.23	2	0.22	70	ND	4.2	1.3	1.9	7.4	6.73
11	5:1	9	Ambient	1.10	2	1.42	85	ND	4.8	0.3	0.0	5.1	4.64
12	5:1	9	Ambient + 0.5	3.81	4	0.53	85	ND	4.4	1.8	1.1	7.3	6.64
13	3:1	9	Ambient	1.00	2	1.52	70	BDL	3.0	0.3	ND	3.3	3.00
14	3:1	9	Ambient + 0.5	3.17	4	0.19	70	ND	3.4	1.2	1.6	6.2	5.64
15	3:1	9	Ambient	2.23	4	1.69	85	ND	4.5	0.4	ND	4.9	4.45
16	3:1	9	Ambient + 0.5	1.67	2	0.15	85	ND	4.1	1.1	2.0	7.2	6.55

DXAA = DCAA + DBAA + BCAA
ND = Not Detected
BDL = Below Detection Limit of 0.5 µg/L
DOC(mg/L) = 1.10

Table A.16
Metedeconk River source hydrophilic fraction (molar concentration)

Trial No.	Cl/N ratio	pH	Br^- (mg/L)	Total Cl Residual ($mg Cl_2/L$) actual	Total Cl Residual ($mg Cl_2/L$) target	Chloramine Demand (mg/L)	Temperature (ºF)	HAA CONCENTRATIONS (µmol/L) TCAA	DCAA	BCAA	DBAA	DXAA	DXAA Yield (DXAA/DOC)	Bromine incorporation factor (n_D')
1	3:1	7	Ambient	1.24	2	1.56	70	BDL	0.03102	ND	ND	0.0310	0.02820	0.000
2	3:1	7	Ambient + 0.5	4.21	4	0.27	70	ND	0.02947	0.0110	0.0073	0.0478	0.04343	0.537
3	3:1	7	Ambient	2.46	4	1.74	85	BDL	0.04421	0.0046	0.0009	0.0497	0.04522	0.130
4	3:1	7	Ambient + 0.5	2.04	2	0.34	85	ND	0.03257	0.0138	0.0133	0.0597	0.05430	0.678
5	5:1	7	Ambient	3.17	4	1.52	70	BDL	0.04498	0.0075	0.0018	0.0543	0.04938	0.206
6	5:1	7	Ambient + 0.5	2.12	2	0.54	70	ND	0.03180	0.0185	0.0193	0.0695	0.06321	0.820
7	5:1	7	Ambient	1.83	2	1.46	85	BDL	0.04964	0.0069	0.0018	0.0584	0.05308	0.181
8	5:1	7	Ambient + 0.5	3.96	4	1.50	85	ND	0.03257	0.0271	0.0230	0.0826	0.07512	0.884
9	5:1	9	Ambient	1.06	4	1.46	70	ND	0.02714	0.0023	0.0009	0.0304	0.02761	0.136
10	5:1	9	Ambient + 0.5	2.23	2	0.22	70	ND	0.03257	0.0075	0.0087	0.0488	0.04436	0.511
11	5:1	9	Ambient	1.10	2	1.42	85	ND	0.03723	0.0017	ND	0.0390	0.03542	0.044
12	5:1	9	Ambient + 0.5	3.81	4	0.53	85	ND	0.03412	0.0104	0.0050	0.0496	0.04505	0.413
13	3:1	9	Ambient	1.78	2	0.93	70	BDL	0.02327	0.0017	ND	0.0250	0.02272	0.069
14	3:1	9	Ambient + 0.5	3.17	4	0.19	70	ND	0.02637	0.0069	0.0073	0.0406	0.03694	0.532
15	3:1	9	Ambient	2.23	4	1.69	85	ND	0.03490	0.0023	ND	0.0372	0.03382	0.062
16	3:1	9	Ambient + 0.5	1.67	2	0.15	85	ND	0.03180	0.0063	0.0092	0.0473	0.04302	0.522

DXAA = DCAA + DBAA + BCAA
ND = Not Detected
BDL = Below Detection Limit of 0.5 µg/L
DOC (mg/L) = 1.10

Table A.17
Metedeconk River source < 3000 AMW fraction (mass concentration)

Trial				Total Cl Residual ($mg\ Cl_2/L$)		Chloramine	Temperature	HAA CONCENTRATIONS (µg/L)					DXAA Yield
No.	Cl/N ratio	pH	Br^- (mg/L)	actual	target	Demand (mg/L)	(ºF)	TCAA	DCAA	BCAA	DBAA	DXAA	(DXAA/DOC)
1	3:1	7	Ambient	1.51	2	3.05	70	0.7	5.30	2.30	0.90	8.50	6.44
2	3:1	7	Ambient + 0.5	3.74	4	1.53	70	0.7	2.40	4.80	4.80	12.00	9.09
3	3:1	7	Ambient	2.84	4	3.90	85	BDL	6.10	3.10	0.90	10.10	7.65
4	3:1	7	Ambient + 0.5	1.91	2	1.44	85	0.5	2.80	4.80	9.20	16.80	12.73
5	5:1	7	Ambient	4.09	4	2.43	70	BDL	7.40	2.30	0.71	10.41	7.89
6	5:1	7	Ambient + 0.5	2.00	2	1.54	70	0.5	2.90	5.10	9.10	17.10	12.95
7	5:1	7	Ambient	1.91	2	2.55	85	0.5	8.30	3.70	1.20	13.20	10.00
8	5:1	7	Ambient + 0.5	1.79	4	4.71	85	0.5	3.40	5.50	10.00	18.90	14.32
9	5:1	9	Ambient	4.23	4	2.23	70	0.6	5.70	0.80	0.50	7.00	5.30
10	5:1	9	Ambient + 0.5	1.87	2	0.35	70	BDL	3.90	1.70	1.60	7.20	5.45
11	5:1	9	Ambient	2.28	2	2.14	85	BDL	6.30	0.60	0.20	7.10	5.38
12	5:1	9	Ambient + 0.5	3.75	4	0.78	85	0.6	4.60	2.80	1.60	9.00	6.82
13	3:1	9	Ambient	1.80	2	1.72	70	ND	3.50	0.40	0.10	4.00	3.03
14	3:1	9	Ambient + 0.5	3.72	4	0.47	70	BDL	2.80	1.90	1.40	6.10	4.62
15	3:1	9	Ambient	4.11	4	2.58	85	BDL	4.00	0.70	0.10	4.80	3.64
16	3:1	9	Ambient + 0.5	1.74	2	1.41	85	BDL	3.50	1.50	1.60	6.60	5.00

DXAA = DCAA + DBAA + BCAA
ND = Not Detected
BDL = Below Detection Limit of 0.5 µg/L
DOC (mg/L) = 1.32

Table A.18
Metedeconk River source < 3000 AMW fraction (molar concentration)

Trial				Total Cl Residual ($mg\ Cl_2/L$)		Chloramine	Temperature	HAA CONCENTRATIONS (µmol/L)					DXAA Yield	Bromine incorporation factor
No.	Cl/N ratio	pH	Br^- (mg/L)	actual	target	Demand (mg/L)	(ºF)	TCAA	DCAA	BCAA	DBAA	DXAA	(DXAA/DOC)	(n_D')
1	3:1	7	Ambient	1.51	2	3.05	70	0.0043	0.0411	0.0133	0.0041	0.0585	0.04432	0.368
2	3:1	7	Ambient + 0.5	3.74	4	1.53	70	0.0043	0.0186	0.0277	0.0220	0.0683	0.05177	1.050
3	3:1	7	Ambient	2.84	4	3.90	85	BDL	0.0473	0.0179	0.0041	0.0693	0.05251	0.377
4	3:1	7	Ambient + 0.5	1.91	2	1.44	85	0.0030	0.0217	0.0277	0.0422	0.0916	0.06942	1.224
5	5:1	7	Ambient	4.09	4	3.51	70	BDL	0.0574	0.0133	0.0033	0.0739	0.05600	0.268
6	5:1	7	Ambient + 0.5	2.00	2	1.54	70	0.0030	0.0225	0.0294	0.0418	0.0937	0.07097	1.206
7	5:1	7	Ambient	1.91	2	2.55	85	0.0030	0.0644	0.0213	0.0055	0.0912	0.06911	0.355
8	5:1	7	Ambient + 0.5	1.79	4	4.71	85	0.0030	0.0264	0.0317	0.0459	0.1040	0.07878	1.188
9	5:1	9	Ambient	4.23	4	2.23	70	0.0036	0.0442	0.0046	0.0023	0.0511	0.03872	0.180
10	5:1	9	Ambient + 0.5	1.87	2	0.35	70	BDL	0.0302	0.0098	0.0073	0.0474	0.03591	0.517
11	5:1	9	Ambient	2.28	2	2.14	85	BDL	0.0489	0.0035	0.0009	0.0532	0.04033	0.099
12	5:1	9	Ambient + 0.5	3.75	4	0.78	85	0.0036	0.0357	0.0161	0.0073	0.0592	0.04482	0.521
13	3:1	9	Ambient	1.80	2	1.72	70	BDL	0.0271	0.0023	0.0005	0.0299	0.02266	0.108
14	3:1	9	Ambient + 0.5	3.72	4	0.47	70	BDL	0.0217	0.0110	0.0064	0.0391	0.02962	0.609
15	3:1	9	Ambient	4.11	4	2.58	85	BDL	0.0310	0.0040	0.0005	0.0355	0.02691	0.140
16	3:1	9	Ambient + 0.5	1.74	2	1.41	85	BDL	0.0271	0.0087	0.0073	0.0431	0.03268	0.541

DXAA = DCAA + DBAA + BCAA
ND = Not Detected
BDL = Below Detection Limit of 0.5 µg/L
DOC (mg/L) = 1.32

APPENDIX B
INITIAL BATCH SCREENING EXPERIMENTS ANOVA ANALYSIS

Table B.1
Summary of ANOVA of various factors influencing the production of DXAA (µg/L) in Lake Austin source water

Parameter	Degrees of Freedom	Mean Squares	F Statistic	Percent Contribution %
pH	1	999.86	***	35.3
Cl_2/N ratio	1	169.63	NS	3.9
Temperature	1	137.08	NS	2.7
Bromide Level	1	568.85	**	19.0
Chloramine Residual	1	110.80	NS	1.7
Error	24	66.20		37.5

Total percent contribution is equal to 100%
***: Significant at 99% Confidence
**: Significant at 95% Confidence
*: Significant at 90% Confidence
NS: Not Significant

Table B.2
Summary of ANOVA of various factors influencing the production of DXAA (µmol/L) in Lake Austin source water

Parameter	Degrees of Freedom	Mean Squares	F Statistic	Percent Contribution %
pH	1	0.0304	***	40.1
Cl_2/N ratio	1	0.0056	*	5.5
Temperature	1	0.0051	NS	4.9
Bromide Level	1	0.0107	**	12.7
Chloramine Residual	1	0.0039	NS	3.2
Error	24	0.0016		33.7

Total percent contribution is equal to 100%
***: Significant at 99% Confidence
**: Significant at 95% Confidence
*: Significant at 90% Confidence
NS: Not Significant

Table B.3
Summary of ANOVA of various factors influencing the production of DXAA (µg/L) in Lake Austin softened water

Parameter	Degrees of Freedom	Mean Squares	F Statistic	Percent Contribution %
pH	1	410.39	***	41.4
Cl_2/N ratio	1	14.13	NS	0.8
Temperature	1	69.21	***	6.5
Bromide Level	1	385.75	***	38.9
Chloramine Residual	1	36.95	**	3.2
Error	24	6.06		9.3

Total percent contribution is equal to 100%
***: Significant at 99% Confidence
**: Significant at 95% Confidence
*: Significant at 90% Confidence
NS: Not Significant

Table B.4
Summary of ANOVA of various factors influencing the production of DXAA (µmol/L) in Lake Austin softened water

Parameter	Degrees of Freedom	Mean Squares	F Statistic	Percent Contribution %
pH	1	0.0104	***	44.0
Cl_2/N ratio	1	0.0006	*	1.8
Temperature	1	0.0025	***	10.1
Bromide Level	1	0.0072	***	30.4
Chloramine Residual	1	0.0013	***	5.2
Error	24	0.0001		8.6

Total percent contribution is equal to 100%
***: Significant at 99% Confidence
**: Significant at 95% Confidence
*: Significant at 90% Confidence
NS: Not Significant

Table B.5
Summary of ANOVA of various factors influencing the production of DXAA (µg/L) in Lake Austin algal source water

Parameter	Degrees of Freedom	Mean Squares	F Statistic	Percent Contribution %
pH	1	742.60	***	36.0
Cl_2/N ratio	1	206.92	***	9.4
Temperature	1	48.51	NS	1.5
Bromide Level	1	727.69	***	35.3
Chloramine Residual	1	107.68	**	4.5
Error	24	17.88		13.3

Total percent contribution is equal to 100%
***: Significant at 99% Confidence
**: Significant at 95% Confidence
*: Significant at 90% Confidence
NS: Not Significant

Table B.6
Summary of ANOVA of various factors influencing the production of DXAA (µmol/L) in Lake Austin algal source water

Parameter	Degrees of Freedom	Mean Squares	F Statistic	Percent Contribution %
pH	1	0.0211	***	39.8
Cl_2/N ratio	1	0.0079	***	14.3
Temperature	1	0.0018	*	2.7
Bromide Level	1	0.0129	***	23.9
Chloramine Residual	1	0.0039	**	6.7
Error	24	0.0004		12.5

Total percent contribution is equal to 100%
***: Significant at 99% Confidence
**: Significant at 95% Confidence
*: Significant at 90% Confidence
NS: Not Significant

Table B.7

Summary of ANOVA of various factors influencing the production of DXAA (µg/L) in Lake Austin hydrophilic fraction

Parameter	Degrees of Freedom	Mean Squares	F Statistic	Percent Contribution %
pH	1	145.63	***	35.4
Cl_2/N ratio	1	81.32	***	19.4
Temperature	1	4.86	NS	0.4
Bromide Level	1	111.66	***	27.0
Chloramine Residual	1	27.29	**	6.0
Error	24	3.13		11.7

Total percent contribution is equal to 100%
***: Significant at 99% Confidence
**: Significant at 95% Confidence
*: Significant at 90% Confidence
NS: Not Significant

Table B.8

Summary of ANOVA of various factors influencing the production of DXAA (µmol/L) in Lake Austin hydrophilic fraction

Parameter	Degrees of Freedom	Mean Squares	F Statistic	Percent Contribution %
pH	1	0.0034	***	31.8
Cl_2/N ratio	1	0.0029	***	26.7
Temperature	1	0.0002	*	1.8
Bromide Level	1	0.0022	***	20.2
Chloramine Residual	1	0.0012	***	10.7
Error	24	0.0001		8.9

Total percent contribution is equal to 100%
***: Significant at 99% Confidence
**: Significant at 95% Confidence
*: Significant at 90% Confidence
NS: Not Significant

Table B.9
Summary of ANOVA of various factors influencing the production of DXAA (μg/L) in Lake Austin hydrophobic fraction

Parameter	Degrees of Freedom	Mean Squares	F Statistic	Percent Contribution %
pH	1	382.32	**	28.5
Cl_2/N ratio	1	16.05	NS	-2.7
Temperature	1	90.30	NS	3.6
Bromide Level	1	176.46	*	11.0
Chloramine Residual	1	28.11	NS	-1.7
Error	24	47.87		61.3

Total percent contribution is equal to 100%
***: Significant at 99% Confidence
**: Significant at 95% Confidence
*: Significant at 90% Confidence
NS: Not Significant

Table B.10
Summary of ANOVA of various factors influencing the production of DXAA (μmol/L) in Lake Austin hydrophobic fraction

Parameter	Degrees of Freedom	Mean Squares	F Statistic	Percent Contribution %
pH	1	0.0135	***	37.0
Cl_2/N ratio	1	0.0004	NS	-2.3
Temperature	1	0.0031	NS	5.7
Bromide Level	1	0.0032	NS	6.1
Chloramine Residual	1	0.0008	NS	-1.3
Error	24	0.0012		54.8

Total percent contribution is equal to 100%
***: Significant at 99% Confidence
**: Significant at 95% Confidence
*: Significant at 90% Confidence
NS: Not Significant

Table B.11
Summary of ANOVA of various factors influencing the production of DXAA (µg/L) in Metedeconk River source water

Parameter	Degrees of Freedom	Mean Squares	F Statistic	Percent Contribution %
pH	1	353.44	***	44.6
Cl_2/N ratio	1	40.32	*	3.8
Temperature	1	68.06	**	7.4
Bromide Level	1	186.32	***	22.8
Chloramine Residual	1	8.12	NS	-0.4
Error	24	11.16		21.8

Total percent contribution is equal to 100%
***: Significant at 99% Confidence
**: Significant at 95% Confidence
*: Significant at 90% Confidence
NS: Not Significant

Table B.12
Summary of ANOVA of various factors influencing the production of DXAA (µmol/L) in Metedeconk River source water

Parameter	Degrees of Freedom	Mean Squares	F Statistic	Percent Contribution %
pH	1	0.0127	***	49.7
Cl_2/N ratio	1	0.0014	*	4.4
Temperature	1	0.0032	**	11.5
Bromide Level	1	0.0040	***	14.8
Chloramine Residual	1	0.0004	NS	0.3
Error	24	0.0003		19.2

Total percent contribution is equal to 100%
***: Significant at 99% Confidence
**: Significant at 95% Confidence
*: Significant at 90% Confidence
NS: Not Significant

Table B.13
Summary of ANOVA of various factors influencing the production of DXAA (μg/L) in Metedeconk River biodegraded water

Parameter	Degrees of Freedom	Mean Squares	F Statistic	Percent Contribution %
pH	1	34.22	**	13.2
Cl_2/N ratio	1	29.70	**	11.2
Temperature	1	46.92	***	18.6
Bromide Level	1	83.72	***	34.5
Chloramine Residual	1	0.00	NS	-1.6
Error	24	3.73		24.1

Total percent contribution is equal to 100%
***: Significant at 99% Confidence
**: Significant at 95% Confidence
*: Significant at 90% Confidence
NS: Not Significant

Table B.14
Summary of ANOVA of various factors influencing the production of DXAA (μmol/L) in Metedeconk River biodegraded water

Parameter	Degrees of Freedom	Mean Squares	F Statistic	Percent Contribution %
pH	1	0.0011	***	13.9
Cl_2/N ratio	1	0.0013	***	16.9
Temperature	1	0.0018	***	23.3
Bromide Level	1	0.0019	***	24.5
Chloramine Residual	1	0.0000	NS	-1.5
Error	24	0.0001		22.9

Total percent contribution is equal to 100%
***: Significant at 99% Confidence
**: Significant at 95% Confidence
*: Significant at 90% Confidence
NS: Not Significant

Table B.15
Summary of ANOVA of various factors influencing the production of DXAA (µg/L) in Metedeconk River hydrophilic fraction water

Parameter	Degrees of Freedom	Mean Squares	F Statistic	Percent Contribution %
pH	1	32.78	***	27.2
Cl_2/N ratio	1	15.41	***	12.1
Temperature	1	7.98	**	5.7
Bromide Level	1	44.56	***	37.4
Chloramine Residual	1	0.23	NS	-1.1
Error	24	1.44		18.8

Total percent contribution is equal to 100%
***: Significant at 99% Confidence
**: Significant at 95% Confidence
*: Significant at 90% Confidence
NS: Not Significant

Table B.16
Summary of ANOVA of various factors influencing the production of DXAA (µmol/L) in Metedeconk River hydrophilic fraction water

Parameter	Degrees of Freedom	Mean Squares	F Statistic	Percent Contribution %
pH	1	0.0011	***	33.3
Cl_2/N ratio	1	0.0006	***	15.6
Temperature	1	0.0004	***	9.8
Bromide Level	1	0.0009	***	26.4
Chloramine Residual	1	0.0000	NS	-0.7
Error	24	0.0000		15.7

Total percent contribution is equal to 100%
***: Significant at 99% Confidence
**: Significant at 95% Confidence
*: Significant at 90% Confidence
NS: Not Significant

Table B.17

Summary of ANOVA of various factors influencing the production of DXAA (µg/L) in Metedeconk River hydrophobic fraction water

Parameter	Degrees of Freedom	Mean Squares	F Statistic	Percent Contribution %
pH	1	170.96	***	50.2
Cl_2/N ratio	1	5.88	NS	-0.2
Temperature	1	29.43	*	6.9
Bromide Level	1	48.65	**	12.8
Chloramine Residual	1	5.64	NS	-0.3
Error	24	6.69		30.6

Total percent contribution is equal to 100%
***: Significant at 99% Confidence
**: Significant at 95% Confidence
*: Significant at 90% Confidence
NS: Not Significant

Table B.18

Summary of ANOVA of various factors influencing the production of DXAA (µmol/L) in Metedeconk River hydrophobic fraction water

Parameter	Degrees of Freedom	Mean Squares	F Statistic	Percent Contribution %
pH	1	0.0062	***	52.6
Cl_2/N ratio	1	0.0002	NS	-0.2
Temperature	1	0.0014	**	10.3
Bromide Level	1	0.0011	*	7.6
Chloramine Residual	1	0.0003	NS	0.5
Error	24	0.0002		29.1

Total percent contribution is equal to 100%
***: Significant at 99% Confidence
**: Significant at 95% Confidence
*: Significant at 90% Confidence
NS: Not Significant

Table B.19
Summary of ANOVA of various factors influencing the production of DXAA (µg/L) in Metedeconk River < 3K AMU water

Parameter	Degrees of Freedom	Mean Squares	F Statistic	Percent Contribution %
pH	1	190.51	***	60.9
Cl_2/N ratio	1	27.59	***	8.1
Temperature	1	12.58	*	3.2
Bromide Level	1	51.09	***	15.7
Chloramine Residual	1	0.30	NS	-0.8
Error	24	2.64		12.9

Total percent contribution is equal to 100%
***: Significant at 99% Confidence
**: Significant at 95% Confidence
*: Significant at 90% Confidence
NS: Not Significant

Table B.20
Summary of ANOVA of various factors influencing the production of DXAA (µmol/L) in Metedeconk River < 3K AMU water

Parameter	Degrees of Freedom	Mean Squares	F Statistic	Percent Contribution %
pH	1	0.0053	***	68.3
Cl_2/N ratio	1	0.0012	***	15.0
Temperature	1	0.0005	***	5.4
Bromide Level	1	0.0004	***	5.2
Chloramine Residual	1	0.0000	NS	-0.4
Error	24	0.0000		6.5

Total percent contribution is equal to 100%
***: Significant at 99% Confidence
**: Significant at 95% Confidence
*: Significant at 90% Confidence
NS: Not Significant

Table B.21
Summary of ANOVA of various factors influencing the production of DXAA (µg/L) in Metedeconk River > 3K AMU water

Parameter	Degrees of Freedom	Mean Squares	F Statistic	Percent Contribution %
pH	1	24.98	NS	7.9
Cl_2/N ratio	1	1.20	NS	-4.2
Temperature	1	22.11	NS	6.5
Bromide Level	1	42.28	*	16.7
Chloramine Residual	1	11.54	NS	1.1
Error	24	9.43		72.0

Total percent contribution is equal to 100%
***: Significant at 99% Confidence
**: Significant at 95% Confidence
*: Significant at 90% Confidence
NS: Not Significant

Table B.22
Summary of ANOVA of various factors influencing the production of DXAA (µmol/L) in Metedeconk River > 3K AMU water

Parameter	Degrees of Freedom	Mean Squares	F Statistic	Percent Contribution %
pH	1	0.0016	*	14.6
Cl_2/N ratio	1	0.0000	NS	-3.9
Temperature	1	0.0012	*	10.4
Bromide Level	1	0.0018	**	17.1
Chloramine Residual	1	0.0005	NS	1.8
Error	24	0.0003		60.0

Total percent contribution is equal to 100%
***: Significant at 99% Confidence
**: Significant at 95% Confidence
*: Significant at 90% Confidence
NS: Not Significant

Table B.23
Summary of ANOVA of various factors influencing the production of DXAA (µg/L) in Metedeconk River alum coagulated water

Parameter	Degrees of Freedom	Mean Squares	F Statistic	Percent Contribution %
pH	1	95.06	***	48.7
Cl_2/N ratio	1	27.56	**	14.7
Temperature	1	1.10	NS	0.0
Bromide Level	1	34.22	***	8.2
Chloramine Residual	1	13.32	*	6.3
Error	24	3.42		22.1

Total percent contribution is equal to 100%
***: Significant at 99% Confidence
**: Significant at 95% Confidence
*: Significant at 90% Confidence
NS: Not Significant

Table B.24
Summary of ANOVA of various factors influencing the production of DXAA (µmol/L) in Metedeconk River alum coagulated water

Parameter	Degrees of Freedom	Mean Squares	F Statistic	Percent Contribution %
pH	1	0.0029	***	48.7
Cl_2/N ratio	1	0.0009	***	14.7
Temperature	1	0.0001	NS	0.0
Bromide Level	1	0.0006	**	8.2
Chloramine Residual	1	0.0005	**	6.3
Error	24	0.0001		22.1

Total percent contribution is equal to 100%
***: Significant at 99% Confidence
**: Significant at 95% Confidence
*: Significant at 90% Confidence
NS: Not Significant

Table B.25
Summary of ANOVA of various factors influencing the production of DXAA (µg/L) in Metedeconk River source water

Parameter	Degrees of Freedom	Mean Squares	F Statistic	Percent Contribution %
pH	1	353.44	***	45.1
Cl_2/N ratio	1	40.32	**	4.3
Temperature	1	68.06	**	8.0
Bromide Level	1	186.32	***	23.4
Chloramine Residual	1	8.12	NS	0.2
Cl_2/N x pH	1	49.00	**	5.5
Error	9	6.95		13.6

Total percent contribution is equal to 100%
***: Significant at 99% Confidence
**: Significant at 95% Confidence
*: Significant at 90% Confidence
NS: Not Significant

Table B.26
Summary of ANOVA of various factors influencing the production of DXAA (µmol/L) in Metedeconk River source water

Parameter	Degrees of Freedom	Mean Squares	F Statistic	Percent Contribution %
pH	1	0.0127	***	50.4
Cl_2/N ratio	1	0.0014	**	5.1
Temperature	1	0.0032	***	12.1
Bromide Level	1	0.0040	***	15.5
Chloramine Residual	1	0.0004	NS	1.0
Cl_2/N x pH	1	0.0018	***	6.8
Error	9	0.0002		9.0

Total percent contribution is equal to 100%
***: Significant at 99% Confidence
**: Significant at 95% Confidence
*: Significant at 90% Confidence
NS: Not Significant

Table B.27
Summary of ANOVA of various factors influencing the production of DXAA (µg/L) in Metedeconk River alum coagulated water

Parameter	Degrees of Freedom	Mean Squares	F Statistic	Percent Contribution %
pH	1	46.10	***	29.0
Cl_2/N ratio	1	27.25	***	16.4
Temperature	1	1.17	NS	-1.0
Bromide Level	1	35.11	***	21.7
Chloramine Residual	1	12.78	*	6.8
Cl_2/N x pH	1	3.44	NS	0.5
Error	9	2.66		26.6

Total percent contribution is equal to 100%
***: Significant at 99% Confidence
**: Significant at 95% Confidence
*: Significant at 90% Confidence
NS: Not Significant

Table B.28
Summary of ANOVA of various factors influencing the production of DXAA (µmol/L) in Metedeconk River alum coagulated water

Parameter	Degrees of Freedom	Mean Squares	F Statistic	Percent Contribution %
pH	1	0.0029	***	49.0
Cl_2/N ratio	1	0.0009	***	15.0
Temperature	1	0.0001	NS	0.4
Bromide Level	1	0.0006	**	8.6
Chloramine Residual	1	0.0005	**	6.7
Cl_2/N x pH	1	0.0003	*	3.3
Error	9	0.0001		17.1

Total percent contribution is equal to 100%
***: Significant at 99% Confidence
**: Significant at 95% Confidence
*: Significant at 90% Confidence
NS: Not Significant

Table B.29

Summary of ANOVA of various factors influencing the production of DXAA (μg/L) in Metedeconk River biodegraded water

Parameter	Degrees of Freedom	Mean Squares	F Statistic	Percent Contribution %
pH	1	34.22	**	13.0
Cl_2/N ratio	1	29.70	**	11.0
Temperature	1	46.92	***	18.5
Bromide Level	1	83.72	***	34.3
Chloramine Residual	1	0.00	NS	-1.8
Cl_2/N x pH	1	0.30	NS	-1.6
Error	9	4.11		26.6

Total percent contribution is equal to 100%
***: Significant at 99% Confidence
**: Significant at 95% Confidence
*: Significant at 90% Confidence
NS: Not Significant

Table B.30

Summary of ANOVA of various factors influencing the production of DXAA (μmol/L) in Metedeconk River biodegraded water

Parameter	Degrees of Freedom	Mean Squares	F Statistic	Percent Contribution %
pH	1	0.0011	**	13.8
Cl_2/N ratio	1	0.0013	***	16.7
Temperature	1	0.0018	***	23.1
Bromide Level	1	0.0019	***	24.3
Chloramine Residual	1	0.0000	NS	-1.7
Cl_2/N x pH	1	0.0000	NS	-1.7
Error	9	0.0001		25.4

Total percent contribution is equal to 100%
***: Significant at 99% Confidence
**: Significant at 95% Confidence
*: Significant at 90% Confidence
NS: Not Significant

Table B.31
Summary of ANOVA of various factors influencing the production of DXAA (µg/L) in Metedeconk River < 3K AMU water

Parameter	Degrees of Freedom	Mean Squares	F Statistic	Percent Contribution %
pH	1	190.51	***	60.8
Cl_2/N ratio	1	27.59	**	8.0
Temperature	1	12.58	*	3.2
Bromide Level	1	51.09	***	15.6
Chloramine Residual	1	0.30	NS	-0.8
Cl_2/N x pH	1	0.73	NS	-0.7
Error	9	2.86		13.9

Total percent contribution is equal to 100%
***: Significant at 99% Confidence
**: Significant at 95% Confidence
*: Significant at 90% Confidence
NS: Not Significant

Table B.32
Summary of ANOVA of various factors influencing the production of DXAA (µmol/L) in Metedeconk River < 3K AMU water

Parameter	Degrees of Freedom	Mean Squares	F Statistic	Percent Contribution %
pH	1	0.0053	***	68.2
Cl_2/N ratio	1	0.0012	***	14.9
Temperature	1	0.0005	***	5.4
Bromide Level	1	0.0004	***	5.2
Chloramine Residual	1	0.0000	NS	-0.4
Cl_2/N x pH	1	0.0000	NS	-0.4
Error	9	0.0000		7.0

Total percent contribution is equal to 100%
***: Significant at 99% Confidence
**: Significant at 95% Confidence
*: Significant at 90% Confidence
NS: Not Significant

Table B.33
Summary of ANOVA of various factors influencing the production of DXAA (μg/L) in Metedeconk River > 3K AMU water

Parameter	Degrees of Freedom	Mean Squares	F Statistic	Percent Contribution %
pH	1	24.98	*	9.5
Cl_2/N ratio	1	1.20	NS	-2.6
Temperature	1	22.11	*	8.1
Bromide Level	1	42.28	**	18.3
Chloramine Residual	1	11.54	NS	2.7
Cl_2/N x pH	1	37.79	**	16.0
Error	9	6.28		47.9

Total percent contribution is equal to 100%
***: Significant at 99% Confidence
**: Significant at 95% Confidence
*: Significant at 90% Confidence
NS: Not Significant

Table B.34
Summary of ANOVA of various factors influencing the production of DXAA (μmol/L) in Metedeconk River > 3K AMU water

Parameter	Degrees of Freedom	Mean Squares	F Statistic	Percent Contribution %
pH	1	0.001584	**	16.2
Cl_2/N ratio	1	1.01E-05	NS	-2.3
Temperature	1	0.001231	**	12.0
Bromide Level	1	0.0018	**	18.7
Chloramine Residual	1	0.000491	NS	3.4
Cl_2/N x pH	1	0.001601	**	16.4
Error	9	0.000203		35.6

Total percent contribution is equal to 100%
***: Significant at 99% Confidence
**: Significant at 95% Confidence
*: Significant at 90% Confidence
NS: Not Significant

Table B.35
Summary of ANOVA of various factors influencing the production of DXAA (µg/L) in Metedeconk River hydrophilic fraction

Parameter	Degrees of Freedom	Mean Squares	F Statistic	Percent Contribution %
pH	1	32.78	***	27.8
Cl_2/N ratio	1	15.41	***	12.7
Temperature	1	7.98	***	6.3
Bromide Level	1	44.56	***	38.0
Chloramine Residual	1	0.23	NS	-0.5
Cl_2/N x pH	1	7.70	***	6.0
Error	9	0.75		9.7

Total percent contribution is equal to 100%
***: Significant at 99% Confidence
**: Significant at 95% Confidence
*: Significant at 90% Confidence
NS: Not Significant

Table B.36
Summary of ANOVA of various factors influencing the production of DXAA (µmol/L) in Metedeconk River hydrophilic fraction

Parameter	Degrees of Freedom	Mean Squares	F Statistic	Percent Contribution %
pH	1	0.0011	***	33.9
Cl_2/N ratio	1	0.0006	***	16.2
Temperature	1	0.0004	***	10.4
Bromide Level	1	0.0009	***	27.0
Chloramine Residual	1	0.0000	NS	-0.1
Cl_2/N x pH	1	0.0002	***	6.1
Error	9	0.0000		6.6

Total percent contribution is equal to 100%
***: Significant at 99% Confidence
**: Significant at 95% Confidence
*: Significant at 90% Confidence
NS: Not Significant

Table B.37
Summary of ANOVA of various factors influencing the production of DXAA (μg/L) in Metedeconk River hydrophobic fraction

Parameter	Degrees of Freedom	Mean Squares	F Statistic	Percent Contribution %
pH	1	170.96	***	50.5
Cl_2/N ratio	1	5.88	NS	0.1
Temperature	1	29.43	**	7.3
Bromide Level	1	48.65	**	13.2
Chloramine Residual	1	5.64	NS	0.1
Cl_2/N x pH	1	17.85	NS	3.8
Error	9	5.45		25.0

Total percent contribution is equal to 100%
***: Significant at 99% Confidence
**: Significant at 95% Confidence
*: Significant at 90% Confidence
NS: Not Significant

Table B.38
Summary of ANOVA of various factors influencing the production of DXAA (μmol/L) in Metedeconk River hydrophobic fraction

Parameter	Degrees of Freedom	Mean Squares	F Statistic	Percent Contribution %
pH	1	0.0062	***	53.2
Cl_2/N ratio	1	0.0002	NS	0.4
Temperature	1	0.0014	**	10.8
Bromide Level	1	0.0011	**	8.2
Chloramine Residual	1	0.0003	NS	1.1
Cl_2/N x pH	1	0.0008	**	5.6
Error	9	0.0002		20.8

Total percent contribution is equal to 100%
***: Significant at 99% Confidence
**: Significant at 95% Confidence
*: Significant at 90% Confidence
NS: Not Significant

Table B.39
Summary of ANOVA of various factors influencing the production of DXAA (µg/L) in Lake Austin source water

Parameter	Degrees of Freedom	Mean Squares	F Statistic	Percent Contribution %
pH	1	999.86	***	35.8
Cl_2/N ratio	1	169.63	NS	4.4
Temperature	1	137.08	NS	3.2
Bromide Level	1	568.85	***	19.5
Chloramine Residual	1	110.80	NS	2.2
Cl_2/N x pH	1	185.35	*	5.0
Error	9	52.97		30.0

Total percent contribution is equal to 100%
***: Significant at 99% Confidence
**: Significant at 95% Confidence
*: Significant at 90% Confidence
NS: Not Significant

Table B.40
Summary of ANOVA of various factors influencing the production of DXAA (µmol/L) in Lake Austin source water

Parameter	Degrees of Freedom	Mean Squares	F Statistic	Percent Contribution %
pH	1	0.0304	***	40.6
Cl_2/N ratio	1	0.0056	*	6.0
Temperature	1	0.0051	*	5.4
Bromide Level	1	0.0107	**	13.2
Chloramine Residual	1	0.0039	NS	3.7
Cl/N*pH	1	0.0048	*	5.0
Error	9	0.0013		26.3

Total percent contribution is equal to 100%
***: Significant at 99% Confidence
**: Significant at 95% Confidence
*: Significant at 90% Confidence
NS: Not Significant

Table B.41

Summary of ANOVA of various factors influencing the production of DXAA (μg/L) in Lake Austin algal source water

Parameter	Degrees of Freedom	Mean Squares	F Statistic	Percent Contribution %
pH	1	742.84	***	36.2
Cl_2/N ratio	1	206.93	***	9.5
Temperature	1	48.51	NS	1.7
Bromide Level	1	727.65	***	35.4
Chloramine Residual	1	107.64	**	4.6
Cl_2/N x pH	1	41.28	NS	1.3
Error	9	15.28		11.4

Total percent contribution is equal to 100%
***: Significant at 99% Confidence
**: Significant at 95% Confidence
*: Significant at 90% Confidence
NS: Not Significant

Table B.42

Summary of ANOVA of various factors influencing the production of DXAA (μmol/L) in Lake Austin algal source water

Parameter	Degrees of Freedom	Mean Squares	F Statistic	Percent Contribution %
pH	1	0.0211	***	39.8
Cl_2/N ratio	1	0.0079	***	14.4
Temperature	1	0.0018	*	2.7
Bromide Level	1	0.0129	***	24.0
Chloramine Residual	1	0.0039	**	6.7
Cl_2/N x pH	1	0.0005	NS	0.2
Error	9	0.0004		12.2

Total percent contribution is equal to 100%
***: Significant at 99% Confidence
**: Significant at 95% Confidence
*: Significant at 90% Confidence
NS: Not Significant

Table B.43
Summary of ANOVA of various factors influencing the production of DXAA (µg/L) in Lake Austin softened water

Parameter	Degrees of Freedom	Mean Squares	F Statistic	Percent Contribution %
pH	1	410.27	***	41.4
Cl_2/N ratio	1	14.14	NS	0.9
Temperature	1	69.22	***	6.5
Bromide Level	1	385.73	***	38.9
Chloramine Residual	1	36.97	**	3.2
Cl_2/N x pH	1	12.60	NS	0.7
Error	9	5.34		8.2

Total percent contribution is equal to 100%
***: Significant at 99% Confidence
**: Significant at 95% Confidence
*: Significant at 90% Confidence
NS: Not Significant

Table B.44
Summary of ANOVA of various factors influencing the production of DXAA (µmol/L) in Lake Austin softened water

Parameter	Degrees of Freedom	Mean Squares	F Statistic	Percent Contribution %
pH	1	0.0104	***	44.0
Cl_2/N ratio	1	0.0006	*	1.9
Temperature	1	0.0025	***	10.2
Bromide Level	1	0.0072	***	30.4
Chloramine Residual	1	0.0013	***	5.2
Cl_2/N x pH	1	0.0003	NS	0.6
Error	9	0.0001		7.6

Total percent contribution is equal to 100%
***: Significant at 99% Confidence
**: Significant at 95% Confidence
*: Significant at 90% Confidence
NS: Not Significant

Table B.45
Summary of ANOVA of various factors influencing the production of DXAA (µg/L) in Lake Austin hydrophilic fraction

Parameter	Degrees of Freedom	Mean Squares	F Statistic	Percent Contribution %
pH	1	145.63	***	35.8
Cl_2/N ratio	1	81.32	***	19.9
Temperature	1	4.86	NS	0.8
Bromide Level	1	111.66	***	27.4
Chloramine Residual	1	27.29	***	6.4
Cl_2/N x pH	1	17.73	***	4.0
Error	9	1.51		5.6

Total percent contribution is equal to 100%
***: Significant at 99% Confidence
**: Significant at 95% Confidence
*: Significant at 90% Confidence
NS: Not Significant

Table B.46
Summary of ANOVA of various factors influencing the production of DXAA (µmol/L) in Lake Austin hydrophilic fraction

Parameter	Degrees of Freedom	Mean Squares	F Statistic	Percent Contribution %
pH	1	0.0034	***	32.1
Cl_2/N ratio	1	0.0029	***	27.0
Temperature	1	0.0002	**	2.1
Bromide Level	1	0.0022	***	20.5
Chloramine Residual	1	0.0012	***	11.0
Cl_2/N x pH	1	0.0004	***	3.1
Error	9	0.0000		4.2

Total percent contribution is equal to 100%
***: Significant at 99% Confidence
**: Significant at 95% Confidence
*: Significant at 90% Confidence
NS: Not Significant

Table B.47
Summary of ANOVA of various factors influencing the production of DXAA (µg/L) in Lake Austin hydrophobic fraction

Parameter	Degrees of Freedom	Mean Squares	F Statistic	Percent Contribution %
pH	1	382.32	**	28.9
Cl_2/N ratio	1	16.05	NS	-2.3
Temperature	1	90.30	NS	4.0
Bromide Level	1	176.46	*	11.4
Chloramine Residual	1	28.11	NS	-1.3
Cl_2/N x pH	1	88.42	NS	3.8
Error	9	43.37		55.5

Total percent contribution is equal to 100%
***: Significant at 99% Confidence
**: Significant at 95% Confidence
*: Significant at 90% Confidence
NS: Not Significant

Table B.48
Summary of ANOVA of various factors influencing the production of DXAA (µmol/L) in Lake Austin hydrophobic fraction

Parameter	Degrees of Freedom	Mean Squares	F Statistic	Percent Contribution %
pH	1	0.0135	***	37.5
Cl_2/N ratio	1	0.0004	NS	-1.9
Temperature	1	0.0031	NS	6.2
Bromide Level	1	0.0032	NS	6.5
Chloramine Residual	1	0.0008	NS	-0.8
Cl_2/N x pH	1	0.0026	NS	4.5
Error	9	0.0011		48.1

Total percent contribution is equal to 100%
***: Significant at 99% Confidence
**: Significant at 95% Confidence
*: Significant at 90% Confidence
NS: Not Significant

Table B.49

Summary of ANOVA of various factors influencing the production of DXAA (µg/L) in Metedeconk River source, alum coagulated, biodegraded, < 3K AMU, and hydrophilic waters

Parameter	Degrees of Freedom	Mean Squares	F Statistic	Percent Contribution %
water	4	196.25	***	31.6
pH	1	581.64	***	23.9
Cl_2/N ratio	1	137.58	***	5.5
Temperature	1	101.45	***	4.0
Bromide Level	1	360.78	***	14.7
Chloramine Residual	1	8.39	NS	0.1
Cl_2/N x pH	1	42.21	**	1.5
Error	69	5.75		18.8

Total percent contribution is equal to 100%
***: Significant at 99% Confidence
**: Significant at 95% Confidence
*: Significant at 90% Confidence
NS: Not Significant

Table B.50

Summary of ANOVA of various factors influencing the production of DXAA (µmol/L) in Metedeconk River source, alum coagulated, biodegraded, < 3K AMU, and hydrophilic waters

Parameter	Degrees of Freedom	Mean Squares	F Statistic	Percent Contribution %
water	4	0.0071	***	35.9
pH	1	0.0189	***	24.1
Cl_2/N ratio	1	0.0053	***	6.6
Temperature	1	0.0044	***	5.4
Bromide Level	1	0.0066	***	8.3
Chloramine Residual	1	0.0004	NS	0.3
Cl_2/N x pH	1	0.0012	**	1.3
Error	69	0.0002		18.1

Total percent contribution is equal to 100%
***: Significant at 99% Confidence
**: Significant at 95% Confidence
*: Significant at 90% Confidence
NS: Not Significant

Table B.51
Summary of ANOVA of various factors influencing the production of DXAA (µg/L) in Lake Austin source, algal source, softened waters, and hydrophilic waters

Parameter	Degrees of Freedom	Mean Squares	F Statistic	Percent Contribution %
water	3	446.56	***	17.2
pH	1	2079.27	***	27.9
Cl_2/N ratio	1	403.75	***	5.2
Temperature	1	213.14	**	2.6
Bromide Level	1	1641.57	***	21.9
Chloramine Residual	1	259.30	***	3.2
Cl_2/N x pH	1	193.21	**	2.3
Error	54	23.14		19.8

Total percent contribution is equal to 100%
***: Significant at 99% Confidence
**: Significant at 95% Confidence
*: Significant at 90% Confidence
NS: Not Significant

Table B.52
Summary of ANOVA of various factors influencing the production of DXAA (µmol/L) in Lake Austin source, algal source, softened waters, and hydrophilic fraction waters

Parameter	Degrees of Freedom	Mean Squares	F Statistic	Percent Contribution %
water	3	0.0149	***	21.2
pH	1	0.0576	***	28.2
Cl_2/N ratio	1	0.0145	***	6.8
Temperature	1	0.0081	***	3.7
Bromide Level	1	0.0304	***	14.7
Chloramine Residual	1	0.0096	***	4.4
Cl_2/N x pH	1	0.0041	**	1.7
Error	54	0.0006		19.3

Total percent contribution is equal to 100%
***: Significant at 99% Confidence
**: Significant at 95% Confidence
*: Significant at 90% Confidence
NS: Not Significant

Table B.53

Summary of ANOVA of various factors influencing the production of DXAA (μg/L) in Metedeconk River source, and alum coagulated waters

Parameter	Degrees of Freedom	Mean Squares	F Statistic	Percent Contribution %
water	1	554.45	***	35.8
pH	1	407.55	***	26.2
Cl_2/N ratio	1	67.28	**	3.9
Temperature	1	43.25	**	2.3
Bromide Level	1	190.13	***	11.9
Chloramine Residual	1	21.13	NS	0.9
Cl_2/N x pH	1	53.56	**	3.0
Error	24	7.93		16.1

Total percent contribution is equal to 100%
***: Significant at 99% Confidence
**: Significant at 95% Confidence
*: Significant at 90% Confidence
NS: Not Significant

Table B.54

Summary of ANOVA of various factors influencing the production of DXAA (μmol/L) in Metedeconk River source, and alum coagulated waters

Parameter	Degrees of Freedom	Mean Squares	F Statistic	Percent Contribution %
water	1	0.0211	***	40.2
pH	1	0.0140	***	26.4
Cl_2/N ratio	1	0.0023	**	4.0
Temperature	1	0.0022	**	3.7
Bromide Level	1	0.0038	***	6.8
Chloramine Residual	1	0.0009	*	1.2
Cl_2/N x pH	1	0.0017	**	2.9
Error	24	0.0002		14.8

Total percent contribution is equal to 100%
***: Significant at 99% Confidence
**: Significant at 95% Confidence
*: Significant at 90% Confidence
NS: Not Significant

Table B.55
Summary of ANOVA of various factors influencing the production of DXAA (µg/L) in Lake Austin source, and softened waters

Parameter	Degrees of Freedom	Mean Squares	F Statistic	Percent Contribution %
water	1	82.80	NS	1.4
pH	1	1345.53	***	35.5
Cl_2/N ratio	1	140.85	*	3.0
Temperature	1	200.56	**	4.6
Bromide Level	1	945.72	***	24.7
Chloramine Residual	1	137.88	*	2.9
Cl_2/N x pH	1	147.30	**	3.2
Error	24	29.47		24.6

Total percent contribution is equal to 100%
***: Significant at 99% Confidence
**: Significant at 95% Confidence
*: Significant at 90% Confidence
NS: Not Significant

Table B.56
Summary of ANOVA of various factors influencing the production of DXAA (µmol/L) in Lake Austin source, and softened waters

Parameter	Degrees of Freedom	Mean Squares	F Statistic	Percent Contribution %
water	1	0.0030	*	2.3
pH	1	0.0382	***	38.1
Cl_2/N ratio	1	0.0048	**	4.1
Temperature	1	0.0074	**	6.7
Bromide Level	1	0.0178	***	17.3
Chloramine Residual	1	0.0049	**	4.2
Cl_2/N x pH	1	0.0037	*	3.0
Error	24	0.0008		24.3

Total percent contribution is equal to 100%
***: Significant at 99% Confidence
**: Significant at 95% Confidence
*: Significant at 90% Confidence
NS: Not Significant

Table B.57

Summary of ANOVA of various factors influencing the production of DXAA (µg/L) in Lake Austin and Metedeconk River source waters

Parameter	Degrees of Freedom	Mean Squares	F Statistic	Percent Contribution %
water	2	438.29	***	13.1
pH	1	2011.16	***	31.5
Cl_2/N ratio	1	379.89	***	5.6
Temperature	1	241.62	**	3.4
Bromide Level	1	1385.70	***	21.6
Chloramine Residual	1	188.04	**	2.6
Cl_2/N x pH	1	243.71	**	3.5
Error	39	25.09		18.7

Total percent contribution is equal to 100%
***: Significant at 99% Confidence
**: Significant at 95% Confidence
*: Significant at 90% Confidence
NS: Not Significant

Table B.58

Summary of ANOVA of various factors influencing the production of DXAA (µmol/L) in Lake Austin and Metedeconk River source waters

Parameter	Degrees of Freedom	Mean Squares	F Statistic	Percent Contribution %
water	2	0.0113	***	12.5
pH	1	0.0624	***	36.0
Cl_2/N ratio	1	0.0135	***	7.5
Temperature	1	0.0097	***	5.3
Bromide Level	1	0.0262	***	14.9
Chloramine Residual	1	0.0070	***	3.7
Cl_2/N x pH	1	0.0061	**	3.2
Error	39	0.0006		16.8

Total percent contribution is equal to 100%
***: Significant at 99% Confidence
**: Significant at 95% Confidence
*: Significant at 90% Confidence
NS: Not Significant

Table B.59
Summary of combined ANOVA of various factors influencing the production of DXAA (μg/L) in all Lake Austin and Metedeconk River waters

Parameter	Degrees of Freedom	Mean Squares	F Statistic	Percent Contribution %
Chloramine Residual	1	137.27	***	0.8
Cl_2/N ratio	1	514.17	***	3.6
pH	1	2288.22	***	16.5
Temperature	1	282.76	***	1.9
Bromide Level	1	1586.50	***	11.4
Water	8	728.40	***	42.2
Error	130			23.7

Total percent contribution is equal to 100%
***: Significant at 99% Confidence
**: Significant at 95% Confidence
*: Significant at 90% Confidence
NS: Not Significant

Table B.60
Summary of combined ANOVA of various factors influencing the production of DXAA (μmol/L) in all Lake Austin and Metedeconk River waters

Parameter	Degrees of Freedom	Mean Squares	F Statistic	Percent Contribution %
Chloramine Residual	1	0.0066	***	1.5
Cl_2/N ratio	1	0.0191	***	4.7
pH	1	0.0669	***	17.0
Temperature	1	0.0113	***	2.8
Bromide Level	1	0.0282	***	7.1
Water	8	0.0218	***	44.4
Error	130			22.5

Total percent contribution is equal to 100%
***: Significant at 99% Confidence
**: Significant at 95% Confidence
*: Significant at 90% Confidence
NS: Not Significant

Table B.61
Summary of combined ANOVA of various factors influencing the production of DXAA yield (μg/mg) in all Lake Austin and Metedeconk River waters

Parameter	Degrees of Freedom	Mean Squares	F Statistic	Percent Contribution %
Chloramine Residual	1	15.44	***	1.2
Cl_2/N ratio	1	89.72	***	7.8
pH	1	368.93	***	32.7
Temperature	1	43.20	***	3.7
Bromide Level	1	243.96	***	21.6
Water	8	15.86	***	11.2
Error	130			21.7

Total percent contribution is equal to 100%
***: Significant at 99% Confidence
**: Significant at 95% Confidence
*: Significant at 90% Confidence
NS: Not Significant

Table B.62
Summary of combined ANOVA of various factors influencing the production of DXAA yield (μmol/mg) in all Lake Austin and Metedeconk River waters

Parameter	Degrees of Freedom	Mean Squares	F Statistic	Percent Contribution %
Chloramine Residual	1	0.0008	***	2.4
Cl_2/N ratio	1	0.0034	***	10.4
pH	1	0.0108	***	33.9
Temperature	1	0.0018	***	5.4
Bromide Level	1	0.0042	***	13.0
Water	8	0.0006	***	15.8
Error	130			19.1

Total percent contribution is equal to 100%
***: Significant at 99% Confidence
**: Significant at 95% Confidence
*: Significant at 90% Confidence
NS: Not Significant

APPENDIX C
FOLLOW-UP BATCH SCREENING EXPERIMENTS

Table C.1
Biscayne Aquifer source water (mass concentration)

Trial No.	Cl/N ratio	pH	Br/C Ratio (ug/mg)	Total Cl Residual ($mg\ Cl_2/L$) actual	target	Chloramine Demand (mg/L)	Temperature (°F)	HAA CONCENTRATIONS (µg/L) TCAA	DCAA	BCAA	DBAA	DXAA	DXAA Yield (DXAA/DOC)
1	4:1	7	<15	2.37	2	1.83	70	0.0	13.0	0.0	0.0	13.0	1.09
2	4:1	7	30-50	2.60	2	2.05	70	0.0	13.0	1.7	0.0	14.7	1.23
3	4:1	7	75-100	2.61	2	2.09	70	0.0	11.0	2.9	0.0	13.9	1.16
4	4:1	8	<15	1.87	2	1.63	70	0.0	7.4	0.0	0.0	7.4	0.62
5	4:1	8	30-50	1.76	2	1.74	70	0.0	7.5	0.0	0.0	7.5	0.63
6	4:1	8	75-100	1.81	2	1.74	70	0.0	7.3	0.0	0.0	7.3	0.61
7	4:1	9	<15	1.63	2	1.59	70	0.0	4.6	0.0	0.0	4.6	0.38
8	4:1	9	30-50	1.82	2	1.55	70	0.0	4.9	0.0	0.0	4.9	0.41
9	4:1	9	75-100	1.84	2	1.61	70	0.0	5.0	0.0	0.0	5.0	0.42

DXAA = DCAA + DBAA + BCAA
DOC (mg/L) = 11.95
SUVA= 3.48

Table C.2
Biscayne Aquifer source water (molar concentration)

Trial No.	Cl/N ratio	pH	Br^- (mg/L)	Total Cl Residual ($mg\ Cl_2/L$) actual	target	Chloramine Demand (mg/L)	Temperature (°F)	HAA CONCENTRATIONS (µmol/L) TCAA	DCAA	BCAA	DBAA	DXAA	DXAA Yield (DXAA/DOC)	Bromine incorporation factor (n_D')
1	4:1	7	<15	2.37	2	1.83	70	0.0000	0.1008	0.0000	0.0000	0.1008	0.00844	0.000
2	4:1	7	30-50	2.60	2	2.05	70	0.0000	0.1008	0.0098	0.0000	0.1106	0.00926	0.089
3	4:1	7	75-100	2.61	2	2.09	70	0.0000	0.0853	0.0167	0.0000	0.1020	0.00854	0.164
4	4:1	8	<15	1.87	2	1.63	70	0.0000	0.0574	0.0000	0.0000	0.0574	0.00480	0.000
5	4:1	8	30-50	1.76	2	1.74	70	0.0000	0.0582	0.0000	0.0000	0.0582	0.00487	0.000
6	4:1	8	75-100	1.81	2	1.74	70	0.0000	0.0566	0.0000	0.0000	0.0566	0.00474	0.000
7	4:1	9	<15	1.63	2	1.59	70	0.0000	0.0357	0.0000	0.0000	0.0357	0.00299	0.000
8	4:1	9	30-50	1.82	2	1.55	70	0.0000	0.0380	0.0000	0.0000	0.0380	0.00318	0.000
9	4:1	9	75-100	1.84	2	1.61	70	0.0000	0.0388	0.0000	0.0000	0.0388	0.00324	0.000

DXAA = DCAA + DBAA + BCAA
DOC (mg/L) = 11.95
SUVA= 3.48

Table C.3
Biscayne Aquifer softened water (mass concentration)

Trial No.	Cl/N ratio	pH	Br/C Ratio (ug/mg)	Total Cl Residual ($mg\ Cl_2/L$) actual	target	Chloramine Demand (mg/L)	Temperature (°F)	HAA CONCENTRATIONS (µg/L) TCAA	DCAA	BCAA	DBAA	DXAA	DXAA Yield (DXAA/DOC)
1	4:1	7	<15	2.08	2	1.92	70	1.00	8.50	1.30	0.40	10.20	1.08
2	4:1	7	75-100	2.06	2	2.04	70	1.00	8.30	3.00	1.60	12.90	1.37
3	4:1	9	<15	1.97	2	1.20	70	0.80	6.10	0.60	0.20	6.90	0.73
4	4:1	9	75-100	1.98	2	1.22	70	0.80	5.50	1.10	1.00	7.60	0.80

DXAA = DCAA + DBAA + BCAA
DOC (mg/L) = 9.45
SUVA= 2.94

Table C.4
Biscayne Aquifer softened water (molar concentration)

Trial No.	Cl/N ratio	pH	Br/C Ratio (ug/mg)	Total Cl Residual ($mg\ Cl_2/L$) actual	target	Chloramine Demand (mg/L)	Temperature (°F)	HAA CONCENTRATIONS (µmol/L) TCAA	DCAA	BCAA	DBAA	DXAA	DXAA Yield (DXAA/DOC)	Bromine incorporation factor (n_D')
1	4:1	7	<15	2.08	2	1.92	70	0.01	0.0659	0.0075	0.0018	0.0753	0.00796	0.148
2	4:1	7	75-100	2.06	2	2.04	70	0.01	0.0644	0.0173	0.0073	0.0890	0.00942	0.359
3	4:1	9	<15	1.97	2	1.20	70	0.00	0.0473	0.0035	0.0009	0.0517	0.00547	0.102
4	4:1	9	75-100	1.98	2	1.22	70	0.00	0.0427	0.0063	0.0046	0.0536	0.00567	0.290

DXAA = DCAA + DBAA + BCAA
DOC (mg/L) = 9.45
SUVA= 2.94

Table C.5
St. Paul source water (mass concentration)

Trial No.	Cl/N ratio	pH	Br/C Ratio (ug/mg)	Total Cl Residual ($mg\ Cl_2/L$) actual	target	Chloramine Demand (mg/L)	Temperature (°F)	HAA CONCENTRATIONS (µg/L) TCAA	DCAA	BCAA	DBAA	DXAA	DXAA Yield (DXAA/DOC)
1	4:1	7	<15	2.08	2	1.97	70	0.0	18.0	3.0	0.0	21.0	1.9
2	4:1	7	30-50	3.04	2	2.20	70	0.0	24.0	10.0	2.7	36.7	3.3
3	4:1	7	75-100	2.41	2	2.19	70	0.0	15.0	9.9	4.9	29.8	2.7
4	4:1	8	<15	2.08	2	1.42	70	0.0	14.0	1.4	0.0	15.4	1.4
5	4:1	8	30-50	2.17	2	1.23	70	0.0	12.0	2.3	0.0	14.3	1.3
6	4:1	8	75-100	2.29	2	1.49	70	0.0	12.0	3.5	1.5	17.0	1.5
7	4:1	9	<15	2.57	2	1.33	70	0.0	14.0	1.5	0.0	15.5	1.4
8	4:1	9	30-50	1.95	2	1.12	70	0.0	7.6	1.3	0.0	8.9	0.8
9	4:1	9	75-100	2.56	2	1.57	70	0.0	10.0	3.5	4.0	17.5	1.6

DXAA = DCAA + DBAA + BCAA
DOC (mg/L) = 11.07
SUVA= 2.49

Table C.6
St. Paul source water (molar concentration)

Trial No.	Cl/N ratio	pH	Br/C Ratio (ug/mg)	Total Cl Residual ($mg\ Cl_2/L$) actual	target	Chloramine Demand (mg/L)	Temperature (°F)	HAA CONCENTRATIONS (µmol/L) TCAA	DCAA	BCAA	DBAA	DXAA	DXAA Yield (DXAA/DOC)	Bromine incorporation factor (n_D')
1	4:1	7	<15	2.08	2	1.97	70	0.00	0.1396	0.0173	0.0000	0.1569	0.01417	0.110
2	4:1	7	30-50	3.04	2	2.20	70	0.00	0.1861	0.0577	0.0124	0.2562	0.02314	0.322
3	4:1	7	75-100	2.41	2	2.19	70	0.00	0.1163	0.0571	0.0225	0.1959	0.01770	0.521
4	4:1	8	<15	2.08	2	1.42	70	0.00	0.1086	0.0081	0.0000	0.1167	0.01054	0.069
5	4:1	8	30-50	2.17	2	1.23	70	0.00	0.0931	0.0133	0.0000	0.1063	0.00961	0.125
6	4:1	8	75-100	2.29	2	1.49	70	0.00	0.0931	0.0202	0.0069	0.1201	0.01085	0.283
7	4:1	9	<15	2.57	2	1.33	70	0.00	0.1086	0.0087	0.0000	0.1172	0.01059	0.074
8	4:1	9	30-50	1.95	2	1.12	70	0.00	0.0589	0.0075	0.0000	0.0664	0.00600	0.113
9	4:1	9	75-100	2.56	2	1.57	70	0.00	0.0776	0.0202	0.0184	0.1161	0.01049	0.490

DXAA = DCAA + DBAA + BCAA
DOC (mg/L) = 11.07
SUVA= 2.49

Table C.7
St. Paul softened water (mass concentration)

Trial No.	Cl/N ratio	pH	Br/C Ratio (ug/mg)	Total Cl Residual ($mg\ Cl_2/L$) actual	target	Chloramine Demand (mg/L)	Temperature (°F)	HAA CONCENTRATIONS (µg/L) TCAA	DCAA	BCAA	DBAA	DXAA	DXAA Yield (DXAA/DOC)
1	4:1	7	<15	2.16	2	1.29	70	0.20	11.60	2.00	0.20	13.80	2.09
2	4:1	7	<15	1.89	2	1.26	70	0.20	7.90	5.70	3.50	17.10	2.59
3	4:1	9	75-100	2.05	2	1.00	70	0.10	6.70	0.90	0.10	7.70	1.17
4	4:1	9	75-100	1.90	2	0.90	70	0.10	5.60	2.00	1.40	9.00	1.36

DXAA = DCAA + DBAA + BCAA
DOC (mg/L) = 6.60
SUVA= 1.59

Table C.8
St. Paul softened water (molar concentration)

Trial No.	Cl/N ratio	pH	Br/C Ratio (ug/mg)	Total Cl Residual ($mg\ Cl_2/L$) actual	target	Chloramine Demand (mg/L)	Temperature (°F)	HAA CONCENTRATIONS (µmol/L) TCAA	DCAA	BCAA	DBAA	DXAA	DXAA Yield (DXAA/DOC)	Bromine incorporation factor (n_D')
1	4:1	7	<15	2.16	2	1.29	70	0.00	0.0900	0.0115	0.0009	0.1024	0.01552	0.131
2	4:1	7	75-100	1.89	2	1.26	70	0.00	0.0613	0.0329	0.0161	0.1102	0.01670	0.590
3	4:1	9	<15	2.05	2	1.00	70	0.00	0.0520	0.0052	0.0005	0.0576	0.00873	0.106
4	4:1	9	75-100	1.90	2	0.90	70	0.00	0.0434	0.0115	0.0064	0.0614	0.00930	0.397

DXAA = DCAA + DBAA + BCAA
DOC (mg/L) = 6.60
SUVA= 1.59

Table C.9
Metedeconk River source water (mass concentration)

Trial No.	Cl/N ratio	pH	Br/C Ratio (μg/mg)	Total Cl Residual ($mg\ Cl_2/L$) actual	target	Chloramine Demand (mg/L)	Temperature (°F)	HAA CONCENTRATIONS (μg/L) TCAA	DCAA	BCAA	DBAA	DXAA	DXAA Yield (DXAA/TOC)
1	4:1	7	<15	1.89	2	2.41	70	0.0	18.0	0.0	0.0	18.0	4.17
2	4:1	7	30-50	1.84	2	1.27	70	0.0	9.9	0.0	0.0	9.9	2.29
3	4:1	7	75-100	2.40	2	1.89	70	0.0	13.0	3.3	1.9	18.2	4.21
4	4:1	8	<15	2.23	2	1.92	70	0.0	15.0	0.0	0.0	15.0	3.47
5	4:1	8	30-50	2.24	2	1.86	70	0.0	13.0	1.7	0.0	14.7	3.40
6	4:1	8	75-100	2.20	2	1.77	70	0.0	11.0	2.8	1.6	15.4	3.56
7	4:1	9	<15	2.13	2	1.87	70	0.0	12.0	0.0	0.0	12.0	2.78
8	4:1	9	30-50	1.80	2	1.80	70	0.0	7.6	0.0	0.0	7.6	1.76
9	4:1	9	75-100	2.14	2	1.60	70	0.0	7.9	1.5	1.0	10.4	2.41

DXAA = DCAA + DBAA + BCAA

TOC (mg/L) = 4.32

SUVA= 3.96

Table C.10
Metedeconk River source water (molar concentration)

Trial No.	Cl/N ratio	pH	Br/C Ratio (μg/mg)	Total Cl Residual ($mg\ Cl_2/L$) actual	target	Chloramine Demand (mg/L)	Temperature (°F)	HAA CONCENTRATIONS (μmol/L) TCAA	DCAA	BCAA	DBAA	DXAA	DXAA Yield (DXAA/TOC)	Bromine incorporation factor (n_D')
1	4:1	7	<15	1.89	2	2.41	70	0.00	0.1396	0.0000	0.0000	0.1396	0.03231	0.000
2	4:1	7	30-50	1.84	2	1.27	70	0.00	0.0768	0.0000	0.0000	0.0768	0.01777	0.000
3	4:1	7	75-100	2.40	2	1.89	70	0.00	0.1008	0.0190	0.0087	0.1286	0.02976	0.284
4	4:1	8	<15	2.23	2	1.92	70	0.00	0.1163	0.0000	0.0000	0.1163	0.02693	0.000
5	4:1	8	30-50	2.24	2	1.86	70	0.00	0.1008	0.0098	0.0000	0.1106	0.02561	0.089
6	4:1	8	75-100	2.20	2	1.77	70	0.00	0.0853	0.0161	0.0073	0.1088	0.02519	0.283
7	4:1	9	<15	2.13	2	1.87	70	0.00	0.0931	0.0000	0.0000	0.0931	0.02154	0.000
8	4:1	9	30-50	1.80	2	1.80	70	0.00	0.0589	0.0000	0.0000	0.0589	0.01364	0.000
9	4:1	9	75-100	2.14	2	1.60	70	0.00	0.0613	0.0087	0.0046	0.0745	0.01725	0.239

DXAA = DCAA + DBAA + BCAA

TOC (mg/L) = 4.32

SUVA= 3.96

Table C.11
Charleston source water (mass concentration)

Trial No.	pH	Alkalinity mg/L CaCO3	DOC mg/L target	DOC mg/L Actual	Br^- µg/L target	Br^- µg/L Actual	Total Chloramine Residual ($mg\ Cl_2/L$) target	Total Chloramine Residual ($mg\ Cl_2/L$) actual	Chloramine Demand (mg/L)	HAA CONCENTRATIONS (µg/L) DCAA	DBAA	BCAA	TCAA	DXAA	DXAA Yield (DXAA/DOC)
1	6	Amb	2.98	3.00	46	29	2.00	2.24	2.16	14.1	0.9	4.0	0.6	19.0	6.33
2	6	200	2.98	3.06	383	417	2.00	3.63	2.66	10.7	9.1	11.4	0.8	31.2	10.18
3	6	Amb	4.98	4.98	183	453	2.00	1.86	3.06	17.9	2.5	8.1	0.6	28.5	5.72
4	6	200	4.98	4.98	583	583	2.00	2.57	2.57	14.1	14.5	16.0	0.7	44.6	8.96
5	6	200	7.00	7.09	183	260	2.00	1.86	3.14	27.1	9.6	19.2	1.7	55.9	7.88
6	6	Amb	7.00	6.92	583	617	2.00	2.14	4.01	26.3	23.2	27.9	1.6	77.4	11.19
7	6	200	1.00	1.09	46	53	2.00	1.90	1.60	5.8	1.1	3.0	0.0	9.9	9.07
8	6	Amb	1.00	1.09	383	453	2.00	2.03	1.92	2.7	8.3	5.0	0.0	16.0	14.65
9	7	200	2.98	3.06	183	217	2.00	2.17	2.39	9.5	2.4	5.1	0.3	17.0	5.55
10	7	Amb	2.98	3.06	583	617	2.00	1.46	1.91	9.1	6.2	6.6	0.0	21.9	7.15
11	7	200	4.98	4.58	46	30	2.00	2.07	1.93	16.9	0.6	2.0	0.3	19.5	4.26
12	7	Amb	4.98	4.98	383	383	2.00	3.30	1.96	18.1	3.5	7.9	0.6	29.5	5.92
13	7	Amb	7.00	7.00	46	110	2.00	2.33	2.27	21.9	0.6	3.8	0.4	26.3	3.75
14	7	200	7.00	7.17	383	417	2.00	2.32	2.48	26.4	3.9	9.4	0.0	39.7	5.54
15	7	Amb	1.00	1.09	183	253	2.00	1.86	1.05	4.2	2.3	3.0	0.0	9.5	8.70
16	7	200	1.00	1.09	583	653	2.00	2.24	1.21	2.8	6.3	4.0	0.0	13.1	12.00
17	8	Amb	2.98	3.06	183	217	2.00	1.87	1.13	8.5	1.1	2.4	0.0	12.0	3.92
18	8	200	2.98	3.06	583	617	2.00	1.62	1.31	7.1	4.2	3.6	0.0	14.9	4.86
19	8	Amb	4.98	4.61	46	30	2.00	2.07	1.73	14.2	0.1	1.3	0.0	15.6	3.39
20	8	200	4.98	4.98	383	383	2.00	1.67	1.43	11.0	2.0	3.4	0.0	16.4	3.29
21	8	200	7.00	7.00	46	110	2.00	2.18	1.92	18.4	0.5	2.8	0.0	21.7	3.10
22	8	Amb	7.00	6.90	383	417	2.00	1.99	1.86	15.5	2.8	4.2	0.0	22.5	3.27
23	8	200	1.00	1.09	183	253	2.00	1.56	0.85	3.6	1.2	1.6	0.0	6.4	5.86
24	8	Amb	1.00	1.09	583	653	2.00	1.71	0.90	3.0	3.8	2.2	0.0	9.0	8.24
25	9	200	2.98	2.87	46	29	2.00	1.88	1.32	7.2	0.1	0.9	0.0	8.2	2.86
26	9	Amb	2.98	3.06	383	417	2.00	1.75	1.18	6.7	4.3	3.1	1.3	14.1	4.60
27	9	200	4.98	4.98	183	453	2.00	2.32	1.80	9.9	0.6	2.2	0.2	12.7	2.55
28	9	Amb	4.98	4.98	583	583	2.00	2.29	1.48	9.9	4.8	4.5	0.0	19.2	3.86
29	9	Amb	7.00	7.09	183	260	2.00	1.9	1.90	12.5	2.3	3.8	0.0	18.6	2.62
30	9	200	7.00	6.94	583	617	2.00	2	1.90	11.9	5.5	4.6	0.0	22.0	3.17
31	9	Amb	1.00	1.09	46	53	2.00	1.57	0.93	3.5	0.3	0.7	0.0	4.5	4.12
32	9	200	1.00	1.09	383	453	2.00	2.12	0.98	3.2	2.4	1.5	0.0	7.1	6.50

DXAA = DCAA + DBAA + BCAA
ND = Not Detected
BDL = Below Detection Limit of 0.5 mg/L
Amb = Ambient

Table C.12
Charleston source water (molar concentration)

Trial No.	pH	Alkalinity mg/L CaCO3	DOC mg/L target	DOC mg/L Actual	Br^- µg/L target	Br^- µg/L Actual	Total Chloramine Residual (mg Cl_2/L) target	Total Chloramine Residual (mg Cl_2/L) actual	Chloramine Demand (mg/L)	HAA CONCENTRATIONS (µmol/L) DCAA	DBAA	BCAA	TCAA	DXAA	DXAA Yield (DXAA/DOC)	DXAA molar bromide incorporation factor (nD')
1	6	Amb	2.98	3.00	46	29	2	2.24	2.16	0.10935	0.00413	0.02307	0.00365	0.1366	0.0455	0.1992
2	6	200	2.98	3.06	383	417	2	3.63	2.66	0.08298	0.04177	0.06575	0.00487	0.1905	0.0622	0.5644
3	6	Amb	4.98	4.98	183	453	2	1.86	3.06	0.13882	0.01148	0.04672	0.00365	0.1970	0.0396	0.2954
4	6	200	4.98	4.98	583	583	2	2.57	2.57	0.10935	0.06656	0.09228	0.00426	0.2682	0.0539	0.5923
5	6	200	7.00	7.09	183	260	2	1.86	3.14	0.21018	0.04407	0.11073	0.01034	0.3650	0.0515	0.4241
6	6	Amb	7.00	6.92	583	617	2	2.14	4.01	0.20397	0.1065	0.16091	0.00973	0.4714	0.0681	0.5673
7	6	200	1.00	1.09	46	53	2	1.9	1.60	0.04498	0.00505	0.0173	0.00000	0.0673	0.0617	0.3320
8	6	Amb	1.00	1.09	383	453	2	2.03	1.92	0.02094	0.0381	0.02884	0.00000	0.0879	0.0805	0.7617
9	7	200	2.98	3.06	183	217	2	2.17	2.39	0.07368	0.01102	0.02941	0.00182	0.1141	0.0372	0.3543
10	7	Amb	2.98	3.06	583	617	2	1.46	1.91	0.07058	0.02846	0.03806	0.00000	0.1371	0.0447	0.4852
11	7	200	4.98	4.58	46	30	2	2.07	1.93	0.13107	0.00275	0.01153	0.00182	0.1454	0.0317	0.0983
12	7	Amb	4.98	4.98	383	383	2	3.3	1.96	0.14038	0.01607	0.04556	0.00365	0.2020	0.0406	0.3051
13	7	Amb	7.00	7.00	46	110	2	2.33	2.27	0.16985	0.00275	0.02192	0.00243	0.1945	0.0278	0.1268
14	7	200	7.00	7.17	383	417	2	2.32	2.48	0.20475	0.0179	0.05421	0.00000	0.2769	0.0386	0.2605
15	7	Amb	1.00	1.09	183	253	2	1.86	1.05	0.03257	0.01056	0.0173	0.00000	0.0604	0.0553	0.4610
16	7	200	1.00	1.09	583	653	2	2.24	1.21	0.02172	0.02892	0.02307	0.00000	0.0737	0.0675	0.7054
17	8	Amb	2.98	3.06	183	217	2	1.87	1.13	0.06592	0.00505	0.01384	0.00000	0.0848	0.0277	0.2227
18	8	200	2.98	3.06	583	617	2	1.62	1.31	0.05506	0.01928	0.02076	0.00000	0.0951	0.0310	0.4210
19	8	Amb	4.98	4.61	46	30	2	2.07	1.73	0.11013	0.00046	0.0075	0.00000	0.1181	0.0256	0.0674
20	8	200	4.98	4.98	383	383	2	1.67	1.43	0.08531	0.00918	0.01961	0.00000	0.1141	0.0229	0.2523
21	8	200	7.00	7.00	46	110	2	2.18	1.92	0.14270	0.0023	0.01615	0.00000	0.1611	0.0230	0.1145
22	8	Amb	7.00	6.90	383	417	2	1.99	1.86	0.12052	0.01285	0.02422	0.00000	0.1576	0.0228	0.2353
23	8	200	1.00	1.09	183	253	2	1.56	0.85	0.02792	0.00551	0.00923	0.00000	0.0427	0.0391	0.3455
24	8	Amb	1.00	1.09	583	653	2	1.71	0.90	0.02327	0.01744	0.01269	0.00000	0.0534	0.0489	0.5643
25	9	200	2.98	2.87	46	29	2	1.88	1.32	0.05584	0.00046	0.00519	0.00000	0.0615	0.0214	0.0919
26	9	Amb	2.98	3.06	383	417	2	1.75	1.18	0.05196	0.01974	0.01788	0.00791	0.0896	0.0292	0.4199
27	9	200	4.98	4.98	183	453	2	2.32	1.80	0.07678	0.00275	0.01269	0.00122	0.0922	0.0185	0.1674
28	9	Amb	4.98	4.98	583	583	2	2.29	1.48	0.07678	0.02203	0.02595	0.00000	0.1248	0.0251	0.3846
29	9	Amb	7.00	7.09	183	260	2	1.9	1.90	0.09694	0.01056	0.02192	0.00000	0.1294	0.0182	0.2509
30	9	200	7.00	6.94	583	617	2	2	1.90	0.09229	0.02525	0.02653	0.00000	0.1441	0.0208	0.3594
31	9	Amb	1.00	1.09	46	53	2	1.57	0.93	0.02714	0.00138	0.00404	0.00000	0.0326	0.0298	0.1663
32	9	200	1.00	1.09	383	453	2	2.12	0.98	0.02482	0.01102	0.00865	0.00000	0.0445	0.0407	0.4421

DXAA = DCAA + DBAA + BCAA
ND = Not Detected
BDL = Below Detection Limit of 0.5 mg/L
Amb = Ambient

APPENDIX D
FOLLOW-UP BATCH SCREENING EXPERIMENTS ANOVA ANALYSIS

Table D.1
Summary of ANOVA of various factors influencing the production of DXAA (µmol/L) in Biscayne Aquifer source water

Parameter	Degrees of Freedom	Mean Squares	F Statistic	Percent Contribution %
pH	1	0.0067	***	93.0
Br/DOC	1	0.0000	NS	-1.0
Error	6	0.00007		8.0

Total percent contribution is equal to 100%
***: Significant at 99% Confidence
**: Significant at 95% Confidence
*: Significant at 90% Confidence
NS: Not Significant

Table D.2
Summary of ANOVA of various factors influencing the production of DXAA (µmol/L) in St. Paul source water

Parameter	Degrees of Freedom	Mean Squares	F Statistic	Percent Contribution %
pH	1	0.0159	**	56.5
Br/DOC	1	0.0003	NS	-5.1
Error	6	0.0015		48.6

Total percent contribution is equal to 100%
***: Significant at 99% Confidence
**: Significant at 95% Confidence
*: Significant at 90% Confidence
NS: Not Significant

Table D.3
Summary of ANOVA of various factors influencing the production of DXAA (µmol/L) in Metedeconk River source water

Parameter	Degrees of Freedom	Mean Squares	F Statistic	Percent Contribution %
pH	1	0.0023	*	31.1
Br/DOC	1	0.0001	NS	-7.6
Error	6	0.0006		76.5

Total percent contribution is equal to 100%
***: Significant at 99% Confidence
**: Significant at 95% Confidence
*: Significant at 90% Confidence
NS: Not Significant

Table D.4
Summary of ANOVA of various factors influencing the production of DXAA (µmol/L) in Biscayne Aquifer softened water

Parameter	Degrees of Freedom	Mean Squares	F Statistic	Percent Contribution %
pH	1	0.0008	NS at 90%	86.5
Br/DOC	1	0.0000	NS	2.7
Error	1	0.0009		10.9

Total percent contribution is equal to 100%
***: Significant at 99% Confidence
**: Significant at 95% Confidence
*: Significant at 90% Confidence
NS: Not Significant

Table D.5
Summary of ANOVA of various factors influencing the production of DXAA (µmol/L) in St. Paul softened water

Parameter	Degrees of Freedom	Mean Squares	F Statistic	Percent Contribution %
pH	1	0.0021	***	98.1
Br/DOC	1	0.00003	NS	1.3
Error	1	0.000004		0.5

Total percent contribution is equal to 100%
***: Significant at 99% Confidence
**: Significant at 95% Confidence
*: Significant at 90% Confidence
NS: Not Significant

Table D.6
Summary of ANOVA of various factors influencing the production of DXAA (μmol/L) in Biscayne Aquifer, St. Paul, and Metedeconk River source waters

Parameter	Degrees of Freedom	Mean Squares	F Statistic	Percent Contribution %
pH	1	0.0002	***	71.1
Br/DOC	1	0.0000	NS	13.8
water	2	0.0007	***	-0.6
Error	22	0.00001		15.7

Total percent contribution is equal to 100%
***: Significant at 99% Confidence
**: Significant at 95% Confidence
*: Significant at 90% Confidence
NS: Not Significant

APPENDIX E
KINETICS EXPERIMENTS

TRADITIONAL CHLORAMINATION EXPERIMENTS

Table E.1
Lake Austin source water pH 7 (mass concentration)

Sample	Time (h)	Cl/N ratio	pH	Br^- (mg/L)	Total Cl Residual ($mg\ Cl_2/L$) actual	target	Cl_2 demand	Temperature (°F)	AVERAGE HAA CONCENTRATIONS (µg/L) MCAA	MBAA	DCAA	BCAA	DBAA	TCAA	BDCAA	DBCAA	TBAA	HAA9	HAA5	DXAA	MXAA	DXAA/HAA9	DXAA Yield (DXAA/DOC)
0.5-1	0.5	4:1	7	Ambient	N/A	N/A	N/A	70	1.73	0.51	5.03	3.51	1.67	BDL	ND	ND	ND	12.45	8.94	10.21	2.24	0.82	2.73
1-1	1	4:1	7	Ambient	N/A	N/A	N/A	70	1.76	0.52	5.17	3.85	1.81	BDL	ND	ND	ND	13.11	9.27	10.84	2.28	0.83	2.90
4-1	4	4:1	7	Ambient	N/A	N/A	N/A	70	1.61	0.61	5.43	3.95	1.85	BDL	ND	ND	ND	13.45	9.51	11.23	2.22	0.83	3.00
8-1	8	4:1	7	Ambient	N/A	N/A	N/A	70	1.51	0.63	5.70	4.15	1.99	BDL	ND	ND	ND	13.98	9.83	11.83	2.15	0.85	3.16
24-1	24	4:1	7	Ambient	N/A	N/A	N/A	70	1.91	0.82	6.52	4.89	2.28	BDL	ND	ND	ND	16.42	11.53	13.68	2.73	0.83	3.66
48-1	48	4:1	7	Ambient	2.21	2	0.64	70	2.10	0.72	7.24	5.55	2.74	BDL	ND	ND	ND	18.35	12.80	15.53	2.82	0.85	4.15
72-1	72	4:1	7	Ambient	2.08	N/A	0.77	70	2.25	0.77	7.89	6.24	3.17	BDL	ND	ND	ND	20.32	14.08	17.30	3.02	0.85	4.63

DXAA = DCAA + DBAA + BCAA
MXAA = MCAA + MBAA
N/A = Not Applicable
ND = Not Detected
BDL = Below Detection Limit of 0.5 µg/L
DOC (mg/L) = 3.74

Table E.2
Lake Austin source water pH 7 (molar concentration)

Sample	Time (h)	Cl/N ratio	pH	Br^- (mg/L)	Total Cl Residual ($mg\ Cl_2/L$) actual	target	Cl_2 demand	Temperature (°F)	AVERAGE HAA CONCENTRATIONS (µmol/L) MCAA	MBAA	DCAA	BCAA	DBAA	TCAA	BDCAA	DBCAA	TBAA	HAA9	HAA5	DXAA	MXAA	DXAA/HAA9	DXAA Yield (DXAA/DOC)	Bromine incorporation factor (n_p')
0.5-1	0.5	4:1	7	Ambient	N/A	N/A	N/A	70	0.0183	0.0036	0.0390	0.0203	0.0077	BDL	ND	ND	ND	0.08892	0.0686	0.0669	0.0220	0.75	0.01790	0.532
1-1	1	4:1	7	Ambient	N/A	N/A	N/A	70	0.0187	0.0037	0.0401	0.0222	0.0083	BDL	ND	ND	ND	0.09302	0.0708	0.0706	0.0224	0.76	0.01889	0.550
4-1	4	4:1	7	Ambient	N/A	N/A	N/A	70	0.0171	0.0044	0.0421	0.0228	0.0085	BDL	ND	ND	ND	0.09485	0.0721	0.0734	0.0215	0.77	0.01962	0.542
8-1	8	4:1	7	Ambient	N/A	N/A	N/A	70	0.0160	0.0046	0.0442	0.0239	0.0091	BDL	ND	ND	ND	0.09779	0.0739	0.0772	0.0206	0.79	0.02065	0.546
24-1	24	4:1	7	Ambient	N/A	N/A	N/A	70	0.0202	0.0059	0.0505	0.0282	0.0105	BDL	ND	ND	ND	0.11533	0.0871	0.0892	0.0261	0.77	0.02385	0.551
48-1	48	4:1	7	Ambient	2.21	2	0.64	70	0.0222	0.0052	0.0561	0.0320	0.0126	BDL	ND	ND	ND	0.12812	0.0961	0.1007	0.0274	0.79	0.02693	0.568
72-1	72	4:1	7	Ambient	2.08	N/A	0.77	70	0.0238	0.0056	0.0612	0.0360	0.0145	BDL	ND	ND	ND	0.14108	0.1051	0.1117	0.0293	0.79	0.02988	0.582

DXAA = DCAA + DBAA + BCAA
MXAA = MCAA + MBAA
N/A = Not Applicable
ND = Not Detected
BDL = Below Detection Limit of 0.5 µg/L
DOC (mg/L) = 3.74

Table E.3
Lake Austin source water pH 9 (mass concentration)

Sample	Time (h)	Cl/N ratio	pH	Br^- (mg/L)	Total Cl Residual ($mg Cl_2/L$) actual	target	Cl_2 demand	Temperature (°F)	AVERAGE HAA CONCENTRATIONS (μg/L) MCAA	MBAA	DCAA	BCAA	DBAA	TCAA	BDCAA	DBCAA	TBAA	HAA9	HAA5	DXAA	MXAA	DXAA/HAA9	DXAA Yield (DXAA/DOC)
0.5-1	0.5	4:1	9	Ambient	N/A	N/A	N/A	70	0.60	BDL	3.29	2.58	0.96	BDL	ND	ND	ND	7.42	4.84	6.83	0.60	0.92	1.83
1-1	1	4:1	9	Ambient	N/A	N/A	N/A	70	0.70	BDL	3.32	2.55	1.68	BDL	ND	ND	ND	8.24	5.69	7.54	0.70	0.91	2.02
4-1	4	4:1	9	Ambient	N/A	N/A	N/A	70	0.93	BDL	3.40	2.58	1.68	BDL	ND	ND	ND	8.59	6.01	7.66	0.93	0.89	2.05
8-1	8	4:1	9	Ambient	N/A	N/A	N/A	70	1.14	BDL	3.74	2.66	1.63	BDL	ND	ND	ND	9.17	6.51	8.03	1.14	0.88	2.15
24-1	24	4:1	9	Ambient	N/A	N/A	N/A	70	1.06	BDL	3.95	2.69	1.65	BDL	ND	ND	ND	9.35	6.66	8.29	1.06	0.89	2.22
48-1	48	4:1	9	Ambient	1.95	2	0.13	70	1.24	BDL	4.81	2.80	1.66	BDL	ND	ND	ND	10.51	7.71	9.27	1.24	0.88	2.48
72-1	72	4:1	9	Ambient	1.93	N/A	0.15	70	1.60	BDL	5.98	2.75	1.64	BDL	ND	ND	ND	11.97	9.22	10.37	1.60	0.87	2.77

DXAA = DCAA + DBAA + BCAA
MXAA = MCAA + MBAA
N/A = Not Applicable
ND = Not Detected
BDL = Below Detection Limit of 0.5 μg/L
DOC (mg/L) = 3.74

Table E.4
Lake Austin source water pH 9 (molar concentration)

Sample	Time (h)	Cl/N ratio	pH	Br^- (mg/L)	Total Cl Residual ($mg Cl_2/L$) actual	target	Cl_2 demand	Temperature (°F)	AVERAGE HAA CONCENTRATIONS (μmol/L) MCAA	MBAA	DCAA	BCAA	DBAA	TCAA	BDCAA	DBCAA	TBAA	HAA9	HAA5	DXAA	MXAA	DXAA/HAA9	DXAA Yield (DXAA/DOC)	Bromine incorporation factor (n_D')
0.5-1	0.5	4:1	9	Ambient	N/A	N/A	N/A	70	0.0063	BDL	0.0255	0.0149	0.0044	BDL	ND	ND	ND	0.0511	0.0362	0.0448	0.0063	0.88	0.01198	0.674
1-1	1	4:1	9	Ambient	N/A	N/A	N/A	70	0.0074	BDL	0.0257	0.0147	0.0077	BDL	ND	ND	ND	0.0555	0.0408	0.0481	0.0074	0.87	0.01286	0.652
4-1	4	4:1	9	Ambient	N/A	N/A	N/A	70	0.0098	BDL	0.0264	0.0149	0.0077	BDL	ND	ND	ND	0.0588	0.0439	0.0490	0.0098	0.83	0.01309	0.651
8-1	8	4:1	9	Ambient	N/A	N/A	N/A	70	0.0121	BDL	0.0290	0.0153	0.0075	BDL	ND	ND	ND	0.0639	0.0486	0.0518	0.0121	0.81	0.01385	0.647
24-1	24	4:1	9	Ambient	N/A	N/A	N/A	70	0.0112	BDL	0.0306	0.0155	0.0076	BDL	ND	ND	ND	0.0650	0.0495	0.0537	0.0112	0.83	0.01437	0.678
48-1	48	4:1	9	Ambient	1.95	2	0.13	70	0.0131	BDL	0.0373	0.0161	0.0076	BDL	ND	ND	ND	0.0741	0.0580	0.0611	0.0131	0.82	0.01633	0.677
72-1	72	4:1	9	Ambient	1.93	N/A	0.15	70	0.0169	BDL	0.0464	0.0159	0.0075	BDL	ND	ND	ND	0.0867	0.0708	0.0698	0.0169	0.80	0.01865	0.644

DXAA = DCAA + DBAA + BCAA
MXAA = MCAA + MBAA
N/A = Not Applicable
ND = Not Detected
BDL = Below Detection Limit of 0.5 μg/L
DOC (mg/L) = 3.74

Table E.5
Lake Austin softened water pH 9 (mass concentration)

Sample	Time (h)	Cl/N ratio	pH	Br^- (mg/L)	Total Cl Residual ($mg Cl_2/L$) actual	target	Cl_2 demand	Temperature (°F)	AVERAGE HAA CONCENTRATIONS (μg/L) MCAA	MBAA	DCAA	BCAA	DBAA	TCAA	BDCAA	DBCAA	TBAA	HAA9	HAA5	DXAA	MXAA	DXAA/HAA9	DXAA Yield (DXAA/DOC)
0.5-1	0.5	4:1	9	Ambient	N/A	N/A	N/A	70	1.31	0.67	2.93	1.93	1.40	BDL	ND	ND	ND	8.25	6.33	6.27	1.99	0.76	1.89
1-1	1	4:1	9	Ambient	N/A	N/A	N/A	70	1.36	0.60	2.76	1.97	1.46	BDL	ND	ND	ND	8.16	6.18	6.20	1.96	0.76	1.87
4-1	4	4:1	9	Ambient	N/A	N/A	N/A	70	1.25	BDL	2.78	1.93	1.48	BDL	ND	ND	ND	7.44	5.51	6.19	1.25	0.83	1.87
8-1	8	4:1	9	Ambient	N/A	N/A	N/A	70	1.28	BDL	2.62	1.91	1.44	BDL	ND	ND	ND	7.25	5.34	5.97	1.28	0.82	1.80
24-1	24	4:1	9	Ambient	N/A	N/A	N/A	70	1.39	BDL	2.75	1.91	1.42	BDL	ND	ND	ND	7.48	5.56	6.09	1.39	0.81	1.83
48-1	48	4:1	9	Ambient	1.93	2	0.57	70	1.34	BDL	3.26	2.04	1.47	BDL	ND	ND	ND	8.11	6.07	6.77	1.34	0.84	2.04
72-1	72	4:1	9	Ambient	1.83	N/A	0.67	70	1.60	BDL	4.75	2.14	1.47	BDL	ND	ND	ND	9.96	7.82	8.37	1.60	0.84	2.52

DXAA = DCAA + DBAA + BCAA
MXAA = MCAA + MBAA
N/A = Not Applicable
ND = Not Detected
BDL = Below Detection Limit of 0.5 μg/L
DOC (mg/L) = 3.32

Table E.6
Lake Austin softened water pH 9 (molar concentration)

Sample	Time (h)	Cl/N ratio	pH	Br^- (mg/L)	Total Cl Residual ($mg\ Cl_2/L$) actual	target	Cl_2 demand	Temperature (°F)	AVERAGE HAA CONCENTRATIONS (µmol/L) MCAA	MBAA	DCAA	BCAA	DBAA	TCAA	BDCAA	DBCAA	TBAA	HAA9	HAA5	DXAA	MXAA	DXAA/HAA9	DXAA Yield (DXAA/DOC)	Bromine incorporation factor (n_D')
0.5-1	0.5	4:1	8	Ambient	N/A	N/A	N/A	70	0.0139	BDL	0.0228	0.0111	0.0064	BDL	ND	ND	ND	0.0542	0.0431	0.0403	0.0139	0.74	0.01214	0.595
1-1	1	4:1	8	Ambient	N/A	N/A	N/A	70	0.0144	BDL	0.0214	0.0114	0.0067	BDL	ND	ND	ND	0.0539	0.0425	0.0395	0.0144	0.73	0.01190	0.628
4-1	4	4:1	8	Ambient	N/A	N/A	N/A	70	0.0132	BDL	0.0215	0.0111	0.0068	BDL	ND	ND	ND	0.0527	0.0416	0.0395	0.0132	0.75	0.01190	0.627
8-1	8	4:1	8	Ambient	N/A	N/A	N/A	70	0.0135	BDL	0.0203	0.0110	0.0066	BDL	ND	ND	ND	0.0515	0.0404	0.0380	0.0135	0.74	0.01143	0.640
24-1	24	4:1	8	Ambient	N/A	N/A	N/A	70	0.0147	BDL	0.0213	0.0110	0.0065	BDL	ND	ND	ND	0.0536	0.0426	0.0389	0.0147	0.73	0.01172	0.620
48-1	48	4:1	8	Ambient	1.93	2	0.57	70	0.0141	BDL	0.0253	0.0118	0.0068	BDL	ND	ND	ND	0.0579	0.0462	0.0438	0.0141	0.76	0.01319	0.577
72-1	72	4:1	8	Ambient	1.83	N/A	0.67	70	0.0169	BDL	0.0368	0.0124	0.0068	BDL	ND	ND	ND	0.0728	0.0605	0.0560	0.0169	0.77	0.01685	0.462

DXAA = DCAA + DBAA + BCAA
MXAA = MCAA + MBAA
N/A = Not Applicable
ND = Not Detected
BDL = Below Detection Limit of 0.5 µg/L
DOC (mg/L) = 3.32

Table E.7
Metedeconk River source water (mass concentration)

Sample	Time (h)	Cl/N ratio	pH	Br^- (mg/L)	Total Cl Residual ($mg\ Cl_2/L$) actual	target	Cl_2 demand	Temperature (°F)	AVERAGE HAA CONCENTRATIONS (µg/L) MCAA	MBAA	DCAA	BCAA	DBAA	TCAA	BDCAA	DBCAA	TBAA	HAA9	HAA5	DXAA	MXAA	DXAA/HAA9	DXAA Yield (DXAA/DOC)
0.5-1	0.5	4:1	8	Ambient	N/A	N/A	N/A	70	4.01	BDL	15.20	1.35	BDL	BDL	ND	ND	ND	20.56	19.21	16.55	4.01	0.81	3.39
1-1	1	4:1	8	Ambient	N/A	N/A	N/A	70	4.16	BDL	17.11	1.48	BDL	0.52	ND	ND	ND	23.26	21.78	18.59	4.16	0.80	3.80
4-1	4	4:1	8	Ambient	N/A	N/A	N/A	70	4.27	BDL	16.97	1.44	BDL	BDL	ND	ND	ND	22.68	21.24	18.41	4.27	0.81	3.76
8-1	8	4:1	8	Ambient	N/A	N/A	N/A	70	4.14	BDL	16.35	1.27	BDL	BDL	ND	ND	ND	21.76	20.49	17.62	4.14	0.81	3.60
24-1	24	4:1	8	Ambient	N/A	N/A	N/A	70	5.06	BDL	18.03	1.51	BDL	BDL	ND	ND	ND	24.60	23.09	19.54	5.06	0.79	4.00
48-1	48	4:1	8	Ambient	1.93	2	0.52	70	5.57	BDL	19.73	1.33	BDL	BDL	ND	ND	ND	26.64	25.31	21.07	5.57	0.79	4.31
72-1	72	4:1	8	Ambient	1.79	N/A	0.66	70	6.39	BDL	21.14	1.55	BDL	BDL	ND	ND	ND	29.08	27.53	22.68	6.39	0.78	4.64

DXAA = DCAA + DBAA + BCAA
MXAA = MCAA + MBAA
N/A = Not Applicable
ND = Not Detected
BDL = Below Detection Limit of 0.5 µg/L
DOC (mg/L) = 4.89

Table E.8
Metedeconk River source water (molar concentration)

Sample	Time (h)	Cl/N ratio	pH	Br^- (mg/L)	Total Cl Residual ($mg\ Cl_2/L$) actual	target	Cl_2 demand	Temperature (°F)	AVERAGE HAA CONCENTRATIONS (µmol/L) MCAA	MBAA	DCAA	BCAA	DBAA	TCAA	BDCAA	DBCAA	TBAA	HAA9	HAA5	DXAA	MXAA	DXAA/HAA9	DXAA Yield (DXAA/DOC)	Bromine incorporation factor (n_D')
0.5-1	0.5	4:1	8	Ambient	N/A	N/A	N/A	70	0.0424	BDL	0.1179	0.0078	BDL	BDL	ND	ND	ND	0.1681	0.1603	0.1257	0.0424	0.75	0.02570	0.062
1-1	1	4:1	8	Ambient	N/A	N/A	N/A	70	0.0440	BDL	0.1327	0.0085	BDL	0.0032	ND	ND	ND	0.1884	0.1798	0.1412	0.0440	0.75	0.02888	0.060
4-1	4	4:1	8	Ambient	N/A	N/A	N/A	70	0.0452	BDL	0.1316	0.0083	BDL	BDL	ND	ND	ND	0.1851	0.1768	0.1399	0.0452	0.76	0.02862	0.059
8-1	8	4:1	8	Ambient	N/A	N/A	N/A	70	0.0438	BDL	0.1268	0.0073	BDL	BDL	ND	ND	ND	0.1779	0.1706	0.1341	0.0438	0.75	0.02743	0.055
24-1	24	4:1	8	Ambient	N/A	N/A	N/A	70	0.0535	BDL	0.1398	0.0087	BDL	BDL	ND	ND	ND	0.2021	0.1934	0.1485	0.0535	0.74	0.03038	0.059
48-1	48	4:1	8	Ambient	1.93	2	0.52	70	0.0590	BDL	0.1530	0.0077	BDL	BDL	ND	ND	ND	0.2197	0.2120	0.1607	0.0590	0.73	0.03287	0.048
72-1	72	4:1	8	Ambient	1.79	N/A	0.66	70	0.0676	BDL	0.1639	0.0089	BDL	BDL	ND	ND	ND	0.2405	0.2316	0.1729	0.0676	0.72	0.03535	0.052

DXAA = DCAA + DBAA + BCAA
MXAA = MCAA + MBAA
N/A = Not Applicable
ND = Not Detected
BDL = Below Detection Limit of 0.5 µg/L
DOC (mg/L) = 4.89

Table E.9

Metedeconk River alum coagulated water (mass concentration)

Sample	Time (h)	Cl/N ratio	pH	Br^- (mg/L)	Total Cl Residual ($mg\ Cl_2/L$) actual	target	Cl_2 demand	Temperature (°F)	AVERAGE HAA CONCENTRATIONS (µg/L) MCAA	MBAA	DCAA	BCAA	DBAA	TCAA	BDCAA	DBCAA	TBAA	HAA9	HAA5	DXAA	MXAA	DXAA/HAA9	DXAA Yield (DXAA/DOC)
0.5-1	0.5	4:1	8	Ambient	N/A	N/A	N/A	70	0.69	BDL	4.18	0.96	BDL	BDL	ND	ND	ND	5.83	4.87	5.14	0.69	0.88	2.22
1-1	1	4:1	8	Ambient	N/A	N/A	N/A	70	0.88	BDL	4.16	0.95	BDL	BDL	ND	ND	ND	6.00	5.05	5.12	0.88	0.85	2.21
4-1	4	4:1	8	Ambient	N/A	N/A	N/A	70	BDL	BDL	3.85	0.81	ND	BDL	ND	ND	ND	4.66	3.85	4.66	0.00	1.00	2.02
8-1	8	4:1	8	Ambient	N/A	N/A	N/A	70	0.58	BDL	3.37	0.76	ND	BDL	ND	ND	ND	4.71	3.95	4.13	0.58	0.88	1.79
24-1	24	4:1	8	Ambient	N/A	N/A	N/A	70	0.77	BDL	4.52	0.77	ND	0.54	ND	ND	ND	6.60	5.83	5.29	0.77	0.80	2.29
48-1	48	4:1	8	Ambient	2.05	2	0.1	70	0.91	BDL	6.32	0.84	ND	BDL	ND	ND	ND	8.06	7.23	7.16	0.91	0.89	3.10
72-1	72	4:1	8	Ambient	1.97	N/A	0.18	70	1.00	BDL	6.96	0.84	ND	BDL	ND	ND	ND	8.80	7.97	7.80	1.00	0.89	3.38

DXAA = DCAA + DBAA + BCAA
MXAA = MCAA + MBAA
N/A = Not Applicable
ND = Not Detected
BDL = Below Detection Limit of 0.5 µg/L
DOC (mg/L) = 2.31

Table E.10

Metedeconk River alum coagulated water (molar concentration)

Sample	Time (h)	Cl/N ratio	pH	Br^- (mg/L)	Total Cl Residual ($mg\ Cl_2/L$) actual	target	Cl_2 demand	Temperature (°F)	AVERAGE HAA CONCENTRATIONS (µmol/L) MCAA	MBAA	DCAA	BCAA	DBAA	TCAA	BDCAA	DBCAA	TBAA	HAA9	HAA5	DXAA	MXAA	DXAA/HAA9	DXAA Yield (DXAA/DOC)	Bromine incorporation factor (n_D')
0.5-1	0.5	4:1	8	Ambient	N/A	N/A	N/A	70	0.0073	BDL	0.0324	0.0055	BDL	BDL	ND	ND	ND	0.0453	0.0398	0.0379	0.0073	0.84	0.01642	0.146
1-1	1	4:1	8	Ambient	N/A	N/A	N/A	70	0.0093	BDL	0.0323	0.0055	BDL	BDL	ND	ND	ND	0.0471	0.0416	0.0378	0.0093	0.80	0.01635	0.145
4-1	4	4:1	8	Ambient	N/A	N/A	N/A	70	BDL	BDL	0.0298	0.0047	ND	BDL	ND	ND	ND	0.0345	0.0298	0.0345	0.0000	1.00	0.01494	0.135
8-1	8	4:1	8	Ambient	N/A	N/A	N/A	70	0.0061	BDL	0.0261	0.0044	ND	BDL	ND	ND	ND	0.0366	0.0322	0.0305	0.0061	0.83	0.01322	0.144
24-1	24	4:1	8	Ambient	N/A	N/A	N/A	70	0.0082	BDL	0.0351	0.0044	ND	0.0033	ND	ND	ND	0.0510	0.0465	0.0395	0.0082	0.78	0.01710	0.112
48-1	48	4:1	8	Ambient	2.05	2	0.1	70	0.0096	BDL	0.0490	0.0048	ND	BDL	ND	ND	ND	0.0634	0.0586	0.0538	0.0096	0.85	0.02330	0.090
72-1	72	4:1	8	Ambient	1.97	N/A	0.18	70	0.0106	BDL	0.0540	0.0048	ND	BDL	ND	ND	ND	0.0695	0.0646	0.0588	0.0106	0.85	0.02547	0.082

DXAA = DCAA + DBAA + BCAA
MXAA = MCAA + MBAA
N/A = Not Applicable
ND = Not Detected
BDL = Below Detection Limit of 0.5 µg/L
DOC (mg/L) = 2.31

Table E.11

Metedeconk River source water + 0.5 mg/L bromide (mass concentration)

Sample	Time (h)	Cl/N ratio	pH	Br^- (mg/L)	Total Cl Residual ($mg\ Cl_2/L$) actual	target	Cl_2 demand	Temperature (°F)	AVERAGE HAA CONCENTRATIONS (µg/L) MCAA	MBAA	DCAA	BCAA	DBAA	TCAA	BDCAA	DBCAA	TBAA	HAA9	HAA5	DXAA	MXAA	DXAA/HAA9	DXAA Yield (DXAA/DOC)
0.5-1	0.5	4:1	8	Ambient + 0.5	N/A	N/A	N/A	70	2.40	0.81	8.45	5.75	5.47	BDL	ND	BDL	BDL	22.89	17.14	19.67	3.21	0.86	4.02
1-1	1	4:1	8	Ambient + 0.5	N/A	N/A	N/A	70	2.64	0.72	9.40	5.59	4.93	BDL	ND	ND	BDL	23.28	17.69	19.92	3.36	0.86	4.07
4-1	4	4:1	8	Ambient + 0.5	N/A	N/A	N/A	70	3.22	0.95	10.40	6.46	4.98	BDL	ND	BDL	BDL	26.02	19.56	21.84	4.17	0.84	4.47
8-1	8	4:1	8	Ambient + 0.5	N/A	N/A	N/A	70	3.33	0.91	10.63	6.23	5.49	BDL	ND	BDL	BDL	26.59	20.37	22.35	4.24	0.84	4.57
24-1	24	4:1	8	Ambient + 0.5	N/A	N/A	N/A	70	3.84	0.89	11.89	6.34	5.26	BDL	ND	ND	BDL	28.22	21.88	23.49	4.73	0.83	4.80
48-1	48	4:1	8	Ambient + 0.5	1.93	2	0.57	70	4.12	0.81	13.23	6.48	5.41	BDL	ND	BDL	BDL	30.05	23.57	25.12	4.93	0.84	5.14
72-1	72	4:1	8	Ambient + 0.5	1.83	N/A	0.67	70	6.18	0.87	15.54	6.57	5.02	BDL	ND	BDL	BDL	34.17	27.60	27.12	7.05	0.79	5.55

DXAA = DCAA + DBAA + BCAA
MXAA = MCAA + MBAA
N/A = Not Applicable
ND = Not Detected
BDL = Below Detection Limit of 0.5 µg/L
DOC (mg/L) = 4.89

Table E.12

Metedeconk River source water + 0.5 mg/L bromide (molar concentration)

Sample	Time (h)	Cl/N ratio	pH	Br^- (mg/L)	Total Cl Residual ($mg Cl_2/L$) actual	Total Cl Residual ($mg Cl_2/L$) target	Cl_2 demand	Temperature (°F)	MCAA	MBAA	DCAA	BCAA	DBAA	TCAA	BDCAA	DBCAA	TBAA	HAA9	HAA5	DXAA	MXAA	DXAA/HAA9	DXAA Yield (DXAA/DOC)	Bromine incorporation factor (n_D')
									AVERAGE HAA CONCENTRATIONS (µmol/L)															
0.5-1	0.5	4:1	8	Ambient + 0.5	N/A	N/A	N/A	70	0.0254	0.0058	0.0655	0.0332	0.0251	BDL	ND	BDL	BDL	0.1551	0.1219	0.1238	0.0313	0.80	0.02532	0.674
1-1	1	4:1	8	Ambient + 0.5	N/A	N/A	N/A	70	0.0279	BDL	0.0729	0.0322	0.0226	BDL	ND	ND	BDL	0.1557	0.1234	0.1278	0.0279	0.82	0.02613	0.606
4-1	4	4:1	8	Ambient + 0.5	N/A	N/A	N/A	70	0.0341	0.0069	0.0807	0.0372	0.0229	BDL	ND	BDL	BDL	0.1817	0.1445	0.1408	0.0409	0.77	0.02879	0.589
8-1	8	4:1	8	Ambient + 0.5	N/A	N/A	N/A	70	0.0352	0.0066	0.0824	0.0359	0.0252	BDL	ND	BDL	BDL	0.1854	0.1495	0.1436	0.0418	0.77	0.02936	0.601
24-1	24	4:1	8	Ambient + 0.5	N/A	N/A	N/A	70	0.0406	0.0064	0.0922	0.0366	0.0241	BDL	ND	ND	BDL	0.1999	0.1634	0.1529	0.0470	0.76	0.03127	0.555
48-1	48	4:1	8	Ambient + 0.5	1.93	2	0.57	70	0.0436	0.0059	0.1026	0.0374	0.0248	BDL	ND	BDL	BDL	0.2142	0.1768	0.1648	0.0494	0.77	0.03370	0.528
72-1	72	4:1	8	Ambient + 0.5	1.83	N/A	0.67	70	0.0654	0.0063	0.1205	0.0379	0.0230	BDL	ND	BDL	BDL	0.2530	0.2152	0.1814	0.0716	0.72	0.03709	0.463

DXAA = DCAA + DBAA + BCAA
MXAA = MCAA + MBAA
N/A = Not Applicable
ND = Not Detected
BDL = Below Detection Limit of 0.5 µg/L
DOC (mg/L) = 4.89

Table E.13

Lake Austin softened water - 85 °F (mass concentration)

Sample	Time (h)	Cl/N ratio	pH	Br^- (mg/L)	Total Cl Residual ($mg Cl_2/L$) actual	Total Cl Residual ($mg Cl_2/L$) target	Cl_2 demand	Temperature (°F)	MCAA	MBAA	DCAA	BCAA	DBAA	TCAA	BDCAA	DBCAA	TBAA	HAA9	HAA5	DXAA	MXAA	DXAA/HAA9	DXAA Yield (DXAA/DOC)
									AVERAGE HAA CONCENTRATIONS (µg/L)														
0.5-1	0.5	4:1	9	Ambient	N/A	N/A	N/A	70	1.23	0.80	2.84	1.71	1.30	BDL	ND	ND	ND	7.88	6.17	5.85	2.03	0.74	1.76
1-1	1	4:1	9	Ambient	N/A	N/A	N/A	70	1.14	0.55	2.83	1.77	1.34	BDL	ND	ND	ND	7.62	5.86	5.94	1.69	0.78	1.79
4-1	4	4:1	9	Ambient	N/A	N/A	N/A	70	0.98	BDL	2.87	1.81	1.39	BDL	ND	ND	ND	7.05	5.24	6.07	0.98	0.86	1.83
8-1	8	4:1	9	Ambient	N/A	N/A	N/A	70	1.14	BDL	3.10	1.85	1.33	BDL	ND	ND	ND	7.42	5.57	6.28	1.14	0.85	1.89
24-1	24	4:1	9	Ambient	N/A	N/A	N/A	70	1.35	BDL	3.84	2.02	1.43	BDL	ND	ND	ND	8.62	6.61	7.28	1.35	0.84	2.19
48-1	48	4:1	9	Ambient	2.06	2	0.14	70	1.56	BDL	4.30	2.00	1.35	BDL	ND	ND	ND	9.20	7.20	7.64	1.56	0.83	2.30
72-1	72	4:1	9	Ambient	1.93	N/A	0.27	70	1.55	BDL	5.04	2.09	1.33	BDL	ND	ND	ND	10.00	7.92	8.45	1.55	0.84	2.55

DXAA = DCAA + DBAA + BCAA
MXAA = MCAA + MBAA
N/A = Not Applicable
ND = Not Detected
BDL = Below Detection Limit of 0.5 µg/L
DOC (mg/L) = 3.32

Table E.14

Lake Austin softened water - 85 °F (molar concentration)

Sample	Time (h)	Cl/N ratio	pH	Br^- (mg/L)	Total Cl Residual ($mg Cl_2/L$) actual	Total Cl Residual ($mg Cl_2/L$) target	Cl_2 demand	Temperature (°F)	MCAA	MBAA	DCAA	BCAA	DBAA	TCAA	BDCAA	DBCAA	TBAA	HAA9	HAA5	DXAA	MXAA	DXAA/HAA9	DXAA Yield (DXAA/DOC)	Bromine incorporation factor (n_D')
									AVERAGE HAA CONCENTRATIONS (µmol/L)															
0.5-1	0.5	4:1	9	Ambient	N/A	N/A	N/A	70	0.0130	0.0058	0.0220	0.0098	0.0060	BDL	ND	ND	ND	0.0566	0.0468	0.0379	0.0188	0.67	0.01140	0.576
1-1	1	4:1	9	Ambient	N/A	N/A	N/A	70	0.0120	0.0040	0.0220	0.0102	0.0061	BDL	ND	ND	ND	0.0543	0.0441	0.0383	0.0160	0.71	0.01154	0.587
4-1	4	4:1	9	Ambient	N/A	N/A	N/A	70	0.0104	BDL	0.0223	0.0105	0.0064	BDL	ND	ND	ND	0.0495	0.0390	0.0391	0.0104	0.79	0.01178	0.593
8-1	8	4:1	9	Ambient	N/A	N/A	N/A	70	0.0121	BDL	0.0241	0.0107	0.0061	BDL	ND	ND	ND	0.0529	0.0422	0.0408	0.0121	0.77	0.01230	0.560
24-1	24	4:1	9	Ambient	N/A	N/A	N/A	70	0.0142	BDL	0.0297	0.0116	0.0065	BDL	ND	ND	ND	0.0622	0.0505	0.0479	0.0142	0.77	0.01443	0.516
48-1	48	4:1	9	Ambient	2.06	2	0.14	70	0.0165	BDL	0.0333	0.0115	0.0062	BDL	ND	ND	ND	0.0675	0.0560	0.0510	0.0165	0.76	0.01537	0.468
72-1	72	4:1	9	Ambient	1.93	N/A	0.27	70	0.0164	BDL	0.0391	0.0120	0.0061	BDL	ND	ND	ND	0.0736	0.0616	0.0572	0.0164	0.78	0.01722	0.424

DXAA = DCAA + DBAA + BCAA
MXAA = MCAA + MBAA
N/A = Not Applicable
ND = Not Detected
BDL = Below Detection Limit of 0.5 µg/L
DOC (mg/L) = 3.32

Table E.15
Charleston coagulated water - 85 °F (mass concentration)

Sample	Time (h)	Cl/N ratio	pH	Br^- (mg/L)	Total Cl Residual ($mg Cl_2/L$) actual	Total Cl Residual ($mg Cl_2/L$) target	Cl_2 demand	Temperature (°F)	MCAA	MBAA	DCAA	BCAA	DBAA	TCAA	BDCAA	DBCAA	TBAA	HAA9	HAA5	DXAA	MXAA	DXAA/HAA9	DXAA Yield (DXAA/DOC)
									AVERAGE HAA CONCENTRATIONS (µg/L)														
0.5-1	0.5	4:1	8	Ambient	N/A	N/A	N/A	85	0.87	BDL	2.65	1.66	0.84	BDL	ND	ND	ND	6.02	4.36	5.15	0.87	0.86	1.74
1-1	1	4:1	8	Ambient	N/A	N/A	N/A	85	0.92	BDL	2.71	1.62	0.85	BDL	ND	ND	ND	6.10	4.48	5.18	0.92	0.85	1.75
4-1	4	4:1	8	Ambient	N/A	N/A	N/A	85	0.96	BDL	3.01	1.73	0.84	BDL	ND	ND	ND	6.55	4.82	5.58	0.96	0.85	1.89
8-1	8	4:1	8	Ambient	N/A	N/A	N/A	85	1.03	BDL	3.07	1.74	0.79	BDL	ND	ND	ND	6.64	4.90	5.60	1.03	0.84	1.89
24-1	24	4:1	8	Ambient	N/A	N/A	N/A	85	0.78	BDL	4.04	1.95	0.80	BDL	ND	ND	ND	7.57	5.62	6.78	0.78	0.90	2.29
48-1	48	4:1	8	Ambient	1.99	2	0.16	85	1.15	BDL	4.72	1.93	0.77	BDL	ND	ND	ND	8.58	6.65	7.43	1.15	0.87	2.51
72-1	72	4:1	8	Ambient	1.92	N/A	0.23	85	1.41	BDL	5.49	2.18	0.80	BDL	ND	ND	ND	9.88	7.70	8.47	1.41	0.86	2.86

DXAA = DCAA + DBAA + BCAA
MXAA = MCAA + MBAA
N/A = Not Applicable
ND = Not Detected
BDL = Below Detection Limit of 0.5 µg/L
DOC (mg/L) = 2.96

Table E.16
Charleston coagulated water - 85 °F (molar concentration)

Sample	Time (h)	Cl/N ratio	pH	Br^- (mg/L)	Total Cl Residual ($mg Cl_2/L$) actual	Total Cl Residual ($mg Cl_2/L$) target	Cl_2 demand	Temperature (°F)	MCAA	MBAA	DCAA	BCAA	DBAA	TCAA	BDCAA	DBCAA	TBAA	HAA9	HAA5	DXAA	MXAA	DXAA/HAA9	DXAA Yield (DXAA/DOC)	Bromine incorporation factor (n_D')
									AVERAGE HAA CONCENTRATIONS (µmol/L)															
0.5-1	0.5	4:1	8	Ambient	N/A	N/A	N/A	85	0.0092	BDL	0.0205	0.0096	0.0039	BDL	ND	ND	ND	0.0432	0.0336	0.0340	0.0092	0.79	0.01148	0.509
1-1	1	4:1	8	Ambient	N/A	N/A	N/A	85	0.0097	BDL	0.0210	0.0093	0.0039	BDL	ND	ND	ND	0.0440	0.0347	0.0343	0.0097	0.78	0.01158	0.500
4-1	4	4:1	8	Ambient	N/A	N/A	N/A	85	0.0102	BDL	0.0234	0.0100	0.0039	BDL	ND	ND	ND	0.0474	0.0374	0.0372	0.0102	0.78	0.01257	0.476
8-1	8	4:1	8	Ambient	N/A	N/A	N/A	85	0.0109	BDL	0.0238	0.0100	0.0036	BDL	ND	ND	ND	0.0484	0.0384	0.0375	0.0109	0.77	0.01267	0.462
24-1	24	4:1	8	Ambient	N/A	N/A	N/A	85	0.0083	BDL	0.0313	0.0113	0.0037	BDL	ND	ND	ND	0.0545	0.0432	0.0462	0.0083	0.85	0.01561	0.402
48-1	48	4:1	8	Ambient	1.99	2	0.16	85	0.0122	BDL	0.0366	0.0112	0.0036	BDL	ND	ND	ND	0.0635	0.0524	0.0513	0.0122	0.81	0.01733	0.356
72-1	72	4:1	8	Ambient	1.92	N/A	0.23	85	0.0149	BDL	0.0426	0.0126	0.0037	BDL	ND	ND	ND	0.0737	0.0611	0.0588	0.0149	0.80	0.01987	0.339

DXAA = DCAA + DBAA + BCAA
MXAA = MCAA + MBAA
N/A = Not Applicable
ND = Not Detected
BDL = Below Detection Limit of 0.5 µg/L
DOC (mg/L) = 2.96

Table E.17
Charleston source water (mass concentration)

Sample	Time (h)	Cl/N ratio	pH	Br^- (mg/L)	Total Cl Residual ($mg Cl_2/L$) actual	Total Cl Residual ($mg Cl_2/L$) target	Cl_2 demand	Temperature (°F)	MCAA	MBAA	DCAA	BCAA	DBAA	TCAA	BDCAA	DBCAA	TBAA	HAA9	HAA5	DXAA	MXAA	DXAA/HAA9	DXAA Yield (DXAA/DOC)
									AVERAGE HAA CONCENTRATIONS (µg/L)														
1-1	1	4:1	8	Ambient	N/A	N/A	N/A	70	1.91	0.86	7.66	3.46	0.82	BDL	ND	ND	ND	14.70	11.24	11.93	2.77	0.81	2.14
4-1	4	4:1	8	Ambient	N/A	N/A	N/A	70	1.90	0.70	7.97	3.41	0.80	BDL	ND	ND	ND	14.78	11.36	12.18	2.60	0.82	2.18
8-1	8	4:1	8	Ambient	N/A	N/A	N/A	70	2.09	0.65	8.77	3.38	0.79	BDL	ND	ND	ND	15.67	12.29	12.93	2.74	0.83	2.32
24-1	24	4:1	8	Ambient	N/A	N/A	N/A	70	3.26	BDL	9.96	3.41	0.75	0.54	ND	ND	ND	17.93	14.52	14.12	3.26	0.79	2.53
48-1	48	4:1	8	Ambient	2.03	2	0.52	70	2.96	BDL	11.53	3.75	0.85	BDL	ND	ND	ND	19.10	15.35	16.14	2.96	0.84	2.89
72-1	72	4:1	8	Ambient	1.89	N/A	0.66	70	3.36	BDL	12.50	3.63	0.76	BDL	ND	ND	ND	20.25	16.62	16.89	3.36	0.83	3.03

DXAA = DCAA + DBAA + BCAA
MXAA = MCAA + MBAA
N/A = Not Applicable
ND = Not Detected
BDL = Below Detection Limit of 0.5 µg/L
DOC (mg/L) = 5.58

Table E.18

Charleston source water (molar concentration)

Sample	Time (h)	Cl/N ratio	pH	Br^- (mg/L)	Total Cl Residual ($mg\ Cl_2/L$) actual	Total Cl Residual ($mg\ Cl_2/L$) target	Cl_2 demand	Temperature (°F)	AVERAGE HAA CONCENTRATIONS (µmol/L) MCAA	MBAA	DCAA	BCAA	DBAA	TCAA	BDCAA	DBCAA	TBAA	HAA9	HAA5	DXAA	MXAA	DXAA/HAA9	DXAA Yield (DXAA/DOC)	Bromine incorporation factor (n_D')
1-1	1	4:1	8	Ambient	N/A	N/A	N/A	70	0.0202	0.0062	0.0594	0.0199	0.0037	BDL	ND	ND	ND	0.1095	0.0895	0.0831	0.0264	0.76	0.01489	0.330
4-1	4	4:1	8	Ambient	N/A	N/A	N/A	70	0.0201	0.0050	0.0618	0.0197	0.0037	BDL	ND	ND	ND	0.1103	0.0906	0.0851	0.0251	0.77	0.01526	0.317
8-1	8	4:1	8	Ambient	N/A	N/A	N/A	70	0.0221	0.0047	0.0680	0.0195	0.0036	BDL	ND	ND	ND	0.1179	0.0984	0.0911	0.0268	0.77	0.01632	0.293
24-1	24	4:1	8	Ambient	N/A	N/A	N/A	70	0.0345	BDL	0.0773	0.0197	0.0035	0.0033	ND	ND	ND	0.1382	0.1186	0.1004	0.0345	0.73	0.01799	0.265
48-1	48	4:1	8	Ambient	2.03	2	0.52	70	0.0314	BDL	0.0894	0.0216	0.0039	BDL	ND	ND	ND	0.1463	0.1247	0.1150	0.0314	0.79	0.02061	0.256
72-1	72	4:1	8	Ambient	1.89	N/A	0.66	70	0.0356	BDL	0.0969	0.0209	0.0035	BDL	ND	ND	ND	0.1569	0.1360	0.1214	0.0356	0.77	0.02175	0.230

DXAA = DCAA + DBAA + BCAA
MXAA = MCAA + MBAA
N/A = Not Applicable
ND = Not Detected
BDL = Below Detection Limit of 0.5 µg/L
DOC (mg/L) = 5.58

Table E.19

Charleston coagulated water (mass concentration)

Sample	Time (h)	Cl/N ratio	pH	Br^- (mg/L)	Total Cl Residual ($mg\ Cl_2/L$) actual	Total Cl Residual ($mg\ Cl_2/L$) target	Cl_2 demand	Temperature (°F)	AVERAGE HAA CONCENTRATIONS (µg/L) MCAA	MBAA	DCAA	BCAA	DBAA	TCAA	BDCAA	DBCAA	TBAA	HAA9	HAA5	DXAA	MXAA	DXAA/HAA9	DXAA Yield (DXAA/DOC)
1-1	1	4:1	8	Ambient	N/A	N/A	N/A	70	1.41	BDL	2.71	1.78	0.89	BDL	ND	ND	ND	6.80	5.02	5.39	1.41	0.79	1.82
4-1	4	4:1	8	Ambient	N/A	N/A	N/A	70	0.90	BDL	2.26	1.87	0.90	BDL	ND	ND	ND	5.93	4.06	5.03	0.90	0.85	1.70
8-1	8	4:1	8	Ambient	N/A	N/A	N/A	70	0.76	BDL	2.15	1.84	0.85	BDL	ND	ND	ND	5.59	3.76	4.84	0.76	0.86	1.63
24-1	24	4:1	8	Ambient	N/A	N/A	N/A	70	1.00	BDL	2.74	1.84	0.82	BDL	ND	ND	ND	6.41	4.56	5.40	1.00	0.84	1.83
48-1	48	4:1	8	Ambient	2.03	2	0.12	70	1.42	BDL	3.73	2.14	0.85	BDL	ND	ND	ND	8.15	6.01	6.73	1.42	0.83	2.27
72-1	72	4:1	8	Ambient	2	N/A	0.15	70	1.02	BDL	4.09	2.02	0.87	BDL	ND	ND	ND	8.00	5.98	6.97	1.02	0.87	2.36

DXAA = DCAA + DBAA + BCAA
MXAA = MCAA + MBAA
N/A = Not Applicable
ND = Not Detected
BDL = Below Detection Limit of 0.5 µg/L
DOC (mg/L) = 2.96

Table E.20

Charleston coagulated water (molar concentration)

Sample	Time (h)	Cl/N ratio	pH	Br^- (mg/L)	Total Cl Residual ($mg\ Cl_2/L$) actual	Total Cl Residual ($mg\ Cl_2/L$) target	Cl_2 demand	Temperature (°F)	AVERAGE HAA CONCENTRATIONS (µmol/L) MCAA	MBAA	DCAA	BCAA	DBAA	TCAA	BDCAA	DBCAA	TBAA	HAA9	HAA5	DXAA	MXAA	DXAA/HAA9	DXAA Yield (DXAA/DOC)	Bromine incorporation factor (n_D')
1-1	1	4:1	8	Ambient	N/A	N/A	N/A	70	0.0150	BDL	0.0211	0.0103	0.0041	BDL	ND	ND	ND	0.0504	0.0401	0.0354	0.0150	0.70	0.01197	0.521
4-1	4	4:1	8	Ambient	N/A	N/A	N/A	70	0.0095	BDL	0.0175	0.0108	0.0041	BDL	ND	ND	ND	0.0419	0.0311	0.0325	0.0095	0.77	0.01097	0.588
8-1	8	4:1	8	Ambient	N/A	N/A	N/A	70	0.0080	BDL	0.0167	0.0106	0.0039	BDL	ND	ND	ND	0.0392	0.0286	0.0312	0.0080	0.80	0.01054	0.589
24-1	24	4:1	8	Ambient	N/A	N/A	N/A	70	0.0106	BDL	0.0213	0.0106	0.0038	BDL	ND	ND	ND	0.0463	0.0356	0.0357	0.0106	0.77	0.01205	0.509
48-1	48	4:1	8	Ambient	2.03	2	0.12	70	0.0150	BDL	0.0290	0.0123	0.0039	BDL	ND	ND	ND	0.0603	0.0479	0.0452	0.0150	0.75	0.01528	0.446
72-1	72	4:1	8	Ambient	2	N/A	0.15	70	0.0108	BDL	0.0317	0.0116	0.0040	BDL	ND	ND	ND	0.0581	0.0465	0.0473	0.0108	0.81	0.01598	0.415

DXAA = DCAA + DBAA + BCAA
MXAA = MCAA + MBAA
N/A = Not Applicable
ND = Not Detected
BDL = Below Detection Limit of 0.5 µg/L
DOC (mg/L) = 2.96

Table E.21
Biscayne Aquifer source water (mass concentration)

Sample	Time (h)	Cl/N ratio	pH	Br⁻ (mg/L)	Total Cl Residual (mg Cl_2/L) actual	target	Cl_2 demand	Temperature (°F)	AVERAGE HAA CONCENTRATIONS (µg/L) MCAA	MBAA	DCAA	BCAA	DBAA	TCAA	BDCAA	DBCAA	TBAA	HAA9	HAA5	DXAA	MXAA	DXAA/HAA9	DXAA Yield (DXAA/DOC)
0.5-1	0.5	4:1	9	Ambient	N/A	N/A	N/A	70	2.02	BDL	8.13	1.46	BDL	BDL	ND	ND	ND	11.61	10.15	9.59	2.02	0.83	0.80
1-1	1	4:1	9	Ambient	N/A	N/A	N/A	70	1.81	BDL	8.41	1.50	BDL	BDL	ND	ND	ND	11.72	10.22	9.91	1.81	0.85	0.83
4-1	4	4:1	9	Ambient	N/A	N/A	N/A	70	1.80	BDL	8.50	1.48	BDL	BDL	ND	ND	ND	11.78	10.30	9.98	1.80	0.85	0.83
8-1	8	4:1	9	Ambient	N/A	N/A	N/A	70	1.84	BDL	8.94	1.57	BDL	BDL	ND	ND	ND	12.36	10.79	10.52	1.84	0.85	0.88
24-1	24	4:1	9	Ambient	N/A	N/A	N/A	70	2.41	BDL	9.36	1.57	BDL	BDL	ND	ND	ND	13.35	11.78	10.94	2.41	0.82	0.92
48-1	48	4:1	9	Ambient	2.11	2	0.94	70	2.95	BDL	10.36	1.65	BDL	BDL	ND	ND	ND	14.95	13.30	12.01	2.95	0.80	1.00
72-1	72	4:1	9	Ambient	1.9	N/A	1.12	70	2.79	BDL	11.14	1.66	BDL	BDL	ND	ND	ND	15.58	13.93	12.79	2.79	0.82	1.07

DXAA = DCAA + DBAA + BCAA
MXAA = MCAA + MBAA
N/A = Not Applicable
ND = Not Detected
BDL = Below Detection Limit of 0.5 µg/L
DOC (mg/L) = 11.95

Table E.22
Biscayne Aquifer source water (molar concentration)

Sample	Time (h)	Cl/N ratio	pH	Br⁻ (mg/L)	Total Cl Residual (mg Cl_2/L) actual	target	Cl_2 demand	Temperature (°F)	AVERAGE HAA CONCENTRATIONS (µmol/L) MCAA	MBAA	DCAA	BCAA	DBAA	TCAA	BDCAA	DBCAA	TBAA	HAA9	HAA5	DXAA	MXAA	DXAA/HAA9	DXAA Yield (DXAA/DOC)	Bromine incorporation factor (n_D')
0.5-1	0.5	4:1	9	Ambient	N/A	N/A	N/A	70	0.0213	BDL	0.0631	0.0084	BDL	BDL	ND	ND	ND	0.0928	0.0844	0.0715	0.0213	0.77	0.00598	0.118
1-1	1	4:1	9	Ambient	N/A	N/A	N/A	70	0.0192	BDL	0.0652	0.0087	BDL	BDL	ND	ND	ND	0.0930	0.0844	0.0739	0.0192	0.79	0.00618	0.117
4-1	4	4:1	9	Ambient	N/A	N/A	N/A	70	0.0191	BDL	0.0659	0.0085	BDL	BDL	ND	ND	ND	0.0935	0.0850	0.0744	0.0191	0.80	0.00623	0.114
8-1	8	4:1	9	Ambient	N/A	N/A	N/A	70	0.0195	BDL	0.0694	0.0091	BDL	BDL	ND	ND	ND	0.0980	0.0889	0.0784	0.0195	0.80	0.00656	0.116
24-1	24	4:1	9	Ambient	N/A	N/A	N/A	70	0.0255	BDL	0.0726	0.0091	BDL	BDL	ND	ND	ND	0.1072	0.0982	0.0817	0.0255	0.76	0.00684	0.111
48-1	48	4:1	9	Ambient	2.11	2	0.94	70	0.0312	BDL	0.0803	0.0095	BDL	BDL	ND	ND	ND	0.1210	0.1115	0.0898	0.0312	0.74	0.00752	0.106
72-1	72	4:1	9	Ambient	1.9	N/A	1.12	70	0.0295	BDL	0.0864	0.0095	BDL	BDL	ND	ND	ND	0.1255	0.1159	0.0959	0.0295	0.76	0.00803	0.100

DXAA = DCAA + DBAA + BCAA
MXAA = MCAA + MBAA
N/A = Not Applicable
ND = Not Detected
BDL = Below Detection Limit of 0.5 µg/L
DOC (mg/L) = 11.95

Table E.23
Biscayne Aquifer softened water (mass concentration)

Sample	Time (h)	Cl/N ratio	pH	Br⁻ (mg/L)	Total Cl Residual (mg Cl_2/L) actual	target	Cl_2 demand	Temperature (°F)	AVERAGE HAA CONCENTRATIONS (µg/L) MCAA	MBAA	DCAA	BCAA	DBAA	TCAA	BDCAA	DBCAA	TBAA	HAA9	HAA5	DXAA	MXAA	DXAA/HAA9	DXAA Yield (DXAA/DOC)
0.5-1	0.5	4:1	9	Ambient	N/A	N/A	N/A	70	0.58	BDL	9.02	1.63	BDL	1.24	BDL	ND	ND	12.47	10.84	10.65	0.58	0.85	1.13
1-1	1	4:1	9	Ambient	N/A	N/A	N/A	70	1.07	BDL	9.84	1.66	BDL	1.21	BDL	ND	ND	13.78	12.12	11.51	1.07	0.83	1.22
4-1	4	4:1	9	Ambient	N/A	N/A	N/A	70	0.60	BDL	9.60	1.69	BDL	1.24	BDL	ND	ND	13.13	11.44	11.29	0.60	0.86	1.19
8-1	8	4:1	9	Ambient	N/A	N/A	N/A	70	0.86	BDL	10.00	1.79	BDL	1.27	0.60	ND	ND	14.52	12.13	11.79	0.86	0.81	1.25
24-1	24	4:1	9	Ambient	N/A	N/A	N/A	70	0.84	BDL	9.84	1.77	BDL	1.25	0.63	ND	ND	14.34	11.94	11.61	0.84	0.81	1.23
48-1	48	4:1	9	Ambient	2.1	2	0.5	70	1.39	BDL	10.05	1.70	BDL	1.20	0.61	ND	ND	14.95	12.64	11.75	1.39	0.79	1.24
72-1	72	4:1	9	Ambient	1.99	N/A	0.61	70	NA	NA	NA	NA	NA	NA	NA	NA	NA	NA	NA	NA	NA	NA	NA

DXAA = DCAA + DBAA + BCAA
MXAA = MCAA + MBAA
N/A = Not Applicable
ND = Not Detected
NA = Not Available
BDL = Below Detection Limit of 0.5 µg/L
DOC (mg/L) = 9.45

Table E.24
Biscayne Aquifer softened water (molar concentration)

Sample	Time (h)	Cl/N ratio	pH	Br^- (mg/L)	Total Cl Residual ($mg Cl_2/L$) actual	Total Cl Residual ($mg Cl_2/L$) target	Cl_2 demand	Temperature (°F)	MCAA	MBAA	DCAA	BCAA	DBAA	TCAA	BDCAA	DBCAA	TBAA	HAA9	HAA5	DXAA	MXAA	DXAA/HAA9	DXAA Yield (DXAA/DOC)	Bromine incorporation factor (n_D')
									AVERAGE HAA CONCENTRATIONS (µmol/L)															
0.5-1	0.5	4:1	9	Ambient	N/A	N/A	N/A	70	0.0061	BDL	0.0699	0.0094	BDL	0.0076	BDL	ND	ND	0.0931	0.0837	0.0794	0.0061	0.85	0.00840	0.119
1-1	1	4:1	9	Ambient	N/A	N/A	N/A	70	0.0113	BDL	0.0763	0.0096	BDL	0.0074	BDL	ND	ND	0.1046	0.0950	0.0859	0.0113	0.82	0.00909	0.112
4-1	4	4:1	9	Ambient	N/A	N/A	N/A	70	0.0063	BDL	0.0744	0.0098	BDL	0.0076	BDL	ND	ND	0.0981	0.0884	0.0842	0.0063	0.86	0.00891	0.116
8-1	8	4:1	9	Ambient	N/A	N/A	N/A	70	0.0091	BDL	0.0775	0.0103	BDL	0.0078	0.0029	ND	ND	0.1076	0.0944	0.0879	0.0091	0.82	0.00930	0.117
24-1	24	4:1	9	Ambient	N/A	N/A	N/A	70	0.0089	BDL	0.0763	0.0102	BDL	0.0077	0.0030	ND	ND	0.1061	0.0929	0.0865	0.0089	0.82	0.00916	0.118
48-1	48	4:1	9	Ambient	2.1	2	0.5	70	0.0147	BDL	0.0779	0.0098	BDL	0.0073	0.0029	ND	ND	0.1128	0.1000	0.0878	0.0147	0.78	0.00929	0.112
72-1	72	4:1	9	Ambient	1.99	N/A	0.61	70	NA	NA	NA	NA	NA	NA	NA	NA	NA	NA	NA	NA	NA	NA	NA	NA

DXAA = DCAA + DBAA + BCAA
MXAA = MCAA + MBAA
N/A = Not Applicable
ND = Not Detected
NA = Not Available
BDL = Below Detection Limit of 0.5 µg/L
DOC (mg/L) = 9.45

CHLORAMINATION EXPERIMENTS WITH PRECHLORINATION

Table E.25
Lake Austin source water - 5 minute prechlorination (mass concentration)

Sample	Time (h)	Cl/N ratio	Actual Cl/N ratio	pH	Br^- (mg/L)	Total Cl Residual ($mg Cl_2/L$) actual	Total Cl Residual ($mg Cl_2/L$) target	Cl_2 demand	Temperature (°F)	MCAA	MBAA	DCAA	BCAA	DBAA	TCAA	BDCAA	DBCAA	TBAA	HAA9	HAA5	DXAA	MXAA	DXAA/HAA9	DXAA Yield (DXAA/DOC)
										AVERAGE HAA CONCENTRATIONS (µg/L)														
5LAW5min	5 min	N/A	N/A	9	Ambient	N/A	N/A	N/A	70	BDL	0.97	7.56	20.96	5.67	2.65	ND	BDL	BDL	37.81	16.84	34.19	0.97	0.90	9.14
5LAW 0.5	0.5	4:1	4.02	9	Ambient	N/A	N/A	N/A	70	BDL	BDL	12.81	6.89	3.50	7.96	1.58	BDL	BDL	32.74	24.27	23.20	0.00	0.71	6.20
5LAW 1	1	4:1	4.02	9	Ambient	N/A	N/A	N/A	70	BDL	BDL	11.88	6.29	3.22	7.26	1.52	BDL	BDL	30.16	22.35	21.39	0.00	0.71	5.72
5LAW4	4	4:1	4.02	9	Ambient	N/A	N/A	N/A	70	BDL	BDL	14.07	7.25	3.93	7.20	1.61	BDL	BDL	34.06	25.20	25.25	0.00	0.74	6.75
5LAW8	8	4:1	4.02	9	Ambient	N/A	N/A	N/A	70	BDL	0.57	12.18	6.29	3.25	7.05	1.40	BDL	BDL	30.74	23.05	21.73	0.57	0.71	5.81
5LAW24	24	4:1	4.02	9	Ambient	N/A	N/A	N/A	70	0.78	0.56	15.87	7.82	4.12	8.51	1.68	BDL	BDL	39.34	29.84	27.81	1.34	0.71	7.44
5LAW48	48	4:1	4.02	9	Ambient	1.83	2	0.57	70	BDL	0.69	13.35	6.83	3.84	6.68	1.52	0.57	0.61	34.11	24.57	24.03	0.69	0.70	6.42
5LAW 72-1	72	4:1	4.02	9	Ambient	1.76	N/A	0.63	70	0.54	0.67	14.47	6.83	3.32	7.42	1.29	BDL	BDL	34.54	26.42	24.62	1.21	0.71	6.58

DXAA = DCAA + DBAA + BCAA
MXAA = MCAA + MBAA
N/A = Not Applicable
ND = Not Detected
BDL = Below Detection Limit of 0.5 µg/L
DOC (mg/L) = 3.74

Table E.26

Lake Austin source water - 5 minute prechlorination (molar concentration)

Sample	Time (h)	Cl/N ratio	Actual Cl/N ratio	pH	Br^- (mg/L)	Total Cl Residual ($mg\ Cl_2/L$) actual	Total Cl Residual ($mg\ Cl_2/L$) target	Cl_2 demand	Temperature (°F)	AVERAGE HAA CONCENTRATIONS (µmol/L) MCAA	MBAA	DCAA	BCAA	DBAA	TCAA	BDCAA	DBCAA	TBAA	HAA9	HAA5	DXAA	MXAA	DXAA/HAA9	DXAA Yield (DXAA/DOC)	Bromine incorporation factor (n_D')
5LAW5min	5 min	N/A	N/A	9	Ambient	N/A	N/A	N/A	70	BDL	0.0070	0.0586	0.1209	0.0260	0.0162	ND	BDL	BDL	0.22870	0.1078	0.2055	0.0070	0.90	0.05495	0.842
5LAW 0.5	0.5	4:1	4.02	9	Ambient	N/A	N/A	N/A	71	BDL	BDL	0.0993	0.0397	0.0161	0.0487	0.0076	BDL	BDL	0.21148	0.1641	0.1551	0.0000	0.73	0.04148	0.463
5LAW 1	1	4:1	4.02	9	Ambient	N/A	N/A	N/A	72	BDL	BDL	0.0921	0.0363	0.0148	0.0444	0.0073	BDL	BDL	0.19488	0.1513	0.1432	0.0000	0.73	0.03828	0.460
5LAW4	4	4:1	4.02	9	Ambient	N/A	N/A	N/A	73	BDL	BDL	0.1092	0.0418	0.0180	0.0441	0.0077	BDL	BDL	0.22081	0.1713	0.1690	0.0000	0.77	0.04519	0.461
5LAW8	8	4:1	4.02	9	Ambient	N/A	N/A	N/A	74	BDL	0.0041	0.0945	0.0363	0.0149	0.0431	0.0067	BDL	BDL	0.19966	0.1566	0.1457	0.0041	0.73	0.03896	0.454
5LAW24	24	4:1	4.02	9	Ambient	N/A	N/A	N/A	75	0.0083	0.0040	0.1231	0.0451	0.0189	0.0521	0.0081	BDL	BDL	0.25956	0.2064	0.1871	0.0123	0.72	0.05003	0.443
5LAW48	48	4:1	4.02	9	Ambient	1.83	2	0.57	76	BDL	0.0050	0.1035	0.0394	0.0176	0.0409	0.0073	0.0023	0.0020	0.21814	0.1671	0.1606	0.0050	0.74	0.04294	0.465
5LAW 72-1	72	4:1	4.02	9	Ambient	1.76	N/A	0.63	77	0.0057	0.0048	0.1122	0.0394	0.0152	0.0454	0.0062	BDL	BDL	0.22900	0.1834	0.1669	0.0105	0.73	0.04461	0.419

DXAA = DCAA + DBAA + BCAA
MXAA = MCAA + MBAA
N/A = Not Applicable
ND = Not Detected
BDL = Below Detection Limit of 0.5 µg/L
DOC (mg/L) = 3.74

Table E.27

Lake Austin source water - 20 minute prechlorination (mass concentration)

Sample	Time (h)	Cl/N ratio	Actual Cl/N ratio	pH	Br^- (mg/L)	Total Cl Residual ($mg\ Cl_2/L$) actual	Total Cl Residual ($mg\ Cl_2/L$) target	Cl_2 demand	Temperature (°F)	AVERAGE HAA CONCENTRATIONS (µg/L) MCAA	MBAA	DCAA	BCAA	DBAA	TCAA	BDCAA	DBCAA	TBAA	HAA9	HAA5	DXAA	MXAA	DXAA/HAA9	DXAA Yield (DXAA/DOC)
20LAW20min	20 min	N/A		9	Ambient	N/A	N/A	N/A	70	BDL	1.0551	11.7417	11.2767	6.0076	5.9647	2.0421	0.8048	BDL	38.8926	24.77	29.0260	1.0551	0.75	7.76
20LAW 0.5	0.5	4:1		9	Ambient	N/A	N/A	N/A	71	0.7759	0.6143	15.9713	11.0351	5.4527	8.5160	2.3320	0.8902	0.6699	46.2575	31.33	32.4592	1.3903	0.70	8.68
20LAW 1	1	4:1		9	Ambient	N/A	N/A	N/A	72	0.5422	0.6176	16.5305	11.2870	5.6408	8.8074	1.8580	0.7334	0.5337	46.5506	32.14	33.4583	1.1598	0.72	8.95
20LAW4	4	4:1		9	Ambient	N/A	N/A	N/A	73	0.6599	0.7648	16.2791	10.3781	5.1580	8.3503	2.1122	0.8526	0.5239	45.0789	31.21	31.8152	1.4248	0.71	8.51
20LAW8	8	4:1		9	Ambient	N/A	N/A	N/A	74	0.5461	0.7585	16.3122	10.9208	5.0296	8.1755	2.1196	0.8644	BDL	44.7267	30.82	32.2626	1.3046	0.72	8.63
20LAW24	24	4:1		9	Ambient	N/A	N/A	N/A	75	0.7783	0.5959	20.5578	10.4639	6.6775	9.0512	2.6732	1.0292	0.5686	52.3958	37.66	37.6993	1.3742	0.72	10.08
20LAW48	48	4:1		9	Ambient	1.63	2	1.07	76	0.8739	0.7107	15.5281	11.2550	5.3416	7.9610	1.9056	0.7421	BDL	44.3180	30.42	32.1247	1.5846	0.72	8.59
20LAW 72-1	72	4:1		9	Ambient	1.54	N/A	1.16	77	1.0710	0.9415	19.1489	13.2342	6.6830	10.4614	2.4639	1.2042	1.0321	56.2402	38.31	39.0661	2.0126	0.69	10.45

DXAA = DCAA + DBAA + BCAA
MXAA = MCAA + MBAA
N/A = Not Applicable
ND = Not Detected
BDL = Below Detection Limit of 0.5 µg/L
DOC (mg/L) = 3.74

Table E.28

Lake Austin source water - 20 minute prechlorination (molar concentration)

Sample	Time (h)	Cl/N ratio	Actual Cl/N ratio	pH	Br^- (mg/L)	Total Cl Residual ($mg\ Cl_2/L$) actual	Total Cl Residual ($mg\ Cl_2/L$) target	Cl_2 demand	Temperature (°F)	AVERAGE HAA CONCENTRATIONS (µmol/L) MCAA	MBAA	DCAA	BCAA	DBAA	TCAA	BDCAA	DBCAA	TBAA	HAA9	HAA5	DXAA	MXAA	DXAA/HAA9	DXAA Yield (DXAA/DOC)	Bromine incorporation factor (n_D')
20LAW20min	20 min	N/A		9	Ambient	N/A	N/A	N/A	70	BDL	0.0076	0.0911	0.0650	0.0276	0.0365	0.0098	0.0032	BDL	0.2408	0.1627	0.1837	0.0076	0.76	0.04911	0.654
20LAW 0.5	0.5	4:1		9	Ambient	N/A	N/A	N/A	71	0.0082	0.0044	0.1239	0.0636	0.0250	0.0521	0.0112	0.0035	0.0023	0.2943	0.2136	0.2125	0.0126	0.72	0.05683	0.535
20LAW 1	1	4:1		9	Ambient	N/A	N/A	N/A	72	0.0057	0.0044	0.1282	0.0651	0.0259	0.0539	0.0089	0.0029	0.0018	0.2969	0.2182	0.2192	0.0102	0.74	0.05861	0.533
20LAW4	4	4:1		9	Ambient	N/A	N/A	N/A	73	0.0070	0.0055	0.1263	0.0599	0.0237	0.0511	0.0102	0.0034	0.0018	0.2887	0.2135	0.2098	0.0125	0.73	0.05609	0.511
20LAW8	8	4:1		9	Ambient	N/A	N/A	N/A	74	0.0058	0.0055	0.1265	0.0630	0.0231	0.0500	0.0102	0.0034	BDL	0.2875	0.2109	0.2126	0.0112	0.74	0.05684	0.513
20LAW24	24	4:1		9	Ambient	N/A	N/A	N/A	75	0.0082	0.0043	0.1594	0.0603	0.0307	0.0554	0.0129	0.0041	0.0019	0.3372	0.2580	0.2504	0.0125	0.74	0.06696	0.486
20LAW48	48	4:1		9	Ambient	1.63	2	1.07	76	0.0092	0.0051	0.1204	0.0649	0.0245	0.0487	0.0092	0.0029	BDL	0.2851	0.2080	0.2099	0.0144	0.74	0.05611	0.543
20LAW 72-1	72	4:1		9	Ambient	1.54	N/A	1.16	77	0.0113	0.0068	0.1485	0.0763	0.0307	0.0640	0.0119	0.0048	0.0035	0.3578	0.2613	0.2555	0.0181	0.71	0.06832	0.539

DXAA = DCAA + DBAA + BCAA
MXAA = MCAA + MBAA
N/A = Not Applicable
ND = Not Detected
BDL = Below Detection Limit of 0.5 µg/L
DOC (mg/L) = 3.74

Table E.29

Meteconk River Coagulated water - 5 minute prechlorination (mass concentration)

Sample	Time (h)	Cl/N ratio	Actual Cl/N ratio	pH	Br^- (mg/L)	Total Cl Residual ($mg Cl_2/L$) actual	Total Cl Residual ($mg Cl_2/L$) target	Cl_2 demand	Temperature (°F)	AVERAGE HAA CONCENTRATIONS (μg/L) MCAA	MBAA	DCAA	BCAA	DBAA	TCAA	BDCAA	DBCAA	TBAA	HAA9	HAA5	DXAA	MXAA	DXAA/HAA9	DXAA Yield (DXAA/DOC)
5MR5min	5 min	N/A	N/A	8	Ambient	N/A	N/A	N/A	70	BDL	BDL	13.99	3.74	1.32	20.33	1.28	BDL	BDL	40.65	35.63	19.04	0.00	0.47	5.77
5MR 0.5	0.5	4:1	4.1	8	Ambient	N/A	N/A	N/A	70	BDL	BDL	13.02	3.56	0.99	18.08	1.24	BDL	BDL	36.89	32.09	17.56	0.00	0.48	5.32
5MR 1	1	4:1	4.1	8	Ambient	N/A	N/A	N/A	70	BDL	BDL	12.81	3.58	0.90	20.98	1.14	BDL	BDL	39.41	34.69	17.29	0.00	0.44	5.24
5MR4	4	4:1	4.1	8	Ambient	N/A	N/A	N/A	70	BDL	BDL	13.54	3.54	1.00	26.40	1.08	BDL	BDL	45.57	40.94	18.08	0.00	0.40	5.48
5MR8	8	4:1	4.1	8	Ambient	N/A	N/A	N/A	70	BDL	BDL	13.45	3.56	1.06	24.91	1.15	BDL	BDL	44.13	39.43	18.07	0.00	0.41	5.48
5MR24	24	4:1	4.1	8	Ambient	N/A	N/A	N/A	70	BDL	BDL	14.23	3.53	1.08	17.63	1.23	BDL	BDL	37.70	32.94	18.84	0.00	0.50	5.71
5MR48	48	4:1	4.1	8	Ambient	1.87	2	0.88	70	BDL	BDL	14.30	3.56	1.09	23.07	1.07	BDL	BDL	43.09	38.45	18.94	0.00	0.44	5.74
5MR 72-1	72	4:1	4.1	9	Ambient	1.81	N/A	0.93	70	BDL	BDL	16.43	3.61	1.00	28.44	0.99	BDL	BDL	50.47	45.87	21.05	0.00	0.42	6.38

DXAA = DCAA + DBAA + BCAA
MXAA = MCAA + MBAA
N/A = Not Applicable
ND = Not Detected
BDL = Below Detection Limit of 0.5 μg/L
DOC (mg/L) = 3.30

Table E.30

Meteconk River alum coagulated water - 5 minute prechlorination (molar concentration)

Sample	Time (h)	Cl/N ratio	Actual Cl/N ratio	pH	Br^- (mg/L)	Total Cl Residual ($mg Cl_2/L$) actual	Total Cl Residual ($mg Cl_2/L$) target	Cl_2 demand	Temperature (°F)	AVERAGE HAA CONCENTRATIONS (μmol/L) MCAA	MBAA	DCAA	BCAA	DBAA	TCAA	BDCAA	DBCAA	TBAA	HAA9	HAA5	DXAA	MXAA	DXAA/HAA9	DXAA Yield (DXAA/DOC)	Bromine incorporation factor (n_D')
5MR5min	5 min	N/A	N/A	8	Ambient	N/A	N/A	N/A	70	BDL	BDL	0.1085	0.0216	0.0060	0.1244	0.0062	BDL	BDL	0.26665	0.2389	0.1361	0.0000	0.51	0.04123	0.247
5MR 0.5	0.5	4:1	4.1	8	Ambient	N/A	N/A	N/A	71	BDL	BDL	0.1010	0.0205	0.0045	0.1107	0.0060	BDL	BDL	0.24267	0.2162	0.1260	0.0000	0.52	0.03819	0.235
5MR 1	1	4:1	4.1	8	Ambient	N/A	N/A	N/A	72	BDL	BDL	0.0994	0.0206	0.0041	0.1284	0.0055	BDL	BDL	0.25799	0.2319	0.1241	0.0000	0.48	0.03761	0.233
5MR4	4	4:1	4.1	8	Ambient	N/A	N/A	N/A	73	BDL	BDL	0.1050	0.0204	0.0046	0.1616	0.0052	BDL	BDL	0.29683	0.2712	0.1300	0.0000	0.44	0.03940	0.228
5MR8	8	4:1	4.1	8	Ambient	N/A	N/A	N/A	74	BDL	BDL	0.1043	0.0205	0.0049	0.1525	0.0055	BDL	BDL	0.28772	0.2617	0.1297	0.0000	0.45	0.03931	0.233
5MR24	24	4:1	4.1	8	Ambient	N/A	N/A	N/A	75	BDL	BDL	0.1104	0.0204	0.0050	0.1079	0.0059	BDL	BDL	0.24951	0.2232	0.1357	0.0000	0.54	0.04112	0.223
5MR48	48	4:1	4.1	8	Ambient	1.87	2	0.88	76	BDL	BDL	0.1109	0.0205	0.0050	0.1412	0.0052	BDL	BDL	0.28274	0.2571	0.1364	0.0000	0.48	0.04133	0.224
5MR 72-1	72	4:1	4.1	8	Ambient	1.81	N/A	0.93	77	BDL	BDL	0.1275	0.0208	0.0046	0.1741	0.0047	BDL	BDL	0.33168	0.3061	0.1529	0.0000	0.46	0.04633	0.196

DXAA = DCAA + DBAA + BCAA
MXAA = MCAA + MBAA
N/A = Not Applicable
ND = Not Detected
BDL = Below Detection Limit of 0.5 μg/L
DOC (mg/L) = 3.30

Table E.31

Meteconk River alum coagulated water - 20 minute prechlorination (mass concentration)

Sample	Time (h)	Cl/N ratio	Actual Cl/N ratio	pH	Br^- (mg/L)	Total Cl Residual ($mg Cl_2/L$) actual	Total Cl Residual ($mg Cl_2/L$) target	Cl_2 demand	Temperature (°F)	AVERAGE HAA CONCENTRATIONS (μg/L) MCAA	MBAA	DCAA	BCAA	DBAA	TCAA	BDCAA	DBCAA	TBAA	HAA9	HAA5	DXAA	MXAA	DXAA/HAA9	DXAA Yield (DXAA/DOC)
20MR20min	20 min	N/A	N/A	8	Ambient	N/A	N/A	N/A	70	BDL	BDL	15.1323	3.8955	1.2851	22.7282	2.0914	BDL	BDL	45.1324	39.15	20.3129	0.0000	0.45	6.16
20MR 0.5	0.5	4:1	4.5	8	Ambient	N/A	N/A	N/A	71	BDL	BDL	14.6850	4.0724	1.2315	23.0891	1.7998	BDL	BDL	44.8779	39.01	19.9889	0.0000	0.45	6.06
20MR 1	1	4:1	4.5	8	Ambient	N/A	N/A	N/A	72	BDL	BDL	15.6966	4.1063	1.2130	35.5930	1.9033	BDL	BDL	58.5122	52.50	21.0160	0.0000	0.36	6.37
20MR4	4	4:1	4.5	8	Ambient	N/A	N/A	N/A	73	BDL	BDL	16.0274	3.9465	1.2243	30.5859	1.7997	BDL	BDL	53.5838	47.84	21.1982	0.0000	0.40	6.42
20MR8	8	4:1	4.5	8	Ambient	N/A	N/A	N/A	74	BDL	BDL	15.8791	3.8261	1.1598	27.8952	1.6623	BDL	BDL	50.4224	44.93	20.8649	0.0000	0.41	6.32
20MR24	24	4:1	4.5	8	Ambient	N/A	N/A	N/A	75	BDL	BDL	15.8315	3.8735	1.1631	26.2468	1.6219	BDL	BDL	48.7368	43.24	20.8681	0.0000	0.43	6.32
20MR48	48	4:1	4.5	8	Ambient	1.75	2	1.5	76	BDL	BDL	16.8398	4.9060	1.1742	26.1653	1.5289	BDL	BDL	50.6142	44.18	22.9200	0.0000	0.45	6.95
20MR 72-1	72	4:1	4.5	8	Ambient	1.7	N/A	1.55	77	BDL	BDL	16.6312	3.8001	1.0990	23.3451	1.3383	BDL	BDL	46.2137	41.08	21.5303	0.0000	0.47	6.52

DXAA = DCAA + DBAA + BCAA
MXAA = MCAA + MBAA
N/A = Not Applicable
ND = Not Detected
BDL = Below Detection Limit of 0.5 μg/L
DOC (mg/L) = 3.30

Table E.32

Meteconk River alum coagulated water - 20 minute prechlorination (molar concentration)

Sample	Time (h)	Cl/N ratio	Actual Cl/N ratio	pH	Br^- (mg/L)	Total Cl Residual ($mg Cl_2/L$) actual	Total Cl Residual ($mg Cl_2/L$) target	Cl_2 demand	Temperature (°F)	MCAA	MBAA	DCAA	BCAA	DBAA	TCAA	BDCAA	DBCAA	TBAA	HAA9	HAA5	DXAA	MXAA	DXAA/HAA9	DXAA Yield (DXAA/DOC)	Bromine incorporation factor (n_D')
										AVERAGE HAA CONCENTRATIONS (μmol/L)															
20MR20min	20 min	N/A	N/A	8	Ambient	N/A	N/A	N/A	70	BDL	BDL	0.1174	0.0225	0.0059	0.1391	0.0101	BDL	BDL	0.2949	0.2624	0.1457	0.0000	0.49	0.04416	0.235
20MR 0.5	0.5	4:1	4.5	8	Ambient	N/A	N/A	N/A	71	BDL	BDL	0.1139	0.0235	0.0057	0.1413	0.0087	BDL	BDL	0.2930	0.2608	0.1430	0.0000	0.49	0.04334	0.243
20MR 1	1	4:1	4.5	8	Ambient	N/A	N/A	N/A	72	BDL	BDL	0.1217	0.0237	0.0056	0.2178	0.0092	BDL	BDL	0.3780	0.3451	0.1510	0.0000	0.40	0.04575	0.231
20MR4	4	4:1	4.5	8	Ambient	N/A	N/A	N/A	73	BDL	BDL	0.1243	0.0228	0.0056	0.1872	0.0087	BDL	BDL	0.3485	0.3171	0.1527	0.0000	0.44	0.04627	0.223
20MR8	8	4:1	4.5	8	Ambient	N/A	N/A	N/A	74	BDL	BDL	0.1232	0.0221	0.0053	0.1707	0.0080	BDL	BDL	0.3293	0.2992	0.1505	0.0000	0.46	0.04562	0.217
20MR24	24	4:1	4.5	8	Ambient	N/A	N/A	N/A	75	BDL	BDL	0.1228	0.0223	0.0053	0.1606	0.0078	BDL	BDL	0.3189	0.2887	0.1505	0.0000	0.47	0.04559	0.219
20MR48	48	4:1	4.5	8	Ambient	1.75	2	1.5	76	BDL	BDL	0.1306	0.0283	0.0054	0.1601	0.0074	BDL	BDL	0.3318	0.2961	0.1643	0.0000	0.50	0.04978	0.238
20MR 72-1	72	4:1	4.5	8	Ambient	1.7	N/A	1.55	77	BDL	BDL	0.1290	0.0219	0.0050	0.1429	0.0064	BDL	BDL	0.3053	0.2769	0.1559	0.0000	0.51	0.04726	0.205

DXAA = DCAA + DBAA + BCAA
MXAA = MCAA + MBAA
N/A = Not Applicable
ND = Not Detected
BDL = Below Detection Limit of 0.5 μg/L
DOC (mg/L) = 3.30

Table E.33

Biscayne Aquifer softened water - 5 minute prechlorination (mass concentration)

Sample	Time (h)	Cl/N ratio	Actual Cl/N ratio	pH	Br^- (mg/L)	Total Cl Residual ($mg Cl_2/L$) actual	Total Cl Residual ($mg Cl_2/L$) target	Cl_2 demand	Temperature (°F)	MCAA	MBAA	DCAA	BCAA	DBAA	TCAA	BDCAA	DBCAA	TBAA	HAA9	HAA5	DXAA	MXAA	DXAA/HAA9	DXAA Yield (DXAA/DOC)
										AVERAGE HAA CONCENTRATIONS (μg/L)														
5BR5min	5 min	N/A	N/A	9	Ambient	N/A	N/A	N/A	70	BDL	0.52	12.69	4.68	2.00	5.21	ND	BDL	BDL	25.10	20.42	19.37	0.52	0.77	2.05
5BR 0.5	0.5	4:1	3.9	9	Ambient	N/A	N/A	N/A	70	0.50	0.52	11.50	4.46	1.77	2.89	ND	BDL	BDL	21.65	17.18	17.73	1.03	0.82	1.88
5BR 1	1	4:1	3.9	9	Ambient	N/A	N/A	N/A	70	0.52	0.52	12.49	4.43	1.81	3.50	ND	BDL	BDL	23.26	18.84	18.73	1.04	0.81	1.98
5BR4	4	4:1	3.9	9	Ambient	N/A	N/A	N/A	70	0.52	BDL	11.99	4.28	1.65	2.94	ND	BDL	BDL	21.38	17.11	17.92	0.52	0.84	1.90
5BR8	8	4:1	3.9	9	Ambient	N/A	N/A	N/A	70	0.60	BDL	12.51	4.54	1.79	3.17	ND	BDL	BDL	22.60	18.06	18.84	0.60	0.83	1.99
5BR24	24	4:1	3.9	9	Ambient	N/A	N/A	N/A	70	0.86	0.52	13.61	4.79	2.02	2.83	ND	BDL	BDL	24.65	19.85	20.43	1.39	0.83	2.16
5BR48	48	4:1	3.9	9	Ambient	1.97	2	0.828	70	0.54	0.75	13.36	5.62	4.35	3.20	ND	1.14	1.26	30.22	22.21	23.33	1.29	0.77	2.47
5BR 72-1	72	4:1	3.9	9	Ambient	1.72	N/A	1.08	70	0.81	BDL	13.95	4.54	1.67	2.88	ND	BDL	BDL	23.85	19.31	20.16	0.81	0.85	2.13

DXAA = DCAA + DBAA + BCAA
MXAA = MCAA + MBAA
N/A = Not Applicable
ND = Not Detected
BDL = Below Detection Limit of 0.5 μg/L
DOC (mg/L) = 9.45

Table E.34

Biscayne Aquifer softened water - 5 minute prechlorination (molar concentration)

Sample	Time (h)	Cl/N ratio	Actual Cl/N ratio	pH	Br^- (mg/L)	Total Cl Residual ($mg Cl_2/L$) actual	Total Cl Residual ($mg Cl_2/L$) target	Cl_2 demand	Temperature (°F)	MCAA	MBAA	DCAA	BCAA	DBAA	TCAA	BDCAA	DBCAA	TBAA	HAA9	HAA5	DXAA	MXAA	DXAA/HAA9	DXAA Yield (DXAA/DOC)	Bromine incorporation factor (n_D')
										AVERAGE HAA CONCENTRATIONS (μmol/L)															
5BR5min	5 min	N/A	N/A	9	Ambient	N/A	N/A	N/A	70	BDL	0.0038	0.0984	0.0270	0.0092	0.0319	ND	BDL	BDL	0.17021	0.1432	0.1346	0.0038	0.79	0.01424	0.337
5BR 0.5	0.5	4:1	3.9	9	Ambient	N/A	N/A	N/A	71	0.0053	0.0038	0.0892	0.0257	0.0081	0.0177	ND	BDL	BDL	0.14981	0.1241	0.1230	0.0091	0.82	0.01302	0.341
5BR 1	1	4:1	3.9	9	Ambient	N/A	N/A	N/A	72	0.0055	0.0037	0.0969	0.0255	0.0083	0.0214	ND	BDL	BDL	0.16135	0.1358	0.1307	0.0092	0.81	0.01383	0.322
5BR4	4	4:1	3.9	9	Ambient	N/A	N/A	N/A	73	0.0055	BDL	0.0930	0.0247	0.0076	0.0180	ND	BDL	BDL	0.14879	0.1241	0.1252	0.0055	0.84	0.01325	0.318
5BR8	8	4:1	3.9	9	Ambient	N/A	N/A	N/A	74	0.0063	BDL	0.0970	0.0262	0.0082	0.0194	ND	BDL	BDL	0.15710	0.1309	0.1314	0.0063	0.84	0.01391	0.324
5BR24	24	4:1	3.9	9	Ambient	N/A	N/A	N/A	75	0.0091	0.0038	0.1056	0.0276	0.0093	0.0173	ND	BDL	BDL	0.17272	0.1451	0.1425	0.0129	0.82	0.01508	0.324
5BR48	48	4:1	3.9	9	Ambient	1.97	2	0.828	76	0.0057	0.0054	0.1036	0.0324	0.0200	0.0196	ND	0.0045	0.0042	0.19548	0.1543	0.1560	0.0111	0.80	0.01651	0.464
5BR 72-1	72	4:1	3.9	9	Ambient	1.72	N/A	1.08	77	0.0086	BDL	0.1082	0.0262	0.0077	0.0176	ND	BDL	BDL	0.16823	0.1421	0.1420	0.0086	0.84	0.01503	0.292

DXAA = DCAA + DBAA + BCAA
MXAA = MCAA + MBAA
N/A = Not Applicable
ND = Not Detected
BDL = Below Detection Limit of 0.5 μg/L
DOC (mg/L) = 9.45

Table E.35

Biscayne Aquifer softened water - 20 minute prechlorination (mass concentration)

Sample	Time (h)	Cl/N ratio	Actual Cl/N ratio	pH	Br^- (mg/L)	Total Cl Residual ($mg\ Cl_2/L$) actual	Total Cl Residual ($mg\ Cl_2/L$) target	Cl_2 demand	Temperature (°F)	MCAA	MBAA	DCAA	BCAA	DBAA	TCAA	BDCAA	DBCAA	TBAA	HAA9	HAA5	DXAA	MXAA	DXAA/HAA9	DXAA Yield (DXAA/DOC)
										AVERAGE HAA CONCENTRATIONS (µg/L)														
20BR20min	20 min	N/A	N/A	9	Ambient	N/A	N/A	N/A	70	BDL	0.6493	14.6569	4.7658	2.0897	5.4013	0.9374	BDL	BDL	28.5004	22.80	21.5124	0.6493	0.75	2.28
20BR 0.5	0.5	4:1	3.8	9	Ambient	N/A	N/A	N/A	71	0.6663	0.6771	15.5904	5.0810	2.0692	5.2055	0.9338	BDL	BDL	30.2233	24.21	22.7407	1.3434	0.75	2.41
20BR 1	1	4:1	3.8	9	Ambient	N/A	N/A	N/A	72	BDL	0.5067	16.6951	5.3263	2.1369	6.7942	1.0208	BDL	BDL	32.4799	26.13	24.1583	0.5067	0.74	2.56
20BR4	4	4:1	3.8	9	Ambient	N/A	N/A	N/A	73	1.3285	0.5031	16.6408	4.9177	2.3106	6.5690	1.0675	BDL	0.5022	33.8394	27.35	23.8691	1.8316	0.71	2.53
20BR8	8	4:1	3.8	9	Ambient	N/A	N/A	N/A	74	1.0174	BDL	16.0549	10.6534	2.2292	5.1611	0.7613	BDL	BDL	35.8772	24.46	28.9375	1.0174	0.81	3.06
20BR24	24	4:1	3.8	9	Ambient	N/A	N/A	N/A	75	1.1006	0.5662	16.4040	5.2594	2.1755	4.2705	0.8282	BDL	BDL	30.6045	24.52	23.8389	1.6668	0.78	2.52
20BR48	48	4:1	3.8	9	Ambient	2.12	2	0.88	76	2.4628	0.6277	17.7761	5.4125	2.1193	5.3470	0.7653	BDL	BDL	34.5106	28.33	25.3078	3.0905	0.73	2.68
20BR 72-1	72	4:1	3.8	9	Ambient	1.83	N/A	1.17	77	1.9409	0.5736	17.8967	5.3335	2.1186	5.2848	0.7558	BDL	BDL	33.9039	27.81	25.3488	2.5145	0.75	2.68

DXAA = DCAA + DBAA + BCAA
MXAA = MCAA + MBAA
N/A = Not Applicable
ND = Not Detected
BDL = Below Detection Limit of 0.5 µg/L
DOC (mg/L) = 9.45

Table E.36

Biscayne Aquifer softened water - 20 minute prechlorination (molar concentration)

Sample	Time (h)	Cl/N ratio	Actual Cl/N ratio	pH	Br^- (mg/L)	Total Cl Residual ($mg\ Cl_2/L$) actual	Total Cl Residual ($mg\ Cl_2/L$) target	Cl_2 demand	Temperature (°F)	MCAA	MBAA	DCAA	BCAA	DBAA	TCAA	BDCAA	DBCAA	TBAA	HAA9	HAA5	DXAA	MXAA	DXAA/HAA9	DXAA Yield (DXAA/DOC)	Bromine incorporation factor (n_0')
										AVERAGE HAA CONCENTRATIONS (µmol/L)															
20BR20min	20 min	N/A	N/A	9	Ambient	N/A	N/A	N/A	70	BDL	0.0047	0.1137	0.0275	0.0096	0.0331	0.0045	BDL	BDL	0.1930	0.1610	0.1507	0.0047	0.78	0.01595	0.310
20BR 0.5	0.5	4:1	3.8	9	Ambient	N/A	N/A	N/A	71	0.0071	0.0049	0.1209	0.0293	0.0095	0.0319	0.0045	BDL	BDL	0.2080	0.1742	0.1597	0.0119	0.77	0.01690	0.302
20BR 1	1	4:1	3.8	9	Ambient	N/A	N/A	N/A	72	BDL	0.0036	0.1295	0.0307	0.0098	0.0416	0.0049	BDL	BDL	0.2201	0.1845	0.1700	0.0036	0.77	0.01799	0.296
20BR4	4	4:1	3.8	9	Ambient	N/A	N/A	N/A	73	0.0141	0.0036	0.1291	0.0284	0.0106	0.0402	0.0051	BDL	0.0017	0.2327	0.1975	0.1680	0.0177	0.72	0.01778	0.295
20BR8	8	4:1	3.8	9	Ambient	N/A	N/A	N/A	74	0.0108	BDL	0.1245	0.0614	0.0102	0.0316	0.0037	BDL	BDL	0.2422	0.1771	0.1962	0.0108	0.81	0.02076	0.417
20BR24	24	4:1	3.8	9	Ambient	N/A	N/A	N/A	75	0.0116	0.0041	0.1272	0.0303	0.0100	0.0261	0.0040	BDL	BDL	0.2134	0.1791	0.1675	0.0157	0.79	0.01773	0.300
20BR48	48	4:1	3.8	9	Ambient	2.12	2	0.88	76	0.0261	0.0045	0.1379	0.0312	0.0097	0.0327	0.0037	BDL	BDL	0.2458	0.2109	0.1788	0.0306	0.73	0.01892	0.283
20BR 72-1	72	4:1	3.8	9	Ambient	1.83	N/A	1.17	77	0.0205	0.0041	0.1388	0.0308	0.0097	0.0323	0.0036	BDL	BDL	0.2399	0.2055	0.1793	0.0247	0.75	0.01897	0.280

DXAA = DCAA + DBAA + BCAA
MXAA = MCAA + MBAA
N/A = Not Applicable
ND = Not Detected
BDL = Below Detection Limit of 0.5 µg/L
DOC (mg/L) = 9.45

APPENDIX F
DRINKING WATER TREATMENT PLANT AND DISTRIBUTION SYSTEM SAMPLING

Table F.1
Davis Drinking Water Treatment Plant (Lake Austin source water) and distribution system (mass concentration)

Sample	Time (h)	pH	Br^- (mg/L)	Cl Residual (mg Cl_2/L)	Temperature (°F)	AVERAGE HAA CONCENTRATIONS (µg/L) MCAA	MBAA	DCAA	BCAA	DBAA	TCAA	BDCAA	DBCAA	TBAA	HAA9	HAA5	TXAA	DXAA	MXAA	DXAA/HAA9	DXAA Yield (DXAA/DOC)
Cl pipes (6.2 min)	0.10	7.7	Ambient	0.83	64	0.95	0.89	9.29	7.81	5.69	6.09	2.79	1.09	BDL	34.60	22.91	9.96	22.80	1.84	0.66	6.68
after rapid mix (8.6 min) N addition	0.14	9.4	Ambient	NA	64	1.18	0.79	19.16	9.55	6.27	16.53	3.61	1.16	BDL	58.25	43.94	21.29	34.98	1.98	0.60	10.26
before filter (5.13 hrs)	5.1	9.2	Ambient	2.04	64	0.89	0.82	20.05	9.38	6.44	19.24	3.50	1.15	BDL	61.48	47.44	23.90	35.87	1.71	0.58	10.52
after filter (5.58 hrs)	5.6	9.2	Ambient	2.12	64	0.98	0.97	18.84	9.46	6.32	20.85	3.45	1.17	BDL	62.03	47.95	25.47	34.62	1.94	0.56	10.15
clearwell (11.56 hrs)	11.6	NA	Ambient	N/A	64	1.05	0.95	19.18	9.76	5.93	19.79	3.55	1.17	BDL	61.39	46.90	24.51	34.87	2.01	0.57	10.22
Jollyville RD (1.1 days)	26.4	9.4	Ambient	2.08	64	1.13	0.97	18.72	9.84	6.36	18.57	3.47	1.23	BDL	60.29	45.74	23.28	34.92	2.09	0.58	10.24
Spicewood Pkwy (2.4 days)	57.6	9.4	Ambient	1.79	66	1.30	1.02	19.84	10.37	6.97	18.30	3.38	1.16	BDL	62.35	47.43	22.84	37.18	2.32	0.60	10.90
Cassandra Dr (3.1 days)	74.4	9.4	Ambient	1.54	66	1.17	0.54	17.32	9.72	6.20	14.85	3.26	1.11	BDL	54.17	40.08	19.22	33.23	1.71	0.61	9.75

DXAA = DCAA + DBAA + BCAA
MXAA = MCAA + MBAA
NA = Not Available
ND = Not Detected
BDL = Below Detection Limit of 0.5 µg/L
DOC (mg/L) = 3.41

Table F.2
Davis Drinking Water Treatment Plant (Lake Austin source water) and distribution system (molar concentration)

Sample	Time (h)	pH	Br^- (mg/L)	Cl Residual (mg Cl_2/L)	Temperature (°F)	AVERAGE HAA CONCENTRATIONS (µmol/L) MCAA	MBAA	DCAA	BCAA	DBAA	TCAA	BDCAA	DBCAA	TBAA	HAA9	HAA5	TXAA	DXAA	MXAA	DXAA/HAA9	DXAA Yield (DXAA/DOC)	Bromine incorporation factor (n_D')
Cl pipes (6.2 min)	0.10	7.7	Ambient	0.83	64	0.0100	0.0064	0.0720	0.0451	0.0261	0.0372	0.0134	0.0043	BDL	0.21468	0.1519	0.0550	0.1432	0.0165	0.67	0.04200	0.680
after rapid mix (8.6 min) N addition	0.14	9.4	Ambient	NA	64	0.0125	0.0057	0.1486	0.0551	0.0288	0.1012	0.0174	0.0046	BDL	0.37380	0.2968	0.1231	0.2325	0.0182	0.62	0.06817	0.485
before filter (5.13 hrs)	5.1	9.2	Ambient	2.04	64	0.0094	0.0059	0.1555	0.0541	0.0295	0.1178	0.0169	0.0046	BDL	0.39364	0.3181	0.1392	0.2392	0.0153	0.61	0.07014	0.473
after filter (5.58 hrs)	5.6	9.2	Ambient	2.12	64	0.0103	0.0069	0.1461	0.0546	0.0290	0.1276	0.0166	0.0046	BDL	0.39579	0.3200	0.1488	0.2297	0.0173	0.58	0.06736	0.490
clearwell (11.56 hrs)	11.6	NA	Ambient	N/A	64	0.0112	0.0069	0.1488	0.0563	0.0272	0.1211	0.0171	0.0046	BDL	0.39312	0.3151	0.1428	0.2322	0.0180	0.59	0.06811	0.477
Jollyville RD (1.1 days)	26.4	9.4	Ambient	2.08	64	0.0119	0.0070	0.1452	0.0568	0.0292	0.1137	0.0167	0.0049	BDL	0.38524	0.3069	0.1352	0.2311	0.0189	0.60	0.06778	0.498
Spicewood Pkwy (2.4 days)	57.6	9.4	Ambient	1.79	66	0.0137	0.0074	0.1539	0.0598	0.0320	0.1120	0.0163	0.0046	BDL	0.39966	0.3189	0.1328	0.2457	0.0211	0.61	0.07205	0.504
Cassandra Dr (3.1 days)	74.4	9.4	Ambient	1.54	66	0.0124	0.0039	0.1343	0.0561	0.0285	0.0909	0.0157	0.0044	BDL	0.34609	0.2699	0.1110	0.2188	0.0163	0.63	0.06417	0.516

DXAA = DCAA + DBAA + BCAA
MXAA = MCAA + MBAA
NA = Not Available
ND = Not Detected
BDL = Below Detection Limit of 0.5 µg/L
DOC (mg/L) = 3.41

Table F.3
Drinking water treatment plant and distribution system analysis for model validation

Parameter	Manatee, FL				Port Orange, FL				Chalmette, Louisiana			
	Source	Treated	Distribution System A	Distribution System B	Source	Softened	Distribution System A	Distribution System B	Source	Coagulated	Distribution System A	Distribution System B
pH	7.23	6.49	7.56	7.34	7.33	9.82	8.03	7.9	7.95	7.56	7.78	7.75
TOC (mg/L)	11.81	3.75	3.10	3.15	8.4	6.43	6.53	6.7	3.9	2.54	3.25	3.29
DOC (mg/L)	11.78	3.93	3.1	3.1	8.17	6.44	6.5	6.68	3.8	2.57	3.32	3.4
Hydrophobic DOC (%)	NR	65	NR	NR	NR	51	NR	NR	NR	73	NR	NR
Hydrophilic (XAD-8) DOC (%)	NR	51	NR	NR	NR	29	NR	NR	NR	62	NR	NR
UV_{254} (cm^{-1})	0.463	0.060	0.067	0.068	0.227	0.140	0.141	0.139	0.116	0.092	0.081	0.090
SUVA (L/mg-m)	3.93	1.53	2.16	2.19	2.78	2.17	2.17	2.08	3.05	3.58	2.44	2.65
Bromide (μg/L)	117	117	87	87	203	195	119	101	78	86.6	52.4	55.4
Ambient Br^-/DOC (μg/mg)	9.89	29.67	28.13	28.13	24.85	30.28	18.31	15.12	20.53	33.70	15.78	16.29
Alkalinity (mg $CaCO_3$/L)	29	3	25	25	251	47	52	50	103	95	92	94
Cl2 Dose (mg/L)	NR	6.0	NR	NR	NR	8.0	NR	NR	NR	6.2	NA	NA
Chloramine Demand (mg/L)	NR	0.9	NR	NR	NR	5.0	NR	NR	NR	2.7	NA	NA
Chloramine Residual (mg/L)	NR	NR	5.1	5.1	NR	NR	3.0	3.0	NR	NR	3.52	3.52
Cl_2/N ratio	NR	4/1	NR	NR	NR	3/1	NR	NR	NR	4/1	NR	NR
Detention Time (h)	NR	NR	0.23	2.1	NR	NR	12	24	NR	NR	20	24
DXAA (μg/L)	NR	NR	13.6	13.7	NR	NR	15.6	15.6	NR	NR	31.5	28.3
MCAA (μg/L)	NR	NR	1.1	1.4	NR	NR	2	1.6	NR	NR	2.8	0
MBAA (μg/L)	NR	NR	ND	ND	NR	NR	BDL	BDL	NR	NR	ND	ND
DCAA (μg/L)	NR	NR	7.9	8.0	NR	NR	12.5	13	NR	NR	27	24
DBAA (μg/L)	NR	NR	1.7	1.7	NR	NR	0.7	0.5	NR	NR	0.7	0.6
TCAA (μg/L)	NR	NR	2.4	2.6	NR	NR	4.8	4.6	NR	NR	3.6	3.8
BCAA (μg/L)	NR	NR	4.0	4.0	NR	NR	2.4	2.1	NR	NR	3.8	3.7

ND = Not Detected
NR = Not Run

REFERENCES

Allgeier, Steven C., R.S. Summers, J.G. Jacangelo, V.A. Hatcher, D.M. Moll, S.M. Hooper, J.W. Swetfeger, and R.B. Green. 1996. In *Proceedings - Water Quality Technology Conference*. Denver, Colo.: AWWA.

APHA, AWWA, and WEF (American Public Health Association, American Water Works Association, and Water Environment Federation). 1998. *Standard Methods for the Examination of Water and Wastewater*. 20th Edition, Washington, DC.

Amy, G., M. Siddiqui, W. Zhai, J. DeBroux, and W. Odem. 1994. *Survey of Bromide in Drinking Water and Impacts on DBP Formation.* American Water Works Association Research Foundation, Denver, CO.

Biber, Madeline V., F.O. Gulacar, and J. Buffle. 1996. Seasonal Variations in Principle Groups of Organic Matter in a Eutrophic Lake Using Pyrolysis/GC/MS. *Environ. Sci. Technol.*, 30(12): 3501-3507.

Bousher, A., P. Brimblecombe, D. Midgley. 1986. Rate of Hypobromite Formation in Chlorinated Seawater. *Water Research* 20(7): 865-870.

Bousher, A., P. Brimblecombe, D. Midgley. 1989. Kinetics of Reactions in Solutions Containing Monochloramine and Bromide. *Water Research* 23(8): 1049-1058.

Carlson, M., D. Hardy, E. Snyder III, C. Ziesemer. 1996. Effect of Water Quality Conditions on Controlling Disinfection Byproducts with Chloramines. *Proceedings – Water Quality Technology Conference,* Volume Date 1995, (Pt 2), 1993-2015.

Carlson, M. and D. Hardy. 1998. Controlling DBPs with Monochloramine. *Jour. AWWA.* 90(2): 95-106.

Collins, M.R., G.L. Amy, and C. Steelink. 1986. Molecular Weight Distribution, Carboxylic Acidity, and Humic Substances Content of Aquatic Organic Matter: Implications for Removal During Water Treatment, *Env. Sci. & Technol.*, 20(10):1028-1032.

Connell. G., B. Macler, J. Routt. 2000. Committee Report: Disinfection at Large and Medium-Size Systems. *Jour. AWWA.* 92(5): 32-43.

Cowman, G.A. and P.C. Singer. 1996. Effect of Bromide Ion on Haloacetic Acid Speciation Resulting from Chlorination and Chloramination of Aquatic Humic Substances. *Environ. Sci & Technol.* 30(1): 16-24.

Croué, J.P., E. Lefebvre, B. Martin, and B. Legube. 1993. Removal of Dissolved Hydrophobic and Hydrophilic Organic Substances During Coagulation/Flocculation of Surface Waters. Submitted to International Laboratory Network on Biodegradable Organic Matter, Ecole Polytechnique, Montréal (Québec).

Diehl, A.C., G.E. Speitel Jr., J.M. Symons, S.W. Krasner, C.J. Hwang, and S.E. Barrett. 2000. Factors Affecting Disinfection By-Product Formation During Chloramination. *Jour. AWWA.* 92(6): 76-90.

Diehl, A.C. 2001. Factors Affecting Disinfection By-Product Formation During Chloramination of Drinking Water, Ph.D. Dissertation, University of Texas at Austin.

Duirk, S.E., B. Gombert, and R.L. Valentine. 2001. Modeling the Reactions of Monochloramine and Free Chlorine with NOM. *Proceedings of 2001 American Water Works Association Annual Conference.* Denver, CO.: AWWA.

Duirk, S.E., J.C. Whitney, and R.L. Valentine. 2002a. Preliminary Investigations into Chloramine Loss and DBP Formation in the Presence of NOM and Bromide. *Proceedings of 2002 American Water Works Association Annual Conference,* New Orleans, LA. AWWA, Denver, CO.

Duirk, S.E., B. Gombert, J. Choi, and R.L. Valentine. 2002b. Monochloramine Loss in the Presence of Humic Acid. *J. Environ. Monit.*, 4: 85-89.

El-Rehaili, A.M. and W.J. Weber. 1987. Correlation of Humic Substances Trihalomethane Formation Potential and Adsorption Behavior to Molecular Weight Distributions in Raw and Chemically Treated Water. *Water Research*, 21(5): 573-582.

EPA. 2000. Internet Site: http:/www.epa.gov/enviro/html/icr_query.html.

Gadza, M., L.E. Dejarmi, T.K. Chowdury, R.G. Cooks, and D.E. Margerum. 1993. *Environ. Sci. Technol.*, 27(3): 557-561.

Gadza, M. and D.E. Margerum. 1994. Reactions of Monochloramine with Br_2, Br_3^-, HOBr, and OBr^-: Formation of Bromochloramines. *Inorganic Chemistry*, 33:118-123.

Ghosh, K and M. Schnitzer. 1980. Macromolecular Structures of Humic Substances. *Soil Science*, 129(5): 266-276.

Hartman, D.J., J. Demarco, D.H. Metz, K.W. Himmelstein, and R. Miller. 1991. Disinfection By-Product Precursor Removal by GAC and Alum Coagulation. In *Proc. of the Annual Conference, American Water Works Association.* Denver, Co.: AWWA.

Hooper, Stuart, Summers, Scott R., Solarik, Gabriele, Owen, Douglas M. 1996. Improving GAC Performance by Optimized Coagulation. *Jour. AWWA*, 88:8:107.

Hwang, C.J., M.J. Sclimenti, S.W. Krasner, J.P. Croue, A. Bruchet, and G.L. Amy. 1999. Characterization and Reactivity of Hydrophilic Natural Organic Matter (NOM) in a Low-Humic Water. In *Proceedings of 1999 Water Quality Technology Conference.* Denver, Colo.: AWWA.

Hwang, C.J., M.J. Sclimenti, and S.W. Krasner. 2000. Disinfection By-Product Formation Potential of Natural Organic Matter Fractions of a Low-Humic Water. In *Natural Organic Matter and Disinfection By-Products: Characterization and Control in Drinking Water.* Edited by S.E. Barrett, S.W. Krasner, and G.L. Amy. Washington, D.C.: ACS.

Hwang, C.J., M.J. Sclimenti, S.W. Krasner, J.P. Croue, A. Bruchet, and G.L. Amy. 2001. DBP Yields of Polar NOM Fractions from Low Humic Waters. *Water Quality Technology Conference Proceedings on CD-ROM,* AWWA, Denver, CO.

Isaac, R.A., and J.C. Morris. 1983. Transfer of Active Chlorine and Chloramine to Nitrogenous Organic Compounds. 1. Kinetics. *Environ. Sci. & Technol.*, 17(12):738-742.

Jacangelo, J.G., N.L. Patania, K.M. Reagan, E.M. Aieta, S.W. Krasner, and M.J. McGuire. 1989. Ozonation: Assessing Its Role in the Formation and Control of Disinfection By-products. *Jour. AWWA*, 81(8): 74-84.

Jafvert, C.T., and R.L. Valentine. 1992. Reaction Scheme for the Chlorination of Ammoniacal Water. *Environ. Sci. & Technol.*, 26:577-586.

Krasner, S.W. and G. Amy. 1995. Jar-Test Evaluations of Enhanced Coagulation. *Jour. AWWA.* 87(10): 93-107.

Krasner, S. W., J.M. Symons, G.E. Speitel Jr., A.C. Diehl, C.J. Hwang, R. Xia, S.E. Barrett. 1996. Effect of Water Quality on DBP Formation During Chloramination. In *Proc. of the AWWA Water Quality Conference.* Denver, Co.: AWWA.

Leenheer, J.A. 1981. Comprehensive Approach to Preparative Isolation and Fractionation of Dissolved Organic Carbon from Natural Waters and Wastewaters. *Environ. Sci. & Technol.,* 15(5): 578-587.

Liao, M.Y. and S.J. Randtke. 1985. Removing Fulvic Acid by Lime Softening. *Jour. AWWA.* 77(10): 78-88.

Logan, B. and Q. Jiang. 1990. Molecular Size Distributions of Dissolved Organic Matter. *J. Environ. Eng.,* 116(6): 1046-1062.

Long, B.W., G.S. Logsdon, and D.G. Neden. 1992. Options for Controlling Disinfection By-Products in Greater Vancouver's Water Supply. In *Proc. Water Quality Technology Conference, Part I.* AWWA, Denver, CO.

Lykins, B.W., W.E. Koffsey, and K.S. Patterson. 1994. Alternative Disinfectants for Drinking Water Treatment. *Jour. of Env. Eng.* 120(4):745-758.

MacLeod, B., Manatee County Public Works Department. personal communication 2003.

Malcolm, R.L. 1991. "Factors to be Considered in the Isolation and Characterization of Aquatic Humic Substances" in *Humic Substances in the Aquatic and Terrestrial Environment.* Boren, H. and Allard, B. editors. Wiley, London, UK 369.

McRae, B.M., T.M. LaPara, and R.M. Hozalski. 2004. Biodegradation of Haloacetic Acids by Bacterial Enrichment Cultures. *Chemosphere.* 55(6): 915-925.

Motosugi, K., Esaki, N., and Soda, K. 1982. Purification and Properties of a New Enzyme, DL-2-Haloacid Dehalogenase, from *Pseudomonas* sp. *Journal of Bacteriology*, May 1982, p. 522-527.

Murphy, E.M., J.M. Zachara, S.C. Smith, J.L. Phillips, and T.W. Wietsma. 1994. Interaction of Hydrophobic Organic Compounds with Mineral-Bound Humic Substances. *Environ. Sci. & Technol.,* Vol. 28: 1291-1299.

Rajbhandari, A.N., M.T. Koechling, and R.S. Summers. 1999. Assessment of Chloramine Decay in Treated Waters; Impact of Treatment and Free Chlorine Contact Time. *Proceedings – Annual Conference, American Water Works Association*, 457-473.

Ralls, S.M. 1999. Enhanced Softening for Disinfection By-Product Precursor Control. Master's thesis. University of Texas, Austin.

Ross, P. 1988. *Taguchi Techniques for Quality Engineering.* McGraw Hill Co., New York.

Roth, J.G., C.L. Ozment. 1999. Taking Control of HAAs. In *Water Quality Technology Conference Proceedings on CD-ROM.* Denver, CO.: AWWA.

SAS Institute Inc., 2001. JMP® IN version 4. SAS Campus Drive, Cary, NC.

Seidel, C., M.M. Frey, R.S. Summers. 2000. Understanding Disinfection By-Product Behavior in Water Distribution Systems. *In Proceedings of the Annual Conference, American Water Works Association*, Denver, CO.: AWWA.

Semmens, M.J. and A.B. Staples. 1986. The Nature of Organics Removed During Treatment of Mississippi River Water. *Jour. AWWA.* 78(2): 76-81.

Singer, P.C., G.W. Harrington, G.A. Cowman, M.E. Smith, D.S. Schecter, L.J. Harrington. 1999. Impacts of Ozonation on the Formation of Chlorination and Chloramination By-Products. Report No. 90766 American Water Works Association Research Foundation, Denver, CO.

Sinha, S. and G.L. Amy. 1997. Reactivity of NOM Fractions in Forming Chlorinated DBPs. In *Proc. of the Annual AWWA Water Quality Conference.* Atlanta, Georgia. Denver, Co.: AWWA.

Smith, K.N., C.E. Gerwe, J.H. Kweon, G.E. Speitel, Jr. and D.F. Lawler. 2004. Enhanced Softening: Effects of Source Water Quality on Natural Organic Matter Removal and Disinfection By-Product Formation, in preparation.

Snyder, M.P., and D.W. Margerum 1982. Kinetics of Chlorine Transfer from Chloramine to Amines, Amino Acids, and Peptides. *Inorganic Chemistry*, 21(7):2545-2550.

Speitel, G.E. Jr. 1999. Control of Disinfection By-Product Formation Using Chloramines. In: *Formation and Control of Disinfection By-Products in Drinking Water*, P.C. Singer (ed.), American Water Works Association, Denver, CO.

Stevens, A.A., L.A. Moore, and R.J. Miltner. 1989. Formation and Control of Non-Trihalomethane Disinfection By-Products. *Jour. AWWA*. 81(8): 54-60.

Stumm, W. and J.J. Morgan. 1996. *Aquatic Chemistry*. John Wiley and Sons, New York, NY.

Sugam, R. and M.H. Helz. Seawater Chlorination: A Description of Chemical Speciation. In *Water Chlorination: Chemistry, Environmental Impacts and Health Effects, Vol. 3*, Edited by R.L. Jolley, W.A. Brungs, R.B. Cumming, & V.A. Jacobs. Lewis Publishers, Chelsea, MI (1980).

Summers, R.S., S.M. Hooper, H.M. Shukairy, G. Solarik, and D. Owen. 1996. Assessing the DBP Yield: Uniform Formation Conditions. *Jour. AWWA* 88(6): 80-93.

Symons, J.M., S.W. Krasner, L.A. Simms, and M Sclimenti. 1993. Measurement of THM and Precursor Concentrations Revisited: The Effect of Bromide Ion. *Jour. AWWA* 85(1): 51-62.

Symons, J.M., S.W. Krasner, M Sclimenti, L.A. Simms, H.W. Sorenson Jr., G.E. Speitel Jr., and A.C. Diehl. 1996. Influence of Bromide Ion on Trihalomethane and Haloacetic Acid Formation. In *Disinfection By-Products In Water Treatment: The Chemistry of their Formation and Control*. Edited by R.A. Minear and G.L. Amy. Boca Raton, Fla.: CRC Press.

Symons, J.M., G.E. Speitel Jr., C.J. Hwang, S.W. Krasner, S.E. Barrett, A.C. Diehl, R. Xia. 1998. Factors Affecting Disinfection By-Product Formation During Chloramination. Report No. 90728 American Water Works Association Research Foundation, Denver, CO.

Teng, H., and J.N. Veenstra. A study of Disinfection By-Products Formed Using Four Alternative Disinfectants as a Function of Precursor Characteristics. 1996. In *Disinfection By-Products in Water Treatment: The Chemistry of their Formation and Control*. Edited by R.A. Minear and G.L. Amy. Boca Raton, Fla.: CRC Press.

Thompson, J.D., M.C. White, G.W. Harrington, and P.C. Singer. 1997. Enhanced Softening: Factors Influencing DBP Precursor Removal. *Jour. AWWA*. 89(6): 94-105.

Trofe, T.W., G.W. Inman, and J.D. Johnson. 1980. Kinetics of Monochloramine Decomposition in the Presence of Bromide. *Environ. Sci. & Technol.*, 14(5): 544-549.

USEPA. 1995. Methods for the Determination of Organic Compounds in Drinking Water, Supplement III. EPA/600/R-95/131.

Valentine, R.L. 1986. Bromochloramine Oxidation of N,N-Diethyl-p-phenylenediamine in the Presence of Monochloramine. *Environ. Sci. & Technol.*, 20(2):166-170.

Valentine, R.L., K. Ozekin, and P.J. Vikesland. 1998. Monochloramine Decomposition in Distribution System and Model Waters. Report No. 90721 American Water Works Association Research Foundation, Denver, CO.

Vikesland, P.J., K. Ozekin, and R.L. Valentine. 1998. Effect of Natural Organic Matter on Monochloramine Decomposition: Pathway Elucidation through the Use of Mass and Redox Balances. *Environ. Sci. & Technol.*, 32(10):1409-1416.

Vikesland, P.J., K. Ozekin, and R.L. Valentine. 2001. Monochloramine Decay in Model and Distribution System Waters. *Water Research* 35(7): 1766-1776.

Volk, Christian, L. Wood, B. Johnson, J. Robinson, H.W. Zhu, and L. Kaplan. 2002. Monitoring Dissolved Organic Carbon in Surface and Drinking Waters. *J. Environ. Monit.*, 4: 43-47.

Xie, Y.G., and H. Zhou. 2002. Use of BAC for HAA Removal – Part 2, Column Study. *Jour. AWWA.* 94(5): 126-134.

Wajon, J.E. and J.C. Morris. 1980. Bromamination Chemistry: Rates of Formation of NH_2Br and Some N-Bromamino Acids. In *Water Chlorination: Chemistry, Environmental Impacts and Health Effects*, Vol. 3, Edited by R.L. Jolley, W.A. Brungs, R.B. Cumming, and V.A. Jacobs. Ann Arbor, Mich.: Ann Arbor Science Pubishers.

Wang, J Z. and R.S. Summers. 1995. Biomass growth and distribution in drinking water biofilters and its impact on natural organic matter removal. In *Proceedings – 1995 Annual Conference, American Water Works Association.* Denver, Colo.: AWWA.

Westerhoff, Paul, Q. He, G. Amy, D. Reckhow, W. Swanson, Z. Chowdbury, M. Addison, and J McClellan. 2001. Comparison of chlorine and DBP predictions by mechanistic and empirical models. In *2001 Annual Conference Proceedings on CD-ROM*, AWWA, Denver, CO.

White, M.C., J.D. Thompson, G.W. Harrington, and P.C. Singer. 1997. Evaluating Criteria for Enhanced Coagulation Compliance. *Jour. AWWA.* 89(5): 64-77.

White, G.C. 1999. *Handbook of Chlorination and Alternative Disinfectants*, 4th Ed., John Wiley.

Williams, Sherry L., R.L. Williams, and J. Yuan. 1998. Bacterial Degradation of Haloacetic Acids in the Distribution System. In *1998 Annual Conference Proceedings on CD-ROM, AWWA*, Denver, CO.

Zhang, X., S. Echigo, R.A. Minear, M.J. Piewa. 1999. Characterization and Comparison of Disinfection By-Products from Using Four Major Disinfectants. 39(1): 251-253. American Chemical Society, Division of Environmental Chemistry, Preprints of Extended Abstracts, 217th ACS National Meeting.

Zhu, Q. and D.A. Reckhow. 1993. Investigation of Pre-Ozonation and Chloramination on Haloacetic Acid Formation. In *Proc. of 1993 AWWA Annual Conference.* AWWA, Denver, CO.

ABBREVIATIONS

ACS	American Chemical Society
amu	average mass units
AMW	apparent molecular weight
ANOVA	analysis of variance
APHA	American Public Health Association
AWWA	American Water Works Association
AwwaRF	Awwa Research Foundation
BA	Biscayne Aquifer
BCAA	bromochloroacetic acid
BDCAA	bromodichloroacetic acid
BDL	below detection limit
BDOC	biodegradeable organic carbon
bio	biodegraded
BrClAA	bromochloraacetic acid
Br_2ClAA	dibromochloroacetic acid
$BrCl_2AA$	bromodichloroacetic acid
°C	degrees Celsius
Cl^+	total chlorine
Chl	Charleston
CNX	cyanogen halide
coag	coagulated
CT	concentration times time
DBAA	dibromoacetic acid
DBCAA	dibromochloroacetic acid
1,2-DBP	1,2-dibromopropane
DBP	disinfection by-product
DCAA	dichloroacetic acid
DI	deionized
DOC	dissolved organic carbon
DOX	dissolved organic halides
DPD	*N*,*N*-diethyl-*p*-phenylenediamine
DXAA	sum of dihalogenated haloacetic acid
EPA	Environmental Protection Agency
°F	degrees Fahrenheit
FC_0	initial free chlorine dose
FCCT	free chlorine contact time
FP	formation potential

g	gram
g/L	grams per liter
GAC	granular activated carbon
GC	gas chromatograph
GF/F	binder-free borosilicate glass microfiber filter
HAA	haloacetic acid
HAAFP	haloacetic acid formaton potential
HAASDS	haloacetic acid simulated distribution system concentration
HAA_5	sum of the mass concentrations of the five regulated haloacetic acids
HAA_6	sum of the mass concentrations of the six commonly measured haloacetic acids
HAA_9	sum of the mass concentrations of all nine haloacetic acids
hr	hour
ICR	Information Collection Rule
JMP IN	statistical software
K	1000
L	liter
LA	Lake Austin water
L/mg-m	Liters per milligram times meter
m	meter
M	molar
MBAA	monobromoacetic acid
MCAA	monochloroacetic acid
MCL	maximum contaminant level
MDL	method detection limit
mg	milligram
MGD	million gallons per day
mg/L	milligrams per liter
mg/mL	milligrams per milliliter
min	minute
mL	milliliter
mm	millimeter
MR	Metedeconk River
MTBE	methyl, *tert*-butyl ether
MXAA	sum of monohalogenated haloacetic acids
n	bromine incorporation factor
n'	bromine incorporation factor for the haloacetic acids
n_D'	bromine incorporation factor for the three bromine-substituted dihalogenated haloacetic acids
NA	not analyzed

n/a	not applicable
ND	not detected or not determined
nm	nanometer
NOM	natural organic matter
NPOC	non-purgeable organic carbon
NS	not significant
NTU	nephelometric turbidity unit
OA L4	four-trial orthoganol array
OA L9	nine-trial orthoganol array
PDS	primary dilution standard
pH	negative logarithm of the effective hydrogen ion concentration
philic	hydrophilic
PVC	polyvinyl chloride
rpm	revolutions per minute
sec	second
SDS	simulated distribution system
SUFC	softening uniform formation conditions
SUVA	specific ultraviolet absorbance
TBAA	tribromoacetic acid
TCAA	trichloroacetic acid
TCE	trichloroethene
THM	trihalomethane
TOC	total organic carbon
TTHM	total trihalomethanes
TXAA	sum of trihalogenated haloacetic acids
UFC	uniform formation conditions
US	United States
USEPA	United States Environmental Protection Agency
UV	ultraviolet
UV_{254}	ultraviolet radiation at 254 nanometers
YC	cellulose acetate
YM	regenerated cellulose
μECD	micro electron capture detector
μg	microgram
μg/L	microgram per liter
μg/mg	micrograms per milligram
μg/mg-m	micrograms per milligram per meter
μL	microliter

μm	micrometer
μmol	micromole
μmol/L	micromoles per liter
μmol/mg	micromoles per milligram

IWA Publishing's authorised EU representative for General Product Safety Regulations is Diane D'Arras, 15 rue Duret, 75116 Paris, France, e-mail: safety@iwap.co.uk.

Printed and bound by CPI Group (UK) Ltd, Croydon, CR0 4YY
17/09/2026
02218810-0004